Applied Probability and Statistics

BAILEY · The Elements of Stochastic Processes with Applications to the Natural Sciences

BAILEY · Mathematics, Statistics and Systems for Health

BARTHOLOMEW · Stochastic Models for Social Processes, *Second Edition*

BECK and ARNOLD · Parameter Estimation in Engineering and Science

BENNETT and FRANKLIN · Statistical Analysis in Chemistry and the Chemical Industry

BHAT · Elements of Applied Stochastic Processes

BLOOMFIELD · Fourier Analysis of Time Series: An Introduction

BOX · The Life of a Scientist

BOX and DRAPER · Evolutionary Operation: A Statistical Method for Process Improvement

BOX, HUNTER, and HUNTER · Statistics for Experimenters: An Introduction to Design, Data Analysis and Model Building

BROWN and HOLLANDER · Statistics: A Biomedical Introduction

BROWNLEE · Statistical Theory and Methodology in Science and Engineering, *Second Edition*

BURY · Statistical Models in Applied Science

CHAMBERS · Computational Methods for Data Analysis

CHATTERJEE and PRICE · Regression Analysis by Example

CHERNOFF and MOSES · Elementary Decision Theory

CHOW · Analysis and Control of Dynamic Economic Systems

CLELLAND, deCANI, BROWN, BURSK, and MURRAY · Basic Statistics with Business Applications, *Second Edition*

COCHRAN · Sampling Techniques, *Third Edition*

COCHRAN and COX · Experimental Designs, *Second Edition*

COX · Planning of Experiments

COX and MILLER · The Theory of Stochastic Processes, *Second Edition*

DANIEL · Application of Statistics to Industrial Experimentation

DANIEL and WOOD · Fitting Equations to Data

DAVID · Order Statistics

DEMING · Sample Design in Business Research

DODGE and ROMIG · Sampling Inspection Tables. *Second Edition*

DRAPER and SMITH · Applied Regression Analysis

DUNN · Basic Statistics: A Primer for the Biomedical Sciences, Second Edition

DUNN and CLARK · Applied Statistics: Analysis of Variance and Regression

ELANDT-JOHNSON · Probability Models and Statistical Methods in Genetics

FLEISS · Statistical Methods for Rates and Proportions

GIBBONS, OLKIN and SOBEL · Selecting and Ordering Populations: A New Statistical Methodology

GNANADESIKAN · Methods for Statistical Data Analysis of Multivariate Observations

GOLDBERGER · Econometric Theory

GROSS and CLARK · Survival Distributions: Reliability Applications in the Biomedical Sciences

GROSS and HARRIS · Fundamentals of Queueing Theory

GUTTMAN, WILKS and HUNTER · Introductory Engineering Statistics, *Second Edition*

continued on back

Fitting Equations
to Data

FITTING EQUATIONS TO DATA

Computer Analysis of Multifactor Data for Scientists and Engineers

CUTHBERT DANIEL
Consultant, Rhinebeck, New York

FRED S. WOOD
American Oil Company, Chicago, Illinois

with the assistance of
JOHN W. GORMAN
American Oil Company, Whiting, Indiana

WILEY-INTERSCIENCE

a Division of John Wiley & Sons, Inc.
New York . London . Sydney . Toronto

Library of Congress Catalogue Card Number: 79-130429

ISBN   0-471-19460-3

Printed in the United States of America.

10   9   8   7

To Janet and Edith

Preface

The best way to summarize a mass of multifactor data is by a simple equation or set of equations. The data, however, must be studied critically, and here the standard texts give little guidance beyond stern warnings to be cautious.

Routine use of standard computer programs to fit equations to data does not usually succeed. A large proportion of the failures is due, not to the programs, computers, or data, but to the analyst's approach. It is not caution that is missing. Boldness in conjecture and persistence in follow-up are much more important. When computer work is a substitute for hard thinking, it is of little use. Its value increases insofar as it leads the analyst to new thoughts about the data and hence about the underlying system.

This book is a summary, mainly by examples, of our efforts to study multifactor data critically. It reflects what we have learned in the course of a decade of work with research engineers and scientists. We know that we have much more to learn, and we look forward to suggestions from our readers.

Nearly all engineering and research data are assembled in unbalanced sets, with little or no attention paid to the standard requirements of statistical design of experiments. This book is entirely devoted to methods of studying such sets. But data analysis is itself a form of experimentation. As the reader will see, we often study the data by the methods of statistical design of experiments.

Our purposes are to help the analyst, scientist, or engineer (1) to recognize the strengths and limitations of his data; (2) to test the assumptions implicit in the least-squares methods used to fit the data; (3) to select appropriate forms of the variables; (4) to judge which combinations of variables are most informative; and (5) to state the conditions under which the equations chosen appear to be applicable.

In trying to cover frequent contingencies not mentioned in any text, we have had to develop many devices, all of unknown power, sensitivity, and robustness. These include ways to detect and handle nested data, ways to spot bad or critical values, and ways to examine and select from large collections of alternative equations, those which are most appropriate. We have made extensive use of indicator or dummy variables to handle several sets of data at once, and also of computer-made residual plots to learn more about the fit of the resulting equations. All of these proposals, not now in the statistical canon, have been usefully discussed in seminars at Florida State, Harvard, New York, North Carolina State, Princeton, Texas A & M, Wisconsin, and Yale Universities, as well as at the Bell Telephone Laboratories and the National Cancer Institute. We have adopted all that has survived, and emerged from, these discussions to avoid the widespread practice of assuming without check that the form of the fitting equation is known, that the observations are independently normal with constant variance, and that there are no bad or critical values.

A major innovation, due to our colleague Colin Mallows (Bell Telephone Laboratories), is the statistic C_p, which we use to judge the total mean square error (bias squared plus random error squared) for each of the whole set of 2^K equations generated by various choices from K candidate independent variables. This approach is particularly useful when the data do not suffice to narrow the choice to a single equation. In such cases it is important to see the whole set of equations that give equally good fits.

To achieve simplicity in presentation, we have relied heavily on references to standard texts, especially those by Brownlee, Davies (Ed.), Bennett and Franklin, Hald, and Anderson and Bancroft, for most statistical derivations.

The two computer programs that we use are based on the usual Gaussian (or Laplacian) least-squares method with the classical assumptions. However, many innovations have been incorporated, as will be apparent. All of the computer work can be done quite economically on an IBM 360-65 or on a CDC 6400. In order that the programs may be easily obtained, they have been placed in the SHARE and VIM libraries, together with full program specifications in FORTRAN IV.

To facilitate translation between the several texts, the computer printout language, and engineering usage, we have included a relatively comprehensive Glossary.

We owe a special debt to K. A. Brownlee. His *Industrial Experimentation* came as a revelation to the senior author when it appeared in 1946. The reader will note that his *Statistical Theory and Methodology in Science and Engineering* is our most frequent reference.

It is with pleasure and gratitude that we acknowledge the detailed criticism of an earlier draft by Harry Smith (Rensselaer Polytechnic Institute). Jerry Warren has used the programs and text in courses at North Carolina State University and has sent us valuable comments by all members of his class.

Gus Haggstrom (University of California, Berkeley), David Jowett (Iowa State), and Michael Godfrey (Massachusetts Institute of Technology) have given us useful criticisms. George Box and William Hunter (University of Wisconsin) have made many helpful comments on the techniques used in the nonlinear example. Harold Steinour (Riverside Cement) and Waldemar Hansen (Cement Consultant) have provided valuable insight into the conditions and results of the Portland cement experiment as well as knowledge of current cement technology. Neil Timm (University of Pittsburgh) has efficiently adapted both the linear and nonlinear computer programs for the CDC 6400. M. W. Hemphill (Health, Education, and Welfare) has performed valuable services in reproducing and circulating early drafts to government agencies. We are thankful for all these favors.

We also are indebted to Hertsell Conway, Elizabeth Grimm, Robert Lyon, and George Schustek (American Oil) for their review of the manuscript, to John Boyle and Edward Sands for the initial drafts of a number of the figures, and to Edith, Kathy, and Chris Wood for their help in typing and assembling the many drafts of the manuscript.

Last, but by no means least, we acknowledge the helpfulness of our colleagues John Gorman and Robert Toman who assembled the reference material used in the first courses given at American Oil on fitting equations to data. Their 1966 paper Selection of Variables bears familial resemblance to our Chapter 6. Gorman has continued to be consistently helpful in reviewing, discussing, and contributing ideas during the development of this manuscript.

CUTHBERT DANIEL
FRED WOOD

Rhinebeck, New York
Valparaiso, Indiana
January 1971

Contents

Fitting Equations
to Data

CHAPTER 1

Introduction

1.1 Flow Diagram of Procedures Used

Fitting equations to data is a step-by-step procedure. Figure 1.1 is a flow diagram of the procedure which we have found helpful in our campaigns; it indicates the work to be done and the decisions to be made at each stage. Usually there are a number of recycles to the computer as defects in the data and in equation form are recognized. The computer programs which we use are designed to help us recognize peculiarities in the data and deficiencies in the fitted equation. Each step is discussed in detail in later chapters when the associated problem first appears, be it with one, two, or more independent variables. As the number of variables increases, the problems become more complex.

1.2 Role of Computer

The computer Linear Least-Squares Curve Fitting Program which is used throughout the book to provide examples has been put into the SHARE Library (Number 360D-13.6.008)* and the VIM Library (Number G2-CAL-LINWOOD)† so that others can obtain keypunched cards or tapes of the program, a FORTRAN listing, a User's Manual, and test problems. Although written for the IBM 360-75 computer, the program is easily adapted to other computers. The terms used in the computer printout have been included in the Glossary for ease of reference.

A multiplicity of synonymous words and symbols has frequently evolved in our field of work. We have tried to cross-reference these variations both in

---

* Available through SHARE Library, Triangle Universities Computation Center, P.O. Box 12175, Research Triangle Park, North Carolina 27709.
† Available through VIM Library, Software Distribution Department, Control Data Corporation, 215 Moffett Park Drive, Sunnyvale, California 94086.
*† Not restricted to SHARE or VIM membership. See page 311 on precision of programs.

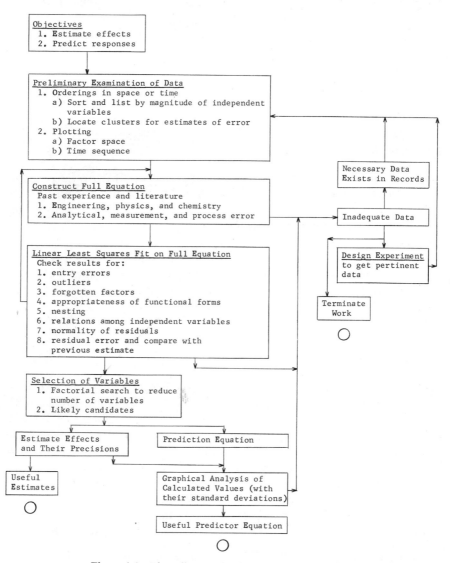

Figure 1.1 Flow diagram for fitting equations to data.

2

the text when the term is first defined and in the Index. The conventions in nomenclature and the symbols that we use are given in the first two sections of the Glossary.

1.3 Sequence of Subjects Discussed

The important but sometimes forgotten assumptions of the least-squares method are discussed in Chapter 2. Difficulties associated with fitting data with *one* independent variable are discussed in Chapter 3. Additional problems associated with *two or more* independent variables are attacked in Chapter 4. Each of these chapters builds on the information of the previous chapters and deals with variables *known* to be influential. In Chapter 5 we give an example of fitting an equation with three independent variables assumed to be known. New ways of studying the data produce conclusions quite different from those obtained by standard analysis.

When there are a number of variables that *may* be influential, a computer search is required to find, with the data at hand, which ones are the most influential. The success of the search may depend on whether or not the influential variables have been observed. If they have, and if their movements have occurred over a range adequate to display their influence, then the suggestions made in Chapter 6 will be useful in making a selection. These proposals *replace* stepwise regression and other one-at-a-time selection methods that use only F- and/or t-tests as the criteria for selection.

In addition to extending the earlier work by treating a ten-factor example, three new devices are introduced in Chapter 7. A new measure of distance between data points is used to spot remote points and to pick out near neighbors. The former are studied to see whether they indicate curvature. The latter are used to develop a new, sometimes less biased estimate of random error. Finally, a table of component effects of each x on each Y is given, which facilitates study of the detailed operation of the fitting equation at each data point.

The widespread presence of *nested* multifactor data is discussed in Chapter 8. So far as we know, this topic has not been treated in the statistical literature.

In Chapter 9 an example of nonlinear least-squares estimation is given. The Nonlinear Least-Squares Curve Fitting Program that we use is a modification of the University of Wisconsin's GAUSHAUS program. The changes introduced make it fairly easy to study a 43-parameter nonlinear equation, which is later simplified to a 17-coefficient equation with no loss in information. This program, together with a FORTRAN listing, User's Manual, and test problems, is also available from the SHARE Library (Number 360D-13.6.007) and the VIM Library (Number G2-CAL-NLWOOD).

For ease of reading, mathematical derivations and examples have been placed in appendices at the end of each chapter. Although each chapter will be self-sufficient for some readers, we believe that even the experienced analyst will find it worthwhile to read the entire text at least once.

Throughout the book we make 51 computer passes on 15 problems, including searches of about 5392 equations to determine the most suitable combinations of variables. To accomplish this, a *total* of $3\frac{1}{2}$ minutes of computer time was used—68 seconds for the linear problems, 144 seconds for the nonlinear problem.

CHAPTER 2

Assumptions and Methods of Fitting Equations

2.1 Assumptions

The purposes of fitting equations are:

a. To summarize a mass of data in order to obtain interpolation formulas or calibration curves.

b. To confirm or refute a theoretical relation; to compare several sets of data in terms of the constants in their representing equations; to aid in the choice of a theoretical model.

We assume that a mass of data is available. A number (N) of runs have been made in a laboratory, pilot plant, or commercial plant. For each run the experimental conditions (the values of each of the independent variables $x_1, x_2, \ldots, x_i, \ldots, x_K$) and the important properties of the system or products produced (the values of the dependent variables $y_1, y_2, \ldots, y_m, \ldots, y_Q$) have been recorded. All the conditions that are thought to be relevant must be included. Some may be *names* of broad classifications or groupings, not measurements. The conditions may have been varied deliberately, or adjustments to changes in raw material or environment may have been unavoidable.

A single equation or set of equivalent equations is desired for each dependent variable that relates its values to the corresponding conditions or *levels* of the independent variables. Statisticians usually call dependent variables *responses* and independent variables *factors*.

A good method of fitting should:

a. Use all the relevant data in estimating each constant.

b. Have reasonable economy in the number of constants required.

c. Provide some estimate of the uncontrolled error in y.

d. Provide some indication of the random error in each constant estimated.

e. Make it possible to find regions of systematic deviations ("function bias") from the equation if any such exist.

f. Show whether the conclusions are unduly sensitive to the results of a small number of runs, perhaps even of one run.

g. Help to spot sets of data that really are not from separate runs, but actually are from parts of one longer run (this defect is called plot-splitting, replication-degeneracy, duplicity, hierarchal arrangement, and nesting by various writers).

h. Give some idea of how well the final equation can be expected to predict the response—both in the overall sense and at important sets of conditions inside the region covered by the data.

In short, we want to obtain all the information possible from the data—both about the equation constants estimated and about the limitations on future use of the equation with these constants. It will be noticed that ease of computation has not been mentioned as a desideratum.

2.2   Methods of Fitting Equations

If the number of runs exceeds the number of constants to be estimated, the equation cannot be expected to fit all the data points exactly. There is then some necessity to choose values for the constants that are somehow "best."

One criterion, in use for over a century, is to choose values for the constants that make the data appear *most likely* when taken as a whole. Note that the data, which may be expected to contain some random variability, are held constant, whereas each of the "constants" which are being estimated is viewed as capable of taking a range of values. This range depends on the distribution of random fluctuation in the dependent variable. If we know both the form of equation to be fitted *and* the form of the distribution of random error in the observations, each independent piece of data has a "probability" that varies with each choice of a set of values for the constants. The *product* of the probabilities of all the data points, for a given set of values for all the constants, is called the *likelihood* of that set. There is usually a particular set that maximizes this product; it is termed the *maximum likelihood set*.

When the distribution of random disturbances is a so-called normal or Gaussian distribution, the maximum likelihood estimates can be found by a simpler procedure called the method of least squares.

2.3   Least Squares

The least-squares (LS) method says: "Find the values of the constants in the chosen equation that minimize the sum of the squared deviations of the observed values from those predicted by the equation."

The justification for this statement is given by the Gauss-Markov theorem (see Hald, Section 18.10), which states that the estimates obtained are the *most precise unbiased estimates* that are *linear* functions of the observations. It is assumed, of course, that the correct form of the equation has been chosen. By "most precise" we mean that the estimates in (imagined) successive sets of data taken under identical circumstances scatter around the true values with *minimum variance* or mean-squared deviation. By "unbiased" we mean that the LS estimates average out in the long run to the true values. By "linear" minimum variance estimates we imply that, although there may be other estimates that have even smaller variances, they will not be linear functions of the observations. This property of linearity is advantageous in its simplicity, but there may be nonlinear estimates with much smaller variance. However, when the random errors are *normally* distributed, the LS estimates are maximum likelihood estimates and are of minimum possible variance.

The first two of the following three key assumptions of the LS method are also crucial assumptions of maximum likelihood.

The first assumption is that we know, or have chosen, *the correct form of the equation*. If we have not used the correct form, some values given by the equation will be biased in the sense that, even if we could take new sets of data, the average of the estimated constants would not converge on the true values. The data will not necessarily give us a clear indication in this regard, but much more can be done than is shown in elementary textbooks. We will see later how we can sometimes use the data to detect systematic bias or lack of fit (also called function bias) in the equations used.

The second assumption is that the data are *typical*, that is, that they are a representative sample from the whole range of situations about which the data analyst wishes to generalize. It should be obvious that an equation which represents the typical behavior of a system cannot be derived from nontypical data. Yet this second assumption is probably the one most often violated. Each reader will have his favorite example. A large collection of data taken on a fluidized-bed catalytic reaction at only one set of temperatures may have no bearing at all on the operation of the system at some widely different set of temperatures. Likewise, data on recipe changes for angel food cake may tell us little about the effects of similar changes in a recipe for chocolate cake.

The third key assumption of the LS method is that the y-observations are *statistically uncorrelated*. Each y is taken to be made up of a "true" value, usually designated by η (eta), plus a random component, e. If we denote any two of these random components by e_j and $e_{j'}$, then this assumption may be stated: The population average value, or expectation, of $(e_j e_{j'})$ is zero. Thus $E(e_j e_{j'}) = 0$.

A more general requirement, which includes the one just mentioned, is *statistical independence.* If the probability of $e_{j'}$ appearing in some range of values is the same, regardless of the magnitude of e_j appearing in some range, then $e_{j'}$ is statistically independent of e_j. Statistical independence implies zero correlation, but the converse is true only when the distribution of the e_j is "normal" (Gaussian). It is pleasant to discover again and again that the random error in many sets of engineering data appears nearly normal. Gauss showed long ago that the presence of a *large number of small independent additive* causes of random disturbance would produce a normal distribution of errors. It is of course logically inadmissible to deduce from the appearance of near normality that we have indeed seen the operation of the four conditions stipulated by the words in italic, but in fact we do often suppose that such is the case.

Three less important assumptions of the LS method have been given extensive attention in the statistical literature. The first assumption is that all observations on y have the *same* (though unknown) *variance.* (When the variance of the dependent variable changes in a known way, then the more general *weighted* least squares method—weighting each point inversely by its variance—can be used. Such cases are discussed later and the computer program contains an option for weighting.) The second is that all the *conditions*, that is, the levels of all of the independent variables ($x_1, x_2, x_3, \ldots, x_i$), are known without error. The third is that the distribution of the uncontrolled error is a *normal* one. The conclusions are not likely to be greatly weakened if the distribution is *nearly* normal or there is *nearly* constant variance. Substantial uncertainty in many of the x_i, however, will bias all the constants toward zero.

An unwritten assumption in all curve fitting is that the data used are "good" data. But most large collections of data and occasionally even small collections contain a few "wild points," sometimes called *mavericks* or *outliers.* What happened to make them nontypical cannot usually be reconstructed. They must be spotted, however, since to retain them may invalidate the judgments we make. Methods of detecting outliers will be discussed in later chapters.

When theory and practical background do not suffice to indicate the general form of the functional relation, the equations must be at least partly empirical. In other cases the data analyst will be able to develop definite forms of the equation to be fitted, basing these forms on theory and on experience.

Polynomial forms are among the easiest empirical equations to fit. The equation chosen is linear in some or all of its K independent variables. The calculated values of the response, Y, may be of the form:

$$Y = b_0 + b_1 x_1 + b_2 x_2 + \cdots b_i x_i + \cdots b_K x_K,$$

instead, the form may be second order:

$$Y = b_0 + b_1 x_1 + b_{11} x_1^2 + b_2 x_2 + b_{22} x_2^2 + b_{12} x_1 x_2 + \cdots,*$$

or it may even contain higher powers of the independent variables. The "coefficients," the b's, are calculated from the data. The independent variables may be the original recorded conditions, x_i, or they may be any *coefficient-free* functions of the x_i. For example, x_i might be $(x_1/x_2)^{1/2}$ or x_1^2, but not $1/(x_1 - c)$ unless c is known.

The ease of fitting comes from the fact that all the above equations are *linear in the coefficients* $b_0, b_1, b_{11}, b_{12}, \ldots$, hence the word linear in the term linear least squares.

2.4 Linear Least-Squares Estimation

By definition the added requirement for *linear* LS estimation is that the equations chosen be *linear in the coefficients*.

The mathematical assumptions and elementary theory of linear LS estimates of equations with one independent variable are given in Appendix 2A. The more general case and a description of the methods used in the computer Linear Least-Squares Curve Fitting Program throughout this book are presented in Appendix 2B.

2.5 Nonlinear Least-Squares Estimation

In addition to the requirements already stated for LS estimation, nonlinear LS estimation requires that:

1. The equation by definition be *nonlinear in some of its coefficients*.
2. Preliminary estimates of all the parameters be available. If these estimates are not close "enough," the later machine-computed estimates will not converge on the best values.

For most nonlinear estimation problems, the services of a competent statistician should be available for a period as short as a day, or as long as several months, depending on the complexity of the equations under study. Able workers in nonlinear LS estimation always emphasize that careful preliminary analysis saves time in the long run. (See Box, 1960, page 803, for an informed discussion of this matter.) Here is one of those cases, actually very common, where hand work and taking thought before going on the machine are major prerequisites. As Caesar is said to have said *Festina lente* ("Make haste slowly").

Three types of nonlinear computer programs are now available: one,

* Here the double subscripts refer to the independent variables in each term. Thus b_{11} is read "b sub one one" and not "b sub eleven."

written by Booth and Peterson and discussed by Hartley, uses a "Taylor-series or Gauss-Newton" method to converge on a solution; another, written by Marquardt (1959), uses a "gradient or steepest-descent" method; the third, developed and discussed by Marquardt (1962, 1963), uses a "maximum neighborhood" method (an interpolation between the other two methods so that both the size and the direction of steps can be determined simultaneously). A mathematical summary of these methods is presented in matrix form by Draper and Smith. Examples of the first two methods have been published by Marquardt (1959) and by Behnken. Recently Wood has revised a program written by Meeter of the University of Wisconsin which utilizes the "maximum neighborhood" method. An analysis of a nonlinear estimation problem using this program is discussed in Chapter 9.

APPENDIX 2A

LINEAR LEAST-SQUARES ESTIMATES— ONE INDEPENDENT VARIABLE

2A.1   Assumptions

We make four assumptions about the relationship between the value x_{1j} of the independent variable in the jth run and the observed value y_j of the dependent variable.

A. There is a linear relationship between the "true" value of a response, η, and the value of the independent variable:

$$\eta = \beta_0 + \beta_1 x_1,$$

where β_0 is the intercept and β_1 the slope.

B. $y_j = \eta_j + e_j = \beta_0 + \beta_1 x_{1j} + e_j$, where e_j is random error.

C. The e_j have the following properties:

1. The expected value of e_j is zero; our observed y_j is an unbiased estimate of η_j.
2. The variance of e_j is $\sigma^2(y)$, which remains constant for all values of x_1.
3. The e_j are statistically uncorrelated, i.e. the expected (population) value of $e_j e_{j'}$ for any pair of points j and j' is zero.

D. The observed values of the independent variable are measured without error. All the error is in the y_j, and none is in the x_{1j}'s.

2A.2   Basic Idea and Derivation

We have the model $\eta = \beta_0 + \beta_1 x_1$ and must somehow estimate the unknown parameters, β_0 and β_1, by statistics (i.e., functions of the data) b_0

and b_1. We will obtain a fitted equation $Y = b_0 + b_1x_1$. The least-squares estimates are those values of b_0 and b_1 which minimize the function

$$Q = \sum_{j=1}^{N}(y_j - Y_j)^2 = \sum_{j=1}^{N^*}(y_j - b_0 - b_1x_{1j})^2, {}^*$$

where Y_j is the fitted value at x_{1j}.

Upon setting

$$\frac{\partial Q}{\partial b_0} = 2\Sigma(y_j - b_0 - b_1x_{1j})(-1) = 0$$

and

$$\frac{\partial Q}{\partial b_1} = 2\Sigma(y_j - b_0 - b_1x_{1j})(-x_{1j}) = 0,$$

we obtain

$$b_0 = \bar{y} - b_1\bar{x}_1$$

Notice that the fitted line must go through the point $(\bar{x}_1, \bar{y})$ and

$$b_1 = \frac{\Sigma(x_{1j} - \bar{x}_1)(y_j - \bar{y})}{\Sigma(x_{1j} - \bar{x}_1)^2} \equiv \frac{[1y]}{[11]}.$$

(For definitions of symbols see the Glossary.)

To obtain the variance of any fitted value Y, Var (Y), we look at our equation $Y = b_0 + b_1x_1$ in a slightly different form, namely,

$$Y = \bar{y} + b_1(x_1 - \bar{x}_1),$$

so that $\bar{y}$ and b_1 are uncorrelated. Since variances under these conditions are additive, we now have

$$\text{Var}(Y_j) = \text{Var}(\bar{y}) + (x_{1j} - \bar{x}_1)^2 \text{Var}(b_1).$$

First we need an estimate for the variance, σ^2. It is

$$s^2(y) = \frac{\Sigma(y_j - Y_j)^2}{N - 2}.$$

Notice that we divide by $(N - 2)$. The degrees of freedom are $(N - 2)$ because we have already used the data to obtain two estimates, namely, b_0 and b_1, from which we calculate the fitted values Y_j. Furthermore,

$$s^2(\bar{y}) = \frac{s^2(y)}{N}$$

$$s^2(b_1) = \frac{s^2(y)}{\Sigma(x_{1j} - \bar{x}_1)^2} \equiv \frac{s^2(y)}{[11]}.$$

---

* Henceforth, since most of the summations are over the total number of observations, from $j = 1$ to N, we will use the summation sign without indices, Σ, to indicate $\sum_{j=1}^{N}$.

(For derivation, see Brownlee, Section 11.2.) Putting the last two expressions into the equation given above for the estimated variance of Y, Est. Var (Y), we obtain some idea of how much uncertainty there is in our fitted values:

$$\text{Est. Var. } (Y) = \frac{s^2(y)}{N} + (x_1 - \bar{x}_1)^2 \frac{s^2(y)}{\Sigma(x_1 - \bar{x}_1)^2}$$

$$= s^2(y)\left[\frac{1}{N} + \frac{(x_1 - \bar{x}_1)^2}{[11]}\right], \quad \text{for any value of } x_1.$$

2A.3 Confidence Regions

To obtain a confidence region for our *line* we must have an appropriate multiplying factor to apply to the standard error of any point on the line. The multiplier is $[2F(0.90, 2, N - 2)]^{1/4}$ for a 90% confidence region (Brownlee, Section 10.3, and Scheffé, Section 3.5 and Problem 2.12). The use of t, rather than $(2F)^{1/2}$, as a multiplier is not appropriate because, when we make a confidence interval statement about the line, *two* parameters rather than one have been estimated, and we want to make a *joint* confidence statement about the two parameters. The first number after the F gives the "level of confidence desired, the second the number of parameters estimated, and the third the number of degrees of freedom for error. The expression

$$Y \pm [2F(0.90, 2, N - 2)]^{1/2}s\left[\frac{1}{N} + \frac{(x_1 - \bar{x}_1)^2}{[11]}\right]^{1/2}$$

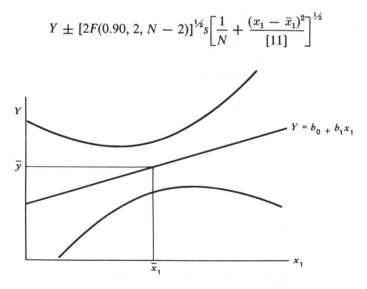

Figure 2A.1 The hyperbolic "90% confidence region" which covers the entire true line.

defines a 90% confidence region about the fitted line, inside which the true line will lie in 90% of the cases. Note that Y and x_1 are continuous variables, whereas [11] depends on the x-values at which the data were taken.

APPENDIX 2B

LINEAR LEAST-SQUARES ESTIMATES—GENERAL CASE

2B.1 Estimates of Coefficients

The coefficients of the equation are obtained by minimizing the sum of squares of the residuals, the differences between each observed value, y_j, and its corresponding fitted value, Y_j. The least-square solution will be illustrated for *three* independent variables. The formulas given are not the most convenient for computational purposes but serve well to illustrate the LS technique.

As in the example with one independent variable, we make four assumptions about the relationship between the jth observed values of the three independent variables, x_{1j}, x_{2j}, and x_{3j}, and the jth observed value of the dependent variable, y_j.

A. There is a linear relationship between the "true" value of a response, η, and the values of the independent variables, that is,

$$\eta = \beta_0 + \beta_1 x_1 + \beta_2 x_2 + \beta_3 x_3.$$

B. $y_j = \eta_j + e_j = \beta_0 + \beta_1 x_{1j} + \beta_2 x_{2j} + \beta_3 x_{3j} + e_j$, where e_j is random error.

C. As in Appendix 2A, the e_j's have the following properties:

1. The expected value of e_j is zero; our observed y_j is an unbiased estimate of η.
2. The variance of e_j is $\sigma^2(y)$, which remains constant for all values of x_i and η.
3. The e_j are statistically uncorrelated, i.e. the expected (population) value of $e_j e_{j'}$ for any pair of points j and j' is zero.

D. Observed values of the independent variables are measured without error. All the error is in the y_j's, and none is in the x_{ij}'s.

The estimated equation is

$$Y = b_0 + b_1 x_1 + b_2 x_2 + b_3 x_3,$$

which can be rewritten as:

$$Y - \bar{y} = b_1(x_1 - \bar{x}_1) + b_2(x_2 - \bar{x}_2) + b_3(x_3 - \bar{x}_3),$$

where $\bar{y}$, $\bar{x}_1$, $\bar{x}_2$, and $\bar{x}_3$ are the means of the variables for the set of N observations. This form takes advantage of the fact that for any LS fit the constant

b_0 is always of the form:

$$b_0 = \bar{y} - \sum_{i=1}^{K} b_i \bar{x}_i, \quad \text{for } K \text{ constants fitted.}$$

Thus in our example we need to find only the coefficients b_1, b_2, and b_3.

The LS solutions for the coefficients are obtained by minimizing the quantity $[rr]$, the sum of squares of the residuals:

$$[rr] \equiv \sum_{j=1}^{N^*} (y_j - Y_j)^2$$

$$= \sum_{j=1}^{N^*} [(y_j - \bar{y}) - b_1(x_{1j} - \bar{x}_1) - b_2(x_{2j} - \bar{x}_2) - b_3(x_{3j} - \bar{x}_3)]^2$$

(For definitions of symbols, see the Glossary.)

Differentiating with respect to each b_i and setting the resultant expressions equal to zero leads to the set of three "normal equations:"

$$b_1[11] + b_2[12] + b_3[13] = [1y]$$

$$b_2[21] + b_2[22] + b_3[23] = [2y]$$

$$b_3[31] + b_2[32] + b_3[33] = [3y]$$

where

$$[11] \equiv \Sigma(x_{1j} - \bar{x}_1)^2;$$

$$[21] \equiv [12] \equiv \Sigma(x_{1j} - \bar{x}_1)(x_{2j} - \bar{x}_2);$$

$$[1y] \equiv \Sigma(x_{1j} - \bar{x}_1)(y_j - \bar{y}); \text{ etc.}$$

The solution for the b's may be obtained by the method of determinants:†

$$b_1 = \frac{\begin{vmatrix} [1y] & [12] & [13] \\ [2y] & [22] & [23] \\ [3y] & [32] & [33] \end{vmatrix}}{\begin{vmatrix} [11] & [12] & [13] \\ [21] & [22] & [23] \\ [31] & [32] & [33] \end{vmatrix}}$$

with similar expressions for b_2 and b_3.

---

* Since most of the summations are over the total number of observations from $j = 1$ to N, we will use the summation sign without indices, Σ, to indicate $\sum_{j=1}^{N}$.

† See any algebra textbook for the solution of sets of linear equations by determinants.

Designating the determinant in the denominator by Δ and expanding the numerator by the first column (expansion by minors), we have

$$b_1 = \frac{1}{\Delta}\left\{[1y]\begin{vmatrix} [22] & [23] \\ [32] & [33] \end{vmatrix} - [2y]\begin{vmatrix} [12] & [13] \\ [32] & [33] \end{vmatrix} + [3y]\begin{vmatrix} [12] & [13] \\ [22] & [23] \end{vmatrix}\right\}.$$

Making similar expansions for b_2 and b_3, we can write the solutions as:

$$b_1 = c_{11}[1y] + c_{12}[2y] + c_{13}[3y]$$
$$b_2 = c_{21}[1y] + c_{22}[2y] + c_{23}[3y]$$
$$b_3 = c_{31}[1y] + c_{32}[2y] + c_{33}[3y]$$

where, for example,

$$c_{12} = -\frac{1}{\Delta}\begin{vmatrix} [12] & [13] \\ [32] & [33] \end{vmatrix} = c_{21}.$$

The elements:

$$
\begin{array}{ccc}
c_{11} & c_{12} & c_{13} \\
c_{21} & c_{22} & c_{23} \\
c_{31} & c_{32} & c_{33}
\end{array}
$$

constitute the inverse of the matrix corresponding to the determinant Δ. To simplify nomenclature, the *off-diagonal* element in the ith row and the i'th column is designated by the symbol $c_{ii'}$, and the ith *diagonal* element is designated by the symbol c_{ii}.

2B.2   Variance and Standard Errors of Coefficients

The b_i are linear functions of the y's. Note that

$$[1y] = \Sigma(x_{1j} - \bar{x}_1)(y_j - \bar{y})$$
$$= \Sigma(x_{1j} - \bar{x}_1)y_j - \bar{y}\Sigma(x_{1j} - \bar{x}_1)$$
$$= \Sigma(x_{1j} - \bar{x}_1)y_j, \quad \text{since } \Sigma(x_{1j} - \bar{x}_1) = 0.$$

Similar relations hold for $[2y]$, $[3y]$, and in general. Hence, the expression for any b_i can be written:

$$b_i = \Sigma y_j\{c_{i1}(x_{1j} - \bar{x}_1) + c_{i2}(x_{2j} - \bar{x}_2) + c_{i3}(x_{3j} - \bar{x}_3)\}.$$

For convenience let the quantity in braces be represented by k_{ij}, so that

$$b_i = \Sigma y_j\{k_{ij}\}.$$

Thus b_i is a linear function of the y_j. (This is the reason for the term linear least-squares estimates.)

The variance of any b_i is obtained directly from the equation given above. The k_{ij} are functions only of the x's, which are assumed to be known without error. The y_j are independent random variables, each with the same variance, $\sigma^2(y)$. Hence the variance of b_i can be written as

$$\text{Var } (b_i) = \sigma^2(y)\Sigma\{k_{ij}^2\}$$

Brownlee (Section 13.6) shows that

$$\sum \{k_{ij}^2\} = c_{ii},$$

the ith diagonal element of the inverse matrix, so that

$$\text{Var. } (b_i) = \sigma^2(y)c_{ii}$$

and the standard error of the coefficient $b_i = \sigma(c_{ii})^{1/2}$.

Exactly analogous equations relate the *estimated* variance of b_i to the *estimated* variance of y; thus the

$$\text{Est. Var. } (b_i) = s^2(y)c_{ii}.$$

2B.3   Computer Computations

The computer Linear Least-Squares Curve Fitting Program given in the User's Manual of this book does not compute elements of the inverse matrix, $c_{ii'}$, but instead computes *multiples* of them, $c'_{ii'}$. The two are related by

$$c'_{ii} = c_{ii}[ii] \quad \text{and} \quad c'_{ii'} = c_{ii'}([ii][i'i'])^{1/2}.$$

The $c'_{ii'}$ are listed by the computer as INVERSE, C(I, I PRIME).* Thus for $s(b_i)$, in terms of the computer inverse elements, we have:

$$s(b_i) = s(y)\left(\frac{c'_{ii}}{[ii]}\right)^{1/2} = \text{S. E. COEF. (computer listing),}$$

where $s(y)$ is the RESIDUAL ROOT MEAN SQUARE value listed by the computer.

2B.4   Partitioning Sums of Squares

Linear LS estimation partitions the total variability in the data (expressed as a total sum of squares) into two parts:

1. The sum of squares due to the fitted equation.
2. The residual sum of squares.

The objective is to account for as much of the total variability as possible by means of the fitted equation. The partitioning is based on the identity:

$$(y_j - \bar{y}) \equiv (Y_j - \bar{y}) + (y_j - Y_j)$$

---

* To conform with computer printout, computer listings are capitalized.

where $Y_j = \bar{y} + \sum\limits_{i=1}^{K} b_i(x_{ij} - \bar{x})$ is the value given for the jth observation by the fitted equation with K independent variables.

The total sum of squares is partitioned thus:

$$\sum (y_j - \bar{y})^2 = \sum (Y_j - \bar{y})^2 + \sum (y_j - Y_j)^2.$$

In the shorter nomenclature of Gauss,

$$[yy] = [YY] + [rr]$$

where (1) $[yy] \equiv \sum (y_j - \bar{y})^2$ is the TOTAL SUM OF SQUARES (computer listing),

(2) $[YY] \equiv \sum (Y_j - \bar{y})^2$ is the reduction in the total sum of squares due to the fitted equation, and

(3) $[rr] \equiv \sum (y_j - Y_j)^2$ is the RESIDUAL SUM OF SQUARES (computer listing), a measure of the squared scatter of the observed values around those calculated by the fitted equation.

2B.5 Multiple Correlation Coefficient Squared R_y^2

The MULT. CORREL. COEF. SQUARED (computer listing), R_y^2, represents the fraction of the total variation accounted for by the fitted equation:

$$R_y^2 = \frac{[YY]}{[yy]} = 1 - \frac{[rr]}{[yy]}.$$

2B.6 F-value

The F-VALUE (computer listing) used to judge the "significance" of the value of R_y^2 is also calculated from the partitioned total sum of squares:

$$\text{F-VALUE} = \frac{[YY]}{[rr]} \cdot \frac{N - K - 1}{K} = \frac{R^2}{1 - R^2} \cdot \frac{N - K - 1}{K}.$$

2B.7 Bias and Random Error

The residual sum of squares (total squared error) is also the sum of two components: (1) bias (lack of fit) and (2) random error:

$$\text{RSS} = \text{SSB} + \text{SSE}.$$

If the fitted equation contains no bias (SSB equals zero), the residual sum of squares will reflect only random error.

2B.8 Residual Mean Square

The RESIDUAL MEAN SQUARE (computer listing) with $(N - K - 1)$ degrees of freedom is:

$$\text{RMS} = \frac{[rr]}{N - K - 1} \equiv \frac{\text{RSS}}{N - K - 1}.$$

If the fitted equation contains no bias, the residual mean square is the estimated variance of y:

$$\mathrm{RMS} = s^2(y).$$

2B.9   Residual Root Mean Square

The RESIDUAL ROOT MEAN SQUARE (computer listing) is by definition the square root of the residual mean square. Again, *if* the fitted equation contains no bias, the residual root mean square is the estimated standard deviation of y:

$$\mathrm{RRMS} = s(y).$$

CHAPTER 3

One Independent Variable

3.1 Plotting Data and Selecting Form of Equation

Plotting is instructive when only one independent variable is thought to be influential. We can *see* whether the data fall on a straight line, show evidence of curvature, or indicate some other form of ill fit.

When the data can be directly represented by a straight line in x and y, we are in luck. There is a simple association between the independent variable x and the dependent variable y, represented by the equation to be fitted:

$$Y = b_0 + b_1 x_1.$$

When there is curvature, we try to identify some form of equation that will fit the data better. For our purposes the fitting equations are of two types: those whose constants are linear or can be linearized by simple transformations, and those whose constants are nonlinear and cannot be easily linearized. To aid in identification, plots of some of the equation forms of each type will be given.

3.2 Plots of Linearizable Equations

The plots in Figures 3.1, 3.2, and 3.3 were made from equations which are linearizable by appropriate transformations of y and x. This system of transformations (Hald, Section 18.7) was produced by combining each of the three functions of y (y, $1/y$, $\ln y$) with each of the corresponding functions of x. Five of the combinations are shown. The asymptotes and intercepts are easily deducible by substitution in the equations.

Another set of linearizable forms is represented by polynomials with integer powers of x:

$$Y = b_0 + b_1 x_1 + b_{11} x_1{}^2 + b_{111} x_1{}^3 + \cdots.$$

19

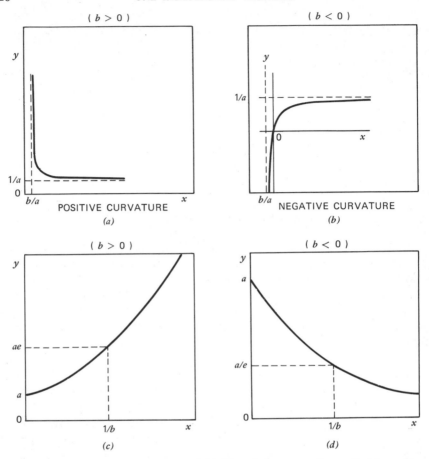

Figure 3.1 Linearizable curves. (*a*) and (*b*) Hyperbolas; $y = x/(ax - b)$. Linear form: $1/y = a - b/x$. (*c*) and (*d*) Exponential functions; $y = ae^{bx}$. Linear form: $\ln y = \ln a + bx$.

We often use a polynomial of such form—rarely going beyond the fourth power, usually not beyond the second—to represent some unknown "true" function. The higher the order of the polynomial, the more precise the data must be. Figure 3.4*a* and 3.4*b* show the diverging curves formed by the positive and negative coefficients of the quadratic term of a second-power polynomial.

Another set of useful shapes compiled by Hoerl (Section 20, page 55) is reproduced, with permission, in Figure 3.4*c*. The equation, $y = ax^be^{cx}$, can be linearized by logs when a and x are greater than zero. Thus

$$\ln y = \ln a + b \ln x + cx$$

is linear in $\ln y$, $\ln x$, and x.

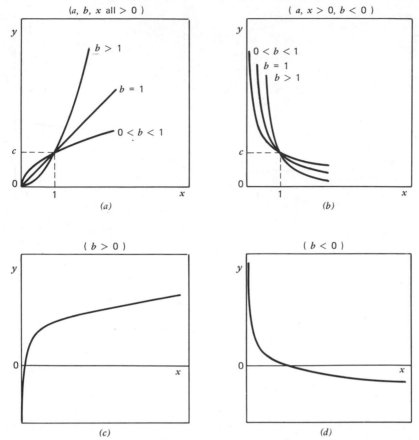

Figure 3.2 Linearizable curves. (*a*) and (*b*) Power functions; $y = ax^b$. Linear form: $\ln y = \ln a + b (\ln x)$. (*c*) and (*d*) Logarithmic functions; $y = a + b (\log x)$. Linear form: $y = a + b (\log x)$.

Although both the polynomials and Hoerl's equation may be identified by x-y plots, each term in the linearized equation must be counted as a separate independent variable in order to estimate the constants. Thus in the polynomial equation we let $x_1 = x_1'$, $x_1^2 = x_2'$, $x_1^3 = x_3'$, ..., and in Hoerl's equation $\ln y = y'$, $\ln x_1 = x_1'$, $x_1 = x_2'$ to obtain the linearized equation form: $y' = b_0 + b_1 x_1' + b_2 x_2' + b_3 x_3' + \cdots$. The added problems associated with fitting equations with two or more variables will be discussed in later chapters.

The distribution of the random errors, e_j, is changed by any nonlinear transformation of the y_j. We will often inspect the "empirical cumulative distribution of the residuals" to see whether y or its transform appears more

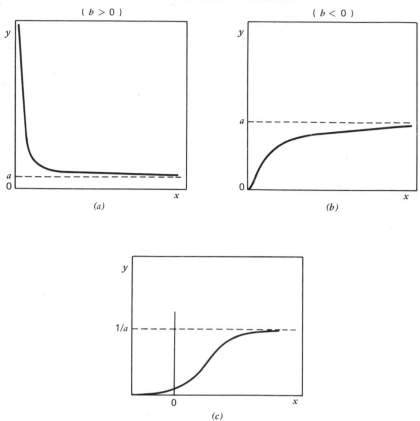

Figure 3.3 Linearizable curves. (a) and (b) Special functions; $y = ae^{b/x}$. Linear form: $\ln y = \ln a + b/x$. (c) $y = 1/(a + be^{-x})$. Linear form: $1/y = a + be^{-x}$.

nearly normal. If the transformation of y chosen for physical or algebraic reasons produces wide changes in the residuals when they are plotted against Y (fitted y-values), we may want to make some guess about how the variance of y is changing with η or with some x. We can then *weight* the y_j *inversely* by their estimated variances.

3.3   Plots of Nonlinearizable Equations

Quite a few commonly occurring shapes are not linearizable; two are shown in Figure 3.5. The S-shaped curve of Figure 3.5a may be fitted by the so-called logistic equation, but most experimenters will want mathematical help at this point. The curve in Figure 3.5b starts at zero, increases to a maximum, and then dies away—either to zero or to a positive asymptote.

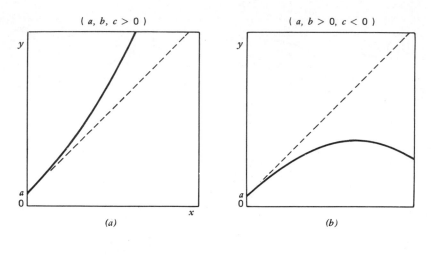

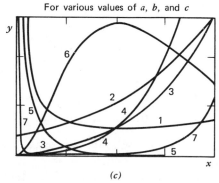

Figure 3.4 Linearizable curves. (*a*) and (*b*) Polynomial functions; $y = a + bx + cx^2$. Linear form: $y = a + bx + cx^2$. (*c*) Hoerl's special functions; $y = ax^b e^{cx}$. Linear form: $\ln y = \ln a + b (\ln x) + cx$.

	Coefficients		
Curve	a	b	c
1	10	−1	0.1
2	2	0	0.1
3	0.1	1	0.1
4	0.01	2	0.1
5	10	−1	−0.2
6	1	2	−0.2
7	0.1	−2	0.5

23

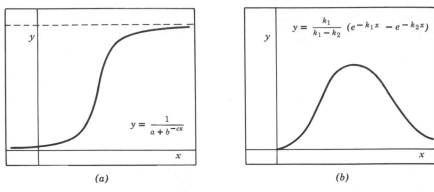

Figure 3.5 Nonlinearizable curves.

The concentration of an intermediate compound, when x represents time (or duration) of a chemical reaction, often follows such a curve.

A useful collection of equations of these shapes has also been made by Hoerl (Section 20, pages 56–77). We advise using the equations in this reference, but not the method of fitting, since results obtained are sensitive to the location of the coordinate points chosen for estimating the constants.

3.4   Statistical Independence and Clusters of Dependent Variable

In any fitted equation the uncertainty in the constants cannot be judged unless one knows or can deduce from the data something about the uncertainty of the y-values. It will not do to assume without check that all y-values have the same independent random variation. For example, each run may produce material which is split into several subbatches for chemical analysis. An x-y plot may then show (Figure 3.6a) several vertical clusters of points, each at a single x-value and each representing a single run. The scatter within each cluster measures only the uncontrolled variation of chemical analysis, perhaps combined with within-run product variability, and hence fails to reflect the variation from one *run* to another even at the same value of x. A statistician would say that, if only r runs were made, the true error for use in judging the precision of the two parameters of the straight line is estimable with $(r - 2)$ degrees of freedom, and not with $(N - 2)$, where N is the number of chemical analyses.

To take another very common case, suppose that a whole sequence of x-values (say, p of them) is whipped through on one batch of raw material. Additional observations may be obtained by running through the whole set of x-settings again, perhaps on other batches of raw material, q times.

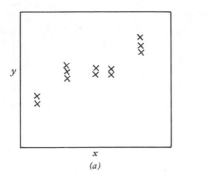 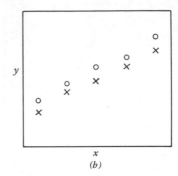

(a)   (b)

Figure 3.6   Randomization patterns. (*a*) Data at each *x*-value taken at one time. (*b*) Each series of five values taken in one run.

The *p* points produced by each set may lie closely on a straight line, but the different sets may lie on different lines (Figure 3.6*b*). From *q* such sequences, we have, at most, (*q* − 1) degrees of freedom for judging the real error in the average values of *y* or the *heights* of the lines at $\bar{x}$, even though all lines are closely parallel, so that the average slope is very precisely determined.

These samples of common randomization patterns are given, not to criticize the way in which the data were taken, but to show that the data cannot be analyzed properly, or valid conclusions drawn, if the analyst has an erroneous view of how they were collected.

3.5   Allocation of Data Points

The allocation of *x*-values needs to be examined. All data points weigh equally in determining the average height of the line, but the slope is influenced more heavily by the extreme values of *x*. Also, a cluster of points is often equivalent to only one point in the determination of the slope. In Figure 3.7*a*, the cluster at low *x* is essentially one point, and the slope is highly dependent on the point at high *x*. Similarly, in Figure 3.7*b* the cluster at the mean of the *x*-values contributes little to determining the slope, which is influenced mainly by the two extreme values. These plots emphasize the value of designed experiments to avoid such unbalanced data.

3.6   Outliers

The data should also be examined for occasional "bad" values (wild points, mavericks, outliers). Depending on their location in "*x*-space," such values can affect the estimate of the average height of the line or the estimate of its slope. Figure 3.8*a* is an example of a bad value at high *x* which will seriously bias the estimated slope. Care is needed in this case, for we may

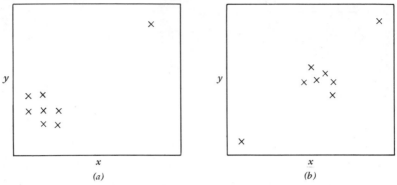

Figure 3.7   Allocation of data points.

not have a bad value; rather, the true equation may be a curve. Figure 3.8*b* is an example of a bad value at the middle of the range of x; it has little effect on the slope but will raise the height of the line, and consequently our error estimate, by a large factor.

3.7   Use of Computer Program

After the data have been examined and the form of the equation to be fit has been tentatively chosen, we are ready to use the computer Linear Least-Squares Program to see whether an equation of the form selected can be made to fit the data well by a suitable choice of coefficients. The key assumptions given in Chapter 2 should always be checked. Some of them may be known to hold. The others can often be supported by critical examination of the data together with the statistics produced by the computer.

When the underlying assumptions are satisfied, the computer printout gives results showing

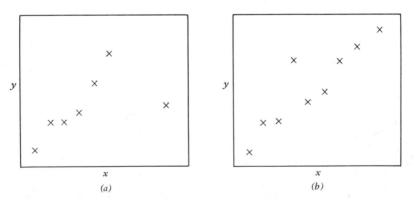

Figure 3.8   Bad values. (*a*) Bad value at high x. (*b*) Bad value at midrange of x.

(a) how well the equation is fitting (MULT. CORREL. COEF. SQUARED),
(b) what the magnitudes of the intercept (COEF. B(0)) and of the coefficients (COEF. B(I)) are,
(c) how well each coefficient is estimated (S. E. COEF.), and
(d) how significant, in the statistical sense, are the coefficients taken all together (F-VALUE).

When the "known' variance of a single observation of y is consistent with the residual mean square value of the fitted equation with degrees of freedom greater than 20, there is little likelihood that any large bias or lack of fit is left in the equation. Under such circumstances we consider ourselves fortunate in having a "good ' fit.

3.8 Study of Residuals

In the event that we have not yet obtained a "good" fit, much can be learned from computer plots of the residuals—the differences between the fitted or calculated value, Y, and the observed value, y. (Residuals are also known as remainders, discrepancies, deviations, and differences.)

The computer plot of the residuals versus their corresponding equation values will sometimes show whether there is some dependence of the magnitude of the residual on the magnitude of the equation value (Anscombe and Tukey, pages 141–160). Four common defects which may be revealed by such a plot are shown in Figure 3.9.

The wedge-shaped plot in Figure 3.9a indicates that the scatter of the residuals increases with Y (fitted Y). The residuals should show roughly equal scatter for all values of Y; otherwise the variance of the points about the line cannot be considered constant. Using log y will often eliminate this type of unequal scatter.

The wedge-shaped plot in Figure 3.9b indicates that the scatter of residuals decreases as the Y-value approaches some value—in this case 100. This suggests that a new fit should be made, but now with each point weighted by $1/(100 - Y_j)^2$. This may give a somewhat different b-value but one with smaller $s(b)$ and hence better fit. Such a weighting produces an equation that is nonlinear in its fitted constants. We defer the treatment of such equations to Chapter 9.

The U-shaped scatter plot (Figure 3.9c) is the result of fitting a straight line to data which are better represented by a simple curve.

The uneven plot (Figure 3.9d) is an example of clustered data. The relatively few observations at the lower values of Y contribute disproportionately to the estimates made and may well determine the slope. Therefore, they should be identified in regard to the time and conditions under which they were obtained. If all of them are judged to be sound random data, then all

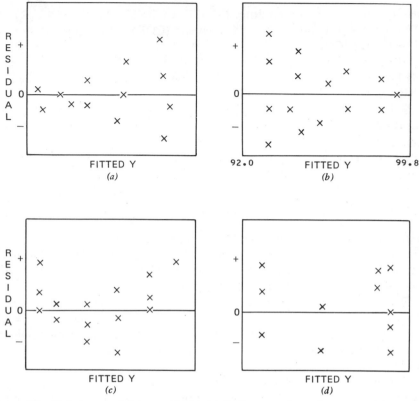

Figure 3.9 Defects revealed by residuals. (*a*) Indicates variance increases with *Y*. (*b*) Indicates variance decreases as *Y* approaches 100. (*c*) Indicates curvature. (*d*) Indicates clustered data.

should be used, but it is clear (is it not?) that we have only three well-separated levels of *x* and hence only 1 degree of freedom for judging curvature, even though we have eleven observations.

Cumulative Distribution of Residuals

The normal distribution is a two-parameter one (μ and σ are the symbols usually employed for its mean and standard deviation, respectively). The *sample* mean and standard deviation are usually denoted by $\bar{y}$ and s_y. When a *K*-parameter equation is fitted to *N* data points (which have normally distributed error), then the *N* residuals have a normal distribution with mean zero and with standard deviation $\sigma\sqrt{(N - K - 1)/(N - 1)}$. There are of course only $(N - K - 1)$ degrees of freedom in these *N* residuals.

The cumulative normal distribution can be linearized so that the height of the straight line through the data estimates the population mean, while the "slope" estimates the standard deviation. The unit for slope is the distance from the 0.1587 (or 16%) to the 0.5000 point. Alternatively, and a little more precisely, the y-distance from the 16% to the 84% point estimates 2σ.

Special grids have been prepared (the best known and in our opinion the best, is by Hazen, Whipple, and Fuller and is sold by Codex Book Co. as their No. 3227) that make it easy to plot the empirical cumulative distribution (usually abbreviated e.c.d.) for any N. The plotting points on the percentage scale are at $(j - \frac{1}{2})/N$, where $j = 1, 2, \ldots, N$, when the residuals are arranged in order of increasing magnitude.

Our purpose in plotting the cumulative distribution of the residuals is not at all to estimate a standard deviation. Rather, we do this to see whether the error structure appears to be roughly normal (Gaussian). It is, then, only the *shape* of the distribution that is of interest. We therefore suppress the actual scale of the residuals. Their actual ordered magnitudes are printed in the table just preceding the residual plot.

Since samples of the cumulative distributions will contain random errors, we need to acquire some feeling for normal departures from normality. Appendix 3A shows some actual plots of random normal deviates with standard deviation 1 and mean 0. They were taken from Rand Corporation tables, where 100,000 such deviates are given. Our sample sizes range from 8 to 384. As might be expected, samples of 8 tell us almost nothing about normality, whereas samples of 384 seem very stable except for their few lowest and highest points. Sets of 16 show shocking wobbles; sets of 32 are visibly better behaved; and sets of 64 nearly always appear straight in their central regions but fluctuate at their ends.

The commonest type of "nonnormality" is a single oversized residual. Figure 3.10 shows an actual example, which indicates too that our computer printout, although admittedly rough (it has 50 steps in each direction), is still amply fine to show curvature or other gross irregularity. Both the cumulative distribution plot and the fitted value plot are needed to identify the point as an outlier. Had the large residual occurred at an extreme Y-value, the point might have implied curvature in the true relation, or a systematic variation of variance with the value of Y. These possibilities should be checked before the point is deemed to be an outlier. Some effort should be made to find the actual cause of its excessive magnitude before it is finally discarded. If the outlier is a high-yield point or a particularly desirable property of a product, knowledge of its cause may be more valuable than all the rest of the data. The experimenter can often make a better guess than the data correlator as to where to look for causes. In many cases such a search has led to the discovery of important variables that had not previously been considered. Numerous patents have resulted from the recognition of outliers.

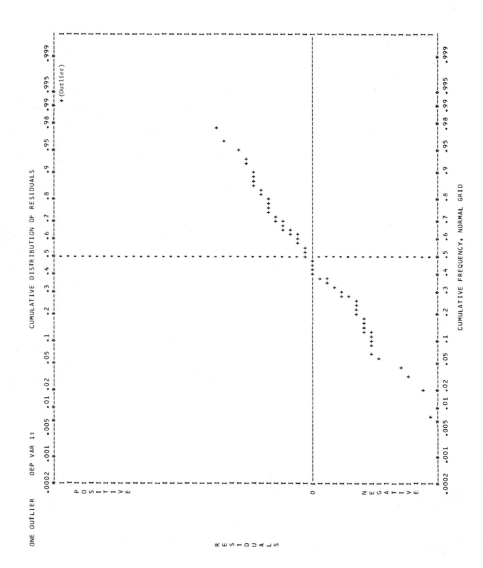

ONE OUTLIER   DEP VAR 1:   CUMULATIVE DISTRIBUTION OF RESIDUALS

CUMULATIVE FREQUENCY, NORMAL GRID

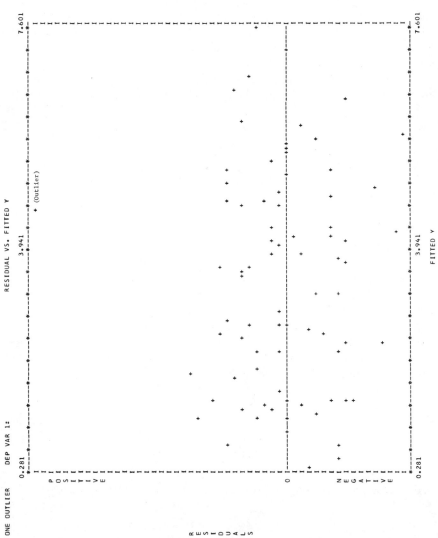

Figure 3.10(a) A single outlier in a set of 58 residuals (b) Same residuals plotted against their *Y*-values.

The time order of the residuals is often informative in judging the quality of data. If observed values are given to the computer in the same order in which they were obtained, the computer program will list them and their residuals in that order. Such a list sometimes reveals sudden jumps, or clusters of points that may be only repeats (and not real replicates) of the same item of information. If the signs of the residuals do not appear to fluctuate from plus to minus in a random order, the independence of the observations should be measured by the runs-of-signs test (Brownlee, Section 6.3). Tables of critical values are given by Owen (Sections 12.4 and 12.5) and by Dixon and Massey (Appendix Table A10).

None of these techniques can be expected to work well with fewer than twenty observations, and all are much more efficient when the number exceeds fifty.

3.9   Dealing with Error in the Independent Variable

Least-squares analysis assumes that the independent variable is measured without error. When the x-values have considerable variance, the estimate of the coefficient (the slope) is biased toward zero. As a rule of thumb, LS analysis can be used safely if the variance of x is less than a tenth of the average scatter of the x's from their mean.

The difficult case of wide uncertainty in several x_i is not discussed in this book. (For details, see the publications by Acton and by Madansky.)

If the variance of x is too large for LS analysis, Bartlett's three-group procedure is recommended. To determine the slope of the fitted line by this procedure, the data are divided into three groups by the observed x-values. The slope of the fitted line is that of the line connecting the average points in the first and the third group. The fitted line must pass through the average point of all the data. Bartlett derives confidence intervals for the slope and height calculated in this manner.

Berkson has shown that when x is a "controlled variable," that is, when each x-value has a random component with population value zero added to a target value, standard LS methods give unbiased estimates of the equation coefficients. For details, the reader should consult Brownlee (Second Edition, page 393) and the additional references that he gives.

3.10   Example of Fitting a Straight Line to Data—One Independent Variable

A numerical example of fitting a straight line to data, using our computer program, is given in Figures 3B.1–3B.5. The example was taken from pilot-plant data. The independent variable, x, is the acid number of a chemical determined by titration, and the dependent variable, y, is its organic acid content determined by extraction and weighing. The aim was to determine how well values obtained by the relatively inexpensive titration method can serve to estimate those obtained by the more expensive extraction and weighing technique. The high values of t and of R^2, the appearance of the x-y graph, the distribution of the residuals, and the even spread of the residuals versus Y plot—all support the conclusion that x is a good estimate of y.

APPENDIX 3A

CUMULATIVE DISTRIBUTION PLOTS OF RANDOM NORMAL DEVIATES

The observed values used in these plots were taken from the Rand Corporation tables of random normal deviates; each set of z values has a standard error of 1 and a mean of 0.

Number of Observations per Set	Figures
8	3A.1–3A.2
16	3A.3–3A.4
32	3A.5–3A.6
64	3A.7–3A.13
384	3A.14

APPENDIX 3B

EXAMPLE OF FITTING A STRAIGHT LINE TO DATA— ONE INDEPENDENT VARIABLE

The objective of this problem is to determine how well values obtained by a relatively inexpensive titration method can serve to estimate those obtained by a more expensive extraction and weighing technique. Twenty samples of a pilot-plant chemical were selected to cover the range of interest. Each sample (chosen from the twenty at random) was thoroughly mixed just before being split and analyzed by both methods.

Figure 3B.1 (page 44) is a plot of the values obtained. There appears to be a reasonably good linear relationship between the acid number (x) determined by titration and the organic acid content (y) determined by extraction and weighing. The error in x is negligible.

Figures 3B.2–3B.5 (pages 45–48) are computer printouts obtained by fitting the equation $y = b_0 + b_1 x$ to the data, using the Linear Least-Squares Curve Fitting Program. Details of the nomenclature used in the printouts are given in the Glossary.

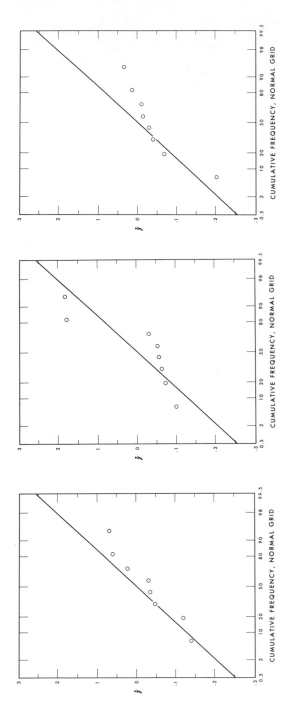

Figure 3A.1 Samples of 8 random normal deviates.

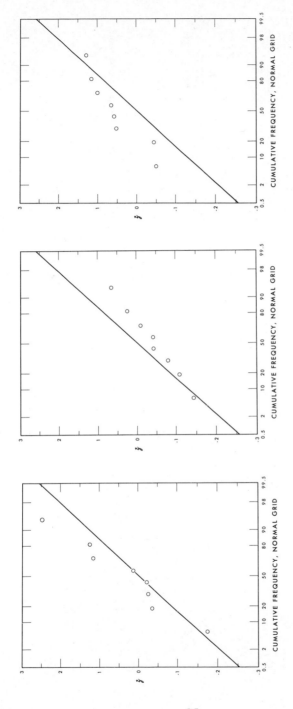

Figure 3A.2   Samples of 8 random normal deviates.

35

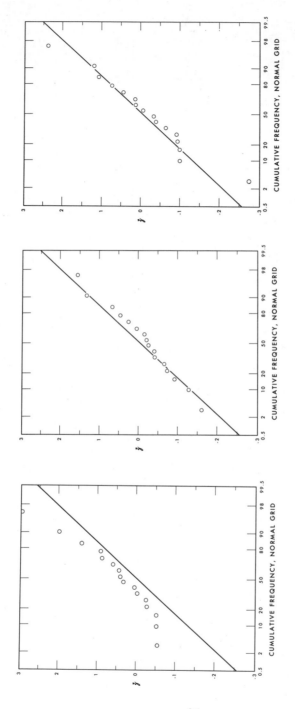

Figure 3A.3 Samples of 16 random normal deviates.

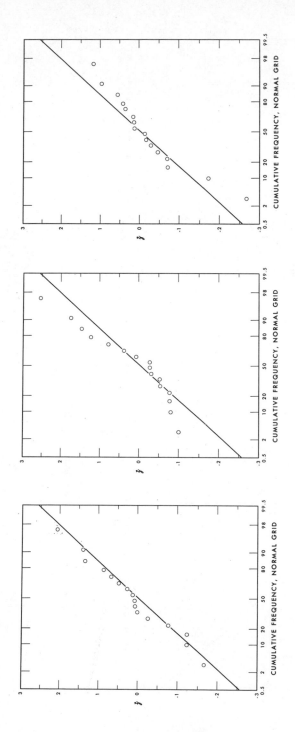

Figure 3A.4 Samples of 16 random normal deviates.

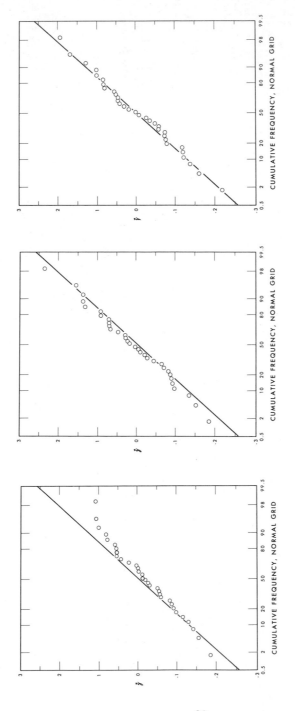

Figure 3A.5 Samples of 32 random normal deviates.

38

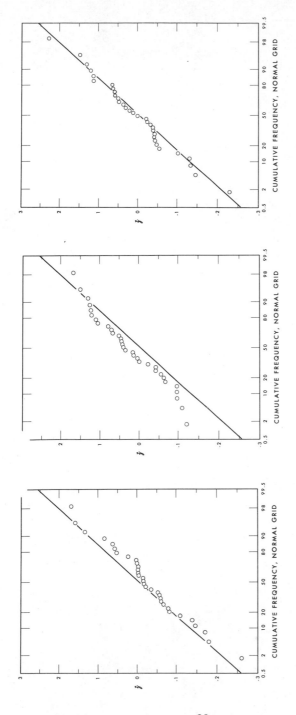

Figure 3A.6 Samples of 32 random normal deviates.

39

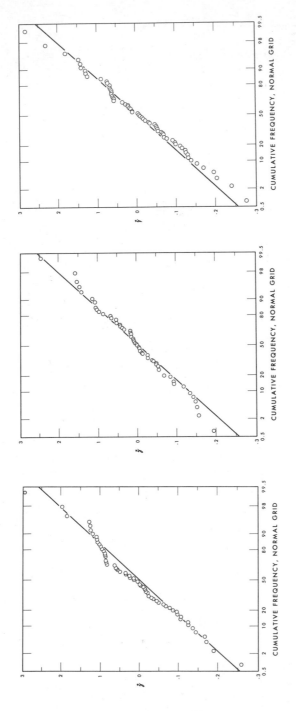

Figure 3A.7 Samples of 64 random normal deviates.

40

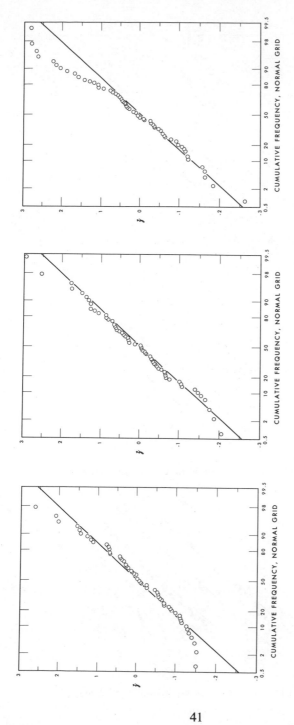

Figure 3A.8 Samples of 64 random normal deviates.

41

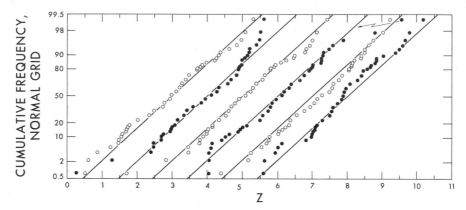

Figure 3A.9 Six samples of 64 random normal deviates.

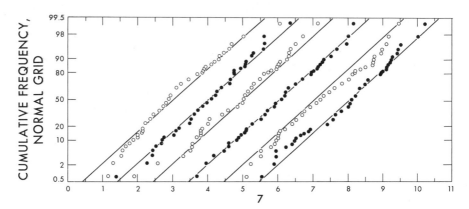

Figure 3A.10 Six samples of 64 random normal deviates.

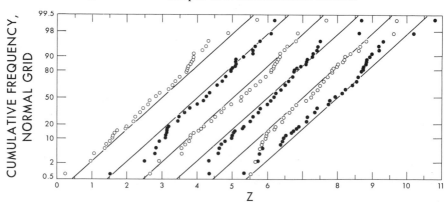

Figure 3A.11 Six samples of 64 random normal deviates.

42

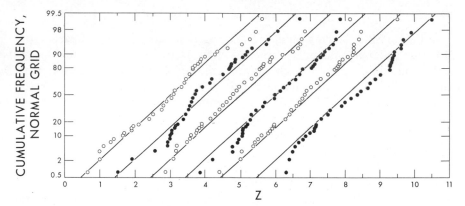

Figure 3A.12 Six samples of 64 random normal deviates.

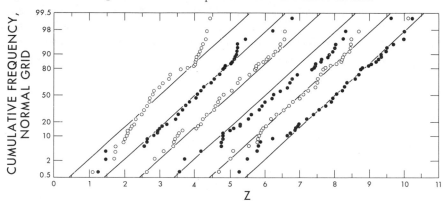

Figure 3A.13 Six samples of 64 random normal deviates.

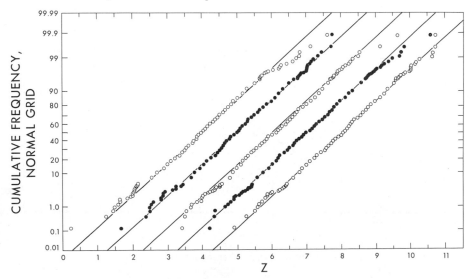

Figure 3A.14 Five samples of 384 random normal deviates.

In Figure 3B.2 the data input is listed together with the sum of each variable, the residual sums of squares and cross products, the calculated mean and standard deviation of each variable, the simple correlation coefficients, and the elements of the inverse matrix. The numbers at the top of each column, such as 1-11-21 and 2-12-22, identify the numbers of the variables listed in that column. In this case, they are 1 and 2. Had there been 21 variables, the 1st, 11th, and 21st variable would have been listed in that order in the first column.

Figure 3B.3 lists many of the important statistical properties of the fit. The b_1 coefficient, given in fixed decimal point form, is 0.32108. Since the standard error of the coefficient is 0.0055, the value of b_1 might well be written as 0.321 ± 0.006. The multiple correlation coefficient squared, R^2, is 0.995; this indicates that 99.5 % of the total sum of squares of y is accounted for by this equation. The residual root mean square is 1.2. Sometimes the residual column allows us to determine that the observations were not statistically independent. There were 10 positive differences where OBS. Y was larger than FITTED Y, and 10 negative differences. In all, there were 13 changes of sign. We see on page 380 of Owen that 15 changes of sign would have been too many and 6 too few. There is in this case, then, no evidence of autocorrelation.

Figure 3B.4 is a computer-made plot of the empirical cumulative distribution of the residuals. The residuals fall, as they should, approximately on a straight line, indicating that they are roughly normally distributed, with no outliers.

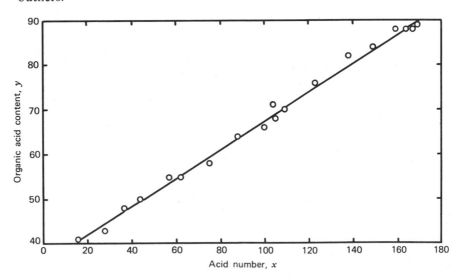

Figure 3B.1 Plot of acid number versus organic acid content.

LINEAR LEAST-SQUARES CURVE FITTING PROGRAM

ACID NUMBER

PROBLEM HAS ONE EQUATION

DATA READ WITH SPECIAL FORMAT

FORMAT CARD 1 (A6,I3,1X,I2, 2F6.3)

BZERO = CALCULATED VALUE

DATA INPUT   1 INDEPENDENT VARIABLES   1 DEPENDENT VARIABLES

OBSV.	SEQ.	1-11-21	2-12-22	3-13-23	4-14-24	5-15-25	6-16-26	7-17-27	8-18-28	9-19-29	10-20-30
1	1	123.000	76.000								
2	1	109.000	70.000								
3	1	62.000	55.000								
4	1	104.000	71.000								
5	1	57.000	55.000								
6	1	37.000	48.000								
7	1	44.000	50.000								
8	1	100.000	66.000								
9	1	16.000	41.000								
10	1	28.000	43.000								
11	1	138.000	82.000								
12	1	105.000	68.000								
13	1	159.000	88.000								
14	1	75.000	58.000								
15	1	88.000	64.000								
16	1	164.000	88.000								
17	1	169.000	89.000								
18	1	167.000	88.000								
19	1	149.000	84.000								
20	1	167.000	88.000								

SUMS OF VARIABLES
   2.06100D 03 1.37200D 03

RESIDUAL SUMS OF SQUARES + CROSS PRODUCTS
   4.90329D 04
   1.57694D 04 5.09880D 03

MEANS OF VARIABLES
   1.03050D 02 6.86000D 01

ROOT MEAN SQUARES OF VARIABLES
   5.08004D 01 1.63816D 01

SIMPLE CORRELATION COEFFICIENTS, R(I,I PRIME)
1   1.000   0.997
2   0.997   1.000

Figure 3B.2

LINEAR LEAST-SQUARES CURVE FITTING PROGRAM

ACID NUMBER   DEP VAR 1: ORG AC   MIN Y = 4.100D 01   MAX Y = 8.900D 01   RANGE Y = 4.800D 01

RESULTANT EQUATION:
$Y = 35.46 + 0.3216X$
Y = ORGANIC ACID CONTENT
X = ACID NUMBER

IND.VAR(I)	NAME	COEF.B(I)	S.E. COEF.	T-VALUE	R(I)SQRD	MIN X(I)	MAX X(I)	RANGE X(I)	REL.INF.X(I)
0		3.54583D 01							
1	ACID N	3.21608D-01	5.55D-03	57.9	0.0	1.600D 01	1.690D 02	1.530D 02	1.03

NO. OF OBSERVATIONS   20
NO. OF IND. VARIABLES   1
RESIDUAL DEGREES OF FREEDOM   18
F-VALUE   3352.3
RESIDUAL ROOT MEAN SQUARE   1.22997881
RESIDUAL MEAN SQUARE   1.51284788
RESIDUAL SUM OF SQUARES   27.23126184
TOTAL SUM OF SQUARES   5098.80000000
MULT. CORREL. COEF. SQUARED   .9947

---ORDERED BY COMPUTER INPUT---

IDENT.	OBSV.	WS DISTANCE	OBS. Y	FITTED Y	RESIDUAL
AN	1	27.	76.000	75.016	0.984
AN	2	2.	70.000	70.514	-0.514
AN	3	115.	55.000	55.398	-0.398
AN	4	0.	71.000	68.906	2.094
AN	5	145.	55.000	53.790	1.210
AN	6	298.	48.000	47.358	0.642
AN	7	238.	50.000	49.609	0.391
AN	8	1.	66.000	67.619	-1.619
AN	9	518.	41.000	40.604	0.396
AN	10	385.	43.000	44.463	-1.463
AN	11	84.	82.000	79.840	2.160
AN	12	0.	68.000	69.227	-1.227
AN	13	214.	88.000	86.594	1.406
AN	14	54.	58.000	59.579	-1.579
AN	15	15.	64.000	63.760	0.240
AN	16	254.	88.000	88.202	-0.202
AN	17	297.	89.000	89.810	-0.810
AN	18	280.	88.000	89.167	-1.167
AN	19	144.	84.000	83.378	0.622
AN	20	280.	88.000	89.167	-1.167

---ORDERED BY RESIDUALS---

OBSV.	OBS. Y	FITTED Y	ORDERED RESID.	SEQ
11	82.000	79.840	2.160	1
4	71.000	68.906	2.094	2
13	88.000	86.594	1.406	3
5	55.000	53.790	1.210	4
1	76.000	75.016	0.984	5
6	48.000	47.358	0.642	6
19	84.000	83.378	0.622	7
9	41.000	40.604	0.396	8
7	50.000	49.609	0.391	9
15	64.000	63.760	0.240	10
16	88.000	88.202	-0.202	11
3	55.000	55.398	-0.398	12
2	70.000	70.514	-0.514	13
17	89.000	89.810	-0.810	14
18	88.000	89.167	-1.167	15
20	88.000	89.167	-1.167	16
12	68.000	69.227	-1.227	17
10	43.000	44.463	-1.463	18
14	58.000	59.579	-1.579	19
8	66.000	67.619	-1.619	20

Figure 3B.3

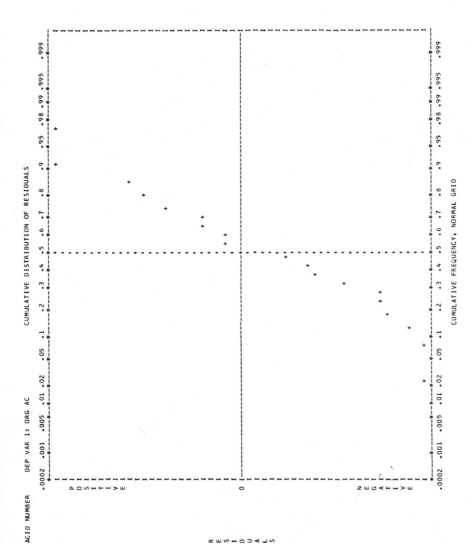

Figure 3B.4

47

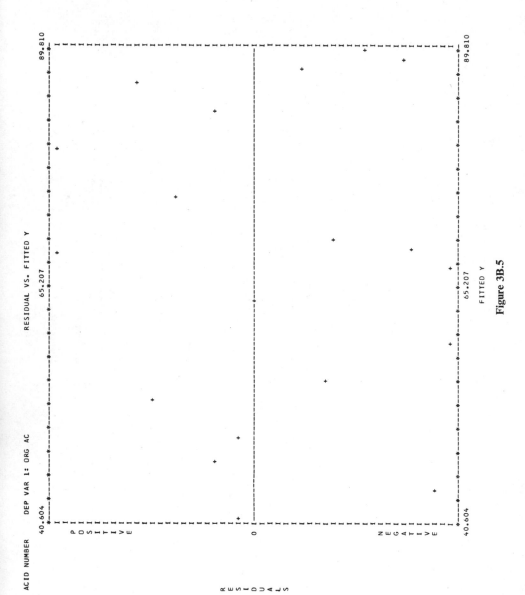

Figure 3B.5

48

Figure 3B.5 is the computer plot of the residuals versus FITTED Y. Since the residuals are evenly distributed, the variance of the points about the line is roughly constant.

We have, then, a good correlation between the two methods of assaying the particular chemical tested. Which one should be used will depend on the relative costs of the two procedures.

CHAPTER 4

Two or More Independent Variables

4.1 Inadequacies of x-y Plots with Two or More Independent Variables

It is natural to think that the graphical methods used with one independent variable will also be useful with two independent variables. This is true only when there are several x_1-values at a fixed x_2-value, or several x_2-values at a fixed x_1-value. Consider the following set of seven points, which were obtained directly from the equation

$$Y = 10 + 7x_1 - 4x_2.$$

x_1	x_2	y
1	3	5
2	1	20
3	3	19
4	5	18
5	7	17
6	4	36
7	7	31

We show in Figure 4.1 the plots of y versus x_1 and of y versus x_2. A line drawn through the x_1-y points suggests a slope of 5 as a value for b_1. No trend can be seen at all in the x_2-y plot.

There are two pairs of points, however, that are matched in their x_2-values (those at $x_2 = 3$ and those at $x_2 = 7$). Thus it is possible to judge the rate of change due to x_1, *at fixed* x_2, and to get a check, from the other pair, on this rate. In this case, the same value, 7, is obtained for b_1. With this value it is now possible to estimate b_2. This set of data is exceptional, however, in having matched points.

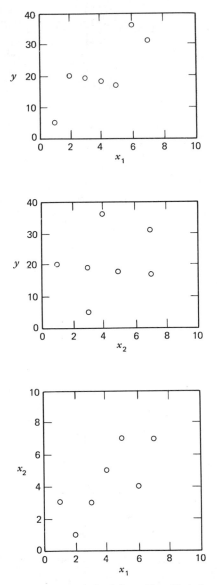

Figure 4.1 Data derived from $Y = 10 + 7x_1 - 4x_2$.

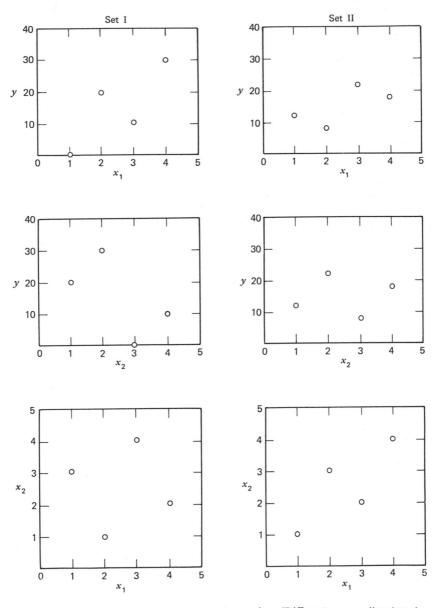

Figure 4.2 Data derived from $Y = b_0 + b_1x_1 + b_2x_2$ (Different x_1, x_2 allocations).

Let us take an even smaller set of points from an equation with different constants, where this matching is not found.

x_1	x_2	y
1	3	0
2	1	20
3	4	10
4	2	30

Here, plotting y versus x_1 (Figure 4.2, Set I), we might imagine that we see a slope, b_1, for x_1 of $+10$. Similarly b_2 might be thought to be -10. But taking another set of data points derived from the same equation, we have:

x_1	x_2	y
1	1	12
2	3	8
3	2	22
4	4	18

Plotting as before (Figure 4.2, Set II), we now find $b_1 = +5$ and $b_2 = 0$. The two sets of data appear to give different values for b_1 and b_2 because of the different allocation of the data points in "x_1-x_2" space. This is shown at the bottom of Figure 4.2, where x_2 is plotted against x_1. Clearly, plotting single independent variables versus y is worthless, or at least hazardous, when only *two* variables operate at the same time, *even though they are operating additively* with no error present. A third variable makes the situation even worse.

It is possible, and indeed advisable, to plot *contours* of constant y on an x_1-x_2 grid when only two independent variables are operating. But this expedient is not extensible to more than two independent variables and so we must draw the following conclusion: plots of x_i versus y when k is greater than 1 are rarely useful. When one independent variable dominates the situation, then of course the corresponding x_i-y plot will show this correctly. But when several x_i are influential, or when several x_i are intercorrelated, confusion is the principal product of such plots.

We must go to more generalizable analytical methods for such cases.

4.2 Equation Forms and Transformations

The general equation, with K independent variables indexed by the subscript i and N observations indexed by the subscript j, has the form:

$$Y_j = b_0 + b_1 x_{1j} + b_2 x_{2j} + \cdots + b_i x_{ij} + \cdots + b_K x_{Kj} = b_0 + \sum_{i=1}^{K} b_i x_{ij}.$$

The constants $b_0, b_1, b_2, \ldots, b_i, \ldots, b_K$ are to be estimated from the data.

A wide variety of shapes can be represented by such an equation. For example,

$$(4.1) \qquad Y = b_0 + b_1\left(\frac{x_1{}^2}{x_2}\right) + b_2 x_1 x_3 + b_3 x_1 \log x_4$$

is linear in the b's and that, as we said in Chapter 1, is all that is required for linear least-squares estimation. The computer program will operate on this equation just as well as on

$$(4.2) \qquad Y = b_0 + b_1 z_1 + b_2 z_2 + b_3 z_3,$$

since equation 4.1 is easily converted to 4.2 by setting $(x_1{}^2/x_2) = z_1$, etc.

We see, then, that many complex equations (or surfaces) are linearizable and should be linearized, when enough is known about the way in which the x_i affect the response.

It may be useful to put equation 4.1 in the form:

$$(4.3) \quad Y = b_0 + b_1 f_1(x_1, x_2, \ldots, x_K) + b_2 f_2(x_1, x_2, \ldots, x_K) + \cdots,$$

where the only restriction on the f's is that they be *known, constant-free* functions of the x_i. Thus $e^{x_1 \sin x_2}$ is acceptable, but $(x_1 - c)/(x_1 - d)$ is not, unless d is known.

If the values of y cover quite a large range, that is, if y has a natural zero and if $y_{\max}/y_{\min}$ is greater than, say, 10, it may be doubted that the residual mean square, $s^2(y)$, is constant for all values of y in the relevant range. If the error variance is not constant, we can weight the y-values. They should be weighted *inversely* as their variances (Brownlee, 1960 ed. only, Section 11.12). It can be proved (but this will not be done here) that the variance of all estimates of constants will be simultaneously minimized by using the correct weights. If guessed weights are used, we may not get minimum variance. If incorrect weights that depend, not on the observed y-values, but only on the x-values are used, we get "unbiased estimates" of the slopes, but not with minimum variance.

When Var. $(x_i) \neq 0$ it can be shown that the estimate of b_i is biased downward from the true slope (Hald, pages 615–616). Since we cannot usually be certain that the variance of x_i is zero, some judgment should be made about the cases in which this variance can be considered "nearly zero" or at least "safely negligible."

Even when some x_i have considerable variances, it is possible to get an unbiased and efficient estimate of each b_i. This case is not discussed here; see Acton, Madansky, or Keeping.

4.3 Variances of Estimated Coefficients, b_i, and Fitted Values, Y_j

With two independent variables the variance of the coefficient b_1 may be calculated from:

$$\text{Est. Var. } (b_1) = \frac{s^2(y)}{[11](1 - r_{12}^2)},$$

where r_{12}^{2*} is the degree of nonorthogonality between x_1 and x_2 *in these data*. It will not have the same value in other sets, but may be only a property (an unfortunate one) of the data before us. If r_{12}^2 is large (i.e., near 1), this is very disadvantageous, since then the variances of the two b_i concerned are larger than they need be, considering only the unavoidable size of $s^2(y)$ and the spread of the x_i, [11], and [22] in these data.

An exactly analogous equation holds for the variance of each b_i in the multifactor case, that is, where there are more than two factors. We need only replace r_{12}^2 by R_i^2, the degree of nonorthogonality of the ith independent variable. Thus the amount of linear dependence of x_i on the other variables determines how much the variance of b_i is inflated. Obviously a similar statement holds for each factor, but a different R_i^2 enters for each. The STANDARD ERROR OF THE COEFFICIENT is obtained by taking the square root of Var. (b_i).

The estimated standard error of a particular Y_j, $s(Y_j)$, depends of course on the variances and covariances of all the b_i, as well as on the actual placement of the point Y_j in x-space.

The equation for computing $s(Y_j)$ is complex but computable. We have not provided a program for this because we find it is rarely needed. The simple bounds given below nearly always suffice.

Minimum $s(Y)$ is always at point $x_i = \bar{x}_i$,

$$\frac{s(y)}{\sqrt{N}} < s(Y) < s(y),$$

$$\text{Average } s(Y) = \sqrt{\frac{K+1}{N}}\, s(y).$$

4.4 Use of Indicator Variables for Discontinuous or Qualitative Classifications

One qualitative factor at two levels

Suppose that our data are divisible into two sets, not necessarily equal in number, and that the distinction between the two sets is not amenable to

* The term $r_{12}^2 = [12]^2/[11][22]$ has the form of a squared correlation coefficient. It measures the fraction of the variation, $[ii]$, in one independent variable that is accounted for by the other in an equation of the form $X_i = A + Bx_{i'}$, $i \neq i'$. It takes the value 1 for exact linear dependence and 0 when the equation gives no improvement over setting $X_i = \bar{x}_i$. When $r_{12}^2 = 0$, we have complete (linear) independence of x_1 and x_2 or "orthogonality." When $r_{12}^2 \neq 0$, we have some degree of dependence and hence a degree of nonorthogonality.

quantitative measure (e.g., day versus night, two suppliers of the same catalyst). We can use an indicator or "dummy" variable, x_1, to make the distinction, *defining* x_1 as 0 for category I, and 1 for category II. No other values of x_1 are to be allowed. Our equation is

$$Y = b_0 + b_1 x_1,$$

which looks rather familiar. For all the data in category I, Y will be b_0; for all in category II, Y will be $b_0 + b_1$. We see that b_1 is simply the difference between the two average values of y.

One qualitative two-level factor and one quantitative (continuous) factor

The equation now reads:

$$Y = b_0 + b_1 x_1 + b_2 x_2.$$

Here x_1 is defined exactly as above, but x_2 may take any value over some range. The data are now represented by two parallel lines with slope b_2, separated by the amount b_1. More than one discrete two-level factor and several continuous factors can be accommodated in this way.

Discrete factors at more than two levels

Suppose that the data produced in *three* shifts must be distinguished in our equation. Here we need *two* fitting constants, since there are 2 degrees of freedom for the three-level factor. Thus:

$$Y = b_0 + b_1 x_1 + b_2 x_2,$$

where now $x_1 = 1$ for shift II,
$\quad\quad\quad = 0$ otherwise;
$\quad\quad x_2 = 1$ for shift III,
$\quad\quad\quad = 0$ otherwise.

In a simple table of translation we have then

	I	II	III
x_1	0	1	0
x_2	0	0	1

and by direct substitution

$$Y_{\mathrm{I}} = b_0,$$
$$Y_{\mathrm{II}} = b_0 + b_1,$$
$$Y_{\mathrm{III}} = b_0 + b_2.$$

Other allocations of the indicator variables to the levels of the discontinuous variable are permissible, but the one shown above gives simple meanings to the three constants.

Discrete factors interacting with continuous factors

If the linear effect (i.e., the slope) of some continuous factor, x_2, is not the same at the two levels of a discontinuous factor, x_1, we can represent this as a product term, $x_1 x_2$:

$$Y = b_0 + b_1 x_1 + b_2 x_2 + b_{12} x_1 x_2.$$

Again $x_1 = 0$ or 1, and x_2 is continuous.

Details of specifying the allocation of indicator variables are discussed in Chapter 8 and illustrated in Figures 8A.1 and 8A.2.

4.5 Allocation of Data in Factor Space

The deliberate allocation of the levels of factors so as to minimize bias and random error is the subject of the branch of statistics called "design of experiments." Statistically designed multifactor experiments are becoming fairly common in some branches of science and engineering research, but the data analyst rarely sees the results of a single balanced experiment. In the first place the analysis and interpretation of such experiments is usually easy and is familiar to the experimenter. A single set of balanced data rarely brings the research work to a satisfactory conclusion, and therefore more experimental work is usually done to clear up ambiguities, to extend the range of some factors, and to include new factors. After all this is done, the entire set is never balanced, and the more general methods of least squares, not analysis of variance, become necessary.

Nearly all of the devices proposed in this work are responses to the fact that most multifactor data are not taken—and often cannot be taken—in balanced arrays. The disposition of the data points in "factor space" then becomes a matter for study. Failure to examine the actual disposition may result in severe misinterpretation.

Statisticians regularly recommend that data would be better if taken in balanced arrays, in accordance with their requirements as to randomization. Doing this is sometimes acceptable but often not practicable. We take the position here that the data have been taken and must be analyzed as they stand. Recommendations to take more data will be more acceptable if preceded by a conscientious study of all that has been done so far. In this way the consumer will see what his data show and what he should do if they do not show enough.

It seems to have escaped the notice of some statisticians that a very large part—perhaps nine tenths—of scientific and engineering research work is carried through to some kind of successful result without any aid from experts in experiment design. There are several reasons for this. Many scientists take their data varying one factor at a time, and so have little trouble analyzing it. Contrariwise, sometimes when an experiment has not come out well, the able researcher will vary three or even eight conditions simultaneously in directions that make sense to him. If the new experiment works, he will hurry on, not going back to find out which of the conditions that he varied was decisive. Moreover some technical work is closely linked to theory and to a wide background of experience. Crucial experiments then permit valid decisions about the part of the theory that is under test.

For most researchers, however, the time comes when they must review and summarize what they have found, perhaps over a long period of time. At this point a mass of data must be studied; it is almost never balanced, is rarely homogeneous, and seldom contains real replicates. It is about these cases and some of their common defects that we write.

Linear dependences among the x_i

One of the x_i, say x_1, may be linearly related to one or more of the other factors. There will then be a linear relationship among some of the x_i, perhaps disturbed by random fluctuations, of the form:

$$x_1 = c_1 + c_2 x_2 + c_3 x_3 + \cdots .$$

In studying two variables we saw that the slopes b_1 and b_2 cannot be estimated *at all* from B_1 and B_2, where these are the constants in the two equations $Y_I = B_I + B_1 x_1$ and $Y_{II} = B_{II} + B_2 x_2$. Even if we look at them two at a time, we can hope for little or no illumination concerning the true linear dependencies among the x_i. We must, in fact, go through the fitting of a full linear equation, using each x_i in turn as the *dependent* variable, in order to detect such relationships.

It is often possible to cut down the full equation to find that only a few of the other factors are linearly related to a particular x_i. It may even happen that only *one* factor is left that is strongly linearly correlated with our present one. But this will not always be the case.

As we have indicated previously, the *variance* of b_i in the equation relating Y to x_i is *inflated* by the factor $1/(1 - R_i^2)$. Thus, if any R_i^2 is as large as 0.9, the variance of the corresponding b_i is inflated by a factor of at least 10.

The outermost points in data space

A process that is well described by a simple equation when running within some standard range of practice may follow a different equation, or be much less stable, when operating outside this range. Therefore we require some means of detecting extreme conditions in multifactor situations. We propose a (squared) measure of "distance from standard" that appears to take satisfactory account of the differing influences of the several factors. It is automatically computed for each data point and is printed along with the fitted values and the residuals. See WSSD in Figures 6A.2, 6B.2, 7A.2, and 7A.9–7A.11 for examples.

When there are no heavy dependencies among the x_i, it is not difficult to locate the points in factor space that are farthest away from the middle of the data. Once again, it is not safe to look at the factors singly, or even in pairs, and to stop there.

Nested data

Plant and pilot-plant data are often taken in naturally grouped sets. Hard-to-change conditions are held constant or nearly constant for considerable periods of time, while those easy to change are varied. If this is done *and recorded*, the analysis of the data, although not completed in a single computer run, is fairly straightforward.

We can expect the quickly changed factors to produce changes in results with small variance. The levels of the hard-to-change factors are run through more slowly and might therefore be expected to reflect weather and other changes with larger random components and therefore greater variance. It will be necessary to run *two* fittings: one on the nested data "within plots" and one "among plots." (The term plots comes from the first appearance of such groupings in agricultural data.)

After the within-plot analysis has been made, we must correct or adjust all plot averages to the same values of the within-plot factors. These revised averages are then treated as new data for the among-plot analysis. All the warnings, pitfalls, and devices indicated above may be expected to apply twice over to "split plot" or "nested" data.

If equations are fitted to such data without taking into account that they are nested, the wrong factors may appear to be significant. The residual mean square will be an unevenly pooled average of the two basic variances; this gives us *too many* whole-plot factors showing "significant" effects, and *too few* (or not any) subplot factors showing "significant" effects.

If the data are recorded helter-skelter, with no record of the time order in which they were taken or of what "nesting" has taken place, then the facts about nesting become more difficult to disentangle. In Chapter 5 we treat a case of nesting by direct inspection, since the situation is quite simple. In Chapter 8 a more complex case is studied, and more general rules are given.

Fitting an Equation in Three Independent Variables

5.1 Introduction

In looking for a multifactor example using real data, we were struck by the fact that no clean-cut, undesigned data were on hand. By "clean-cut" we mean "meeting all the necessary statistical requirements without any complicating circumstances." When we tried to analyze textbook examples, we found unforeseen complications in each. One of the more interesting problems is presented here, not only as an example of fitting an equation with three independent variables, but also to point out some of the precautions that are necessary.

The example is from Brownlee (Section 13.12). It is also presented in Draper and Smith (Chapter 6). The data, given in Table 5.1, represent 21 successive days of operation of a plant oxidizing ammonia to nitric acid. Factor x_1 is the flow of air to the plant. Factor x_2 is the temperature of the cooling water entering the countercurrent nitric oxide absorption tower. Factor x_3 is the concentration of nitric acid in the absorbing liquid (coded for ease of hand calculation by subtracting 50 and then multiplying by 10). The response, y, is 10 times the percentage of the ingoing ammonia that is lost as unabsorbed nitric oxides; it is an indirect measure of the yield of nitric acid.

5.2 First Trials

Figures 5.1–5.3 show summary computer printouts after fitting the equation

$$Y = b_0 + b_1x_1 + b_2x_2 + b_3x_3.$$

Although the standard statistics—the t-values, the F-value, and the multiple correlation coefficient squared—all seem acceptably significant, the

TABLE 5.1

DATA FROM OPERATION OF A PLANT FOR THE OXIDATION OF AMMONIA
TO NITRIC ACID

Observation Number	Air Flow, x_1	Cooling Water Inlet Temperature, x_2	Acid Concentration, x_3	Stack Loss, y
1	80	27	89	42
2	80	27	88	37
3	75	25	90	37
4	62	24	87	28
5	62	22	87	18
6	62	23	87	18
7	62	24	93	19
8	62	24	93	20
9	58	23	87	15
10	58	18	80	14
11	58	18	89	14
12	58	17	88	13
13	58	18	82	11
14	58	19	93	12
15	50	18	89	8
16	50	18	86	7
17	50	19	72	8
18	50	19	79	8
19	50	20	80	9
20	56	20	82	15
21	70	20	91	15

normal plot (Figure 5.2) indicates one very low point (21), and the residuals versus fitted Y plot (Figure 5.3) is irregular in an unfamiliar way. There appear to be five points (1, 2, 3, 4, and 21) whose residuals are widely separated from the others, and the remaining sixteen residuals seem to show some sort of subsidiary trends in sets.

5.3 Possible Causes of Disturbance

In reflecting on possible causes of—or at least names for—these oddities, we list four conceivable sorts of disturbance:

1. A small minority of points, perhaps only one or two, may be entirely erroneous.

LINEAR LEAST-SQUARES CURVE FITTING PROGRAM

STACK LOSS, PASS 1   DEP VAR 1: S.LOSS   MIN Y = 7.000D 00   MAX Y = 4.200D 01   RANGE Y = 3.500D 01

Y = B(0) + B(1)X1 + B(2)X2 + B(3)X3
Y = STACK LOSS
X1 = AIR FLOW
X2 = COOLING WATER TEMPERATURE
X3 = NITRIC ACID CONCENTRATION
ALL DATA USED

IND.VAR(I)	NAME	COEF.B(I)	S.E. COEF.	T-VALUE	R(I)SQRD	MIN X(I)	MAX X(I)	RANGE X(I)	REL.INF.X(I)
0		-3.991970 01							
1	A.FLOW	7.156400-01	1.350-01	5.3	0.6559	5.000D 01	8.000D 01	3.000D 01	0.61
2	W.TEMP	1.295290 00	3.680-01	3.5	0.6113	1.700D 01	2.700D 01	1.000D 01	0.37
3	A.CONC	-1.521230-01	1.560-01	1.0	0.2501	7.200D 01	9.300D 01	2.100D 01	0.09

NO. OF OBSERVATIONS	21
NO. OF IND. VARIABLES	3
RESIDUAL DEGREES OF FREEDOM	17
F-VALUE	59.9
RESIDUAL ROOT MEAN SQUARE	3.24336392
RESIDUAL MEAN SQUARE	10.51940951
RESIDUAL SUM OF SQUARES	178.82996160
TOTAL SUM OF SQUARES	2069.23809524
MULT. CORREL. COEF. SQUARED	.9136

------ORDERED BY COMPUTER INPUT------

IDENT.	OBSV.	WS DISTANCE	OBS. Y	FITTED Y	RESIDUAL
S LOSS	1	24.	42.000	38.765	3.235
S LOSS	2	24.	37.000	38.917	-1.917
S LOSS	3	13.	37.000	32.444	4.556
S LOSS	4	1.	28.000	22.302	5.698
S LOSS	5	0.	18.000	19.712	-1.712
S LOSS	6	1.	18.000	21.007	-3.007
S LOSS	7	2.	19.000	21.389	-2.389
S LOSS	8	2.	20.000	21.389	-1.389
S LOSS	9	1.	15.000	18.144	-3.144
S LOSS	10	2.	14.000	12.733	1.267
S LOSS	11	2.	14.000	11.364	2.636
S LOSS	12	3.	13.000	10.221	2.779
S LOSS	13	1.	11.000	12.429	-1.429
S LOSS	14	1.	12.000	12.050	-0.050
S LOSS	15	7.	8.000	5.639	2.361
S LOSS	16	7.	7.000	6.095	0.905
S LOSS	17	6.	8.000	9.520	-1.520
S LOSS	18	6.	8.000	8.455	-0.455
S LOSS	19	6.	9.000	9.598	-0.598
S LOSS	20	1.	15.000	13.588	1.412
S LOSS	21	5.	15.000	22.238	-7.238

------ORDERED BY RESIDUALS------

OBSV.	OBS. Y	FITTED Y	ORDERED RESID.	SEQ
4	28.000	22.302	5.698	1
2	37.000	32.444	4.556	2
1	42.000	38.765	3.235	3
12	13.000	10.221	2.779	4
11	14.000	11.364	2.636	5
15	8.000	5.639	2.361	6
20	15.000	13.588	1.412	7
10	14.000	12.733	1.267	8
16	7.000	6.095	0.905	9
14	12.000	12.050	-0.050	10
18	8.000	8.455	-0.455	11
19	9.000	9.598	-0.598	12
8	20.000	21.389	-1.389	13
13	11.000	12.429	-1.429	14
17	8.000	9.520	-1.520	15
5	18.000	19.712	-1.712	16
2	37.000	38.917	-1.917	17
7	19.000	21.389	-2.389	18
6	18.000	21.007	-3.007	19
9	15.000	18.144	-3.144	20
21	15.000	22.238	-7.238	21

Figure 5.1

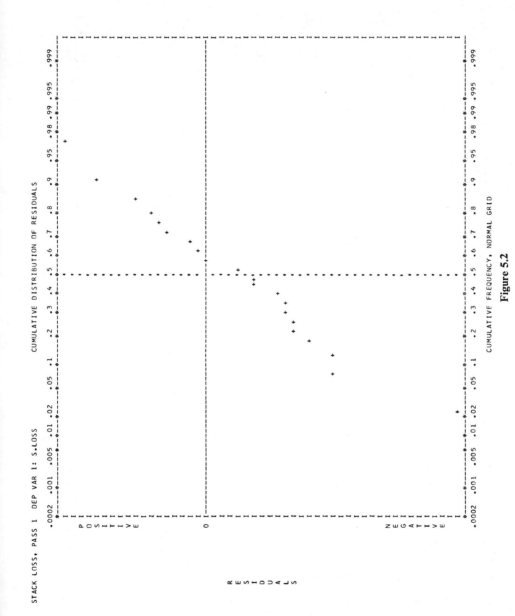

Figure 5.2

63

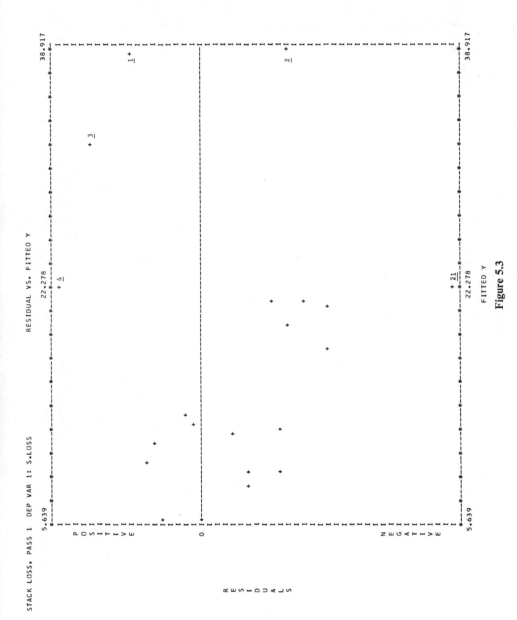

Figure 5.3

2. We may not have the right functional form. The large residuals may be reflecting "function bias."

3. The random error may be quite irregular, not at all "normal."

4. Some points may reflect changing conditions, soon after large changes in some influential factor. The system may not yet have come to equilibrium under the new conditions.

We focus our attention on possibilities 1 and 2. We ask first whether perhaps observation *21* alone was disturbing the equation, and, second, whether there might not be a trend in error increasing with the true magnitude of the response. The latter possibility could be accommodated by using log y instead of y as the dependent variable.

In Pass 2 we drop observation *21*. Figures 5.4–5.6 show the results. We see from the t-values, the F-value, the multiple correlation coefficient squared (R_y^2), and the residual mean square (RMS) that we have a noticeably better fit. But the normal plot of residuals is now skew, and the residuals versus fitted Y plot shows the same undesirable trends in sixteen points that it did in Pass 1. Point *21* was the last piece of data to be taken and reflects a jump in x_1 (from 56 to 70) that might signal lack of equilibrium, but removal of *21* has not sufficed to make everything satisfactory.

5.4 Outlier or Logged Response or Squared Independent Variable

Figures 5.7–5.9 show the computer summary and plots from Pass 3, using log y instead of y. We have no better fit than with y, but the normal plot is greatly improved and the queer trends in the residuals have largely disappeared.

Since x_3 has shown negligible influence in the three passes just made, we drop it and run through the same alternatives (the full set of data, dropping point *21*, using log y for response). The first six lines of Table 5.2 summarize our work so far.

We make two more passes (7 and 8 in Table 5.2) so that all eight alternatives are before us. These may be viewed as a 2^3 factorial experiment:*

Factor A has two levels: observation *21* in or out,

Factor B has two levels: y or log y used as response,

Factor C has two levels: x_3 in or out.

Thus each equation studied can be specified by a simple symbol, for example, *abc* means "*21* out, log y used, x_3 out." We resist the temptation to carry

---

* The language and nomenclature of 2^K factorial designs will be used in more detail later. Here we only require representation of an equation by indicating the presence or absence of a small letter, one for each change in the equation. The reader can easily see this by study of columns 2–5 in Table 5.2.

LINEAR LEAST-SQUARES CURVE FITTING PROGRAM

STACK LOSS, PASS 2   DEP VAR 1: S.LOSS   MIN Y = 7.000D 00   MAX Y = 4.200D 01   RANGE Y = 3.500D 01

```
Y = B(0) + B(1)X1 + B(2)X2 + B(3)X3
Y  = STACK LOSS
X1 = AIR FLOW
X2 = COOLING WATER TEMPERATURE
X3 = NITRIC ACID CONCENTRATION
OBSERVATION 21 OMITTED
```

IND.VAR(I)	NAME	COEF.B(I)	S.E. COEF.	T-VALUE	R(I)SQRD	MIN X(I)	MAX X(I)	RANGE X(I)	REL.INF.X(I)
0		-4.37040D 01							
1	A.FLOW	8.89108D-01	1.19D-01	7.5	0.7052	5.000D 01	8.000D 01	3.000D 01	0.76
2	W.TEMP	8.16620D-01	3.25D-01	2.5	0.6853	1.700D 01	2.700D 01	1.000D 01	0.23
3	A.CONC	-1.07141D-01	1.25D-01	0.9	0.2276	7.200D 01	9.300D 01	2.100D 01	0.06

```
NO. OF OBSERVATIONS          20
NO. OF IND. VARIABLES         3
RESIDUAL DEGREES OF FREEDOM  16
F-VALUE                      98.8
RESIDUAL ROOT MEAN SQUARE    2.56920122
RESIDUAL MEAN SQUARE         6.60079490
RESIDUAL SUM OF SQUARES      105.61271844
TOTAL SUM OF SQUARES         2062.55000000
MULT. CORREL. COEF. SQUARED   .9488
```

-----ORDERED BY COMPUTER INPUT-----

IDENT.	OBSV.	WS DISTANCE	OBS. Y	FITTED Y	RESIDUAL
S LOSS	1	52.	42.000	39.938	2.062
S LOSS	2	52.	37.000	40.045	-3.045
S LOSS	3	29.	37.000	33.752	3.248
S LOSS	4	1.	28.000	21.698	6.302
S LOSS	5	1.	18.000	20.065	-2.065
S LOSS	6	1.	18.000	20.882	-2.882
S LOSS	7	1.	19.000	21.055	-2.055
S LOSS	8	1.	20.000	21.055	-1.055
S LOSS	9	1.	15.000	17.325	-2.325
S LOSS	10	2.	14.000	13.992	0.008
S LOSS	11	1.	14.000	13.028	0.972
S LOSS	12	2.	13.000	12.318	0.682
S LOSS	13	1.	11.000	13.778	-2.778
S LOSS	14	1.	12.000	13.416	-1.416
S LOSS	15	13.	8.000	5.915	2.085
S LOSS	16	13.	7.000	6.236	0.764
S LOSS	17	13.	8.000	8.553	-0.553
S LOSS	18	12.	8.000	7.803	0.197
S LOSS	19	12.	9.000	8.512	0.488
S LOSS	20	2.	15.000	13.633	1.367

-----ORDERED BY RESIDUALS-----

OBSV.	OBS. Y	FITTED Y	ORDERED RESID.	SEQ
4	28.000	21.698	6.302	1
3	37.000	33.752	3.248	2
15	8.000	5.915	2.085	3
1	42.000	39.938	2.062	4
20	15.000	13.633	1.367	5
11	14.000	13.028	0.972	6
16	7.000	6.236	0.764	7
12	13.000	12.318	0.682	8
19	9.000	8.512	0.488	9
18	8.000	7.803	0.197	10
10	14.000	13.992	-0.008	11
17	8.000	8.553	-0.553	12
8	20.000	21.055	-1.055	13
14	12.000	13.416	-1.416	14
7	19.000	21.055	-2.055	15
5	18.000	20.065	-2.065	16
9	15.000	17.325	-2.325	17
13	11.000	13.778	-2.778	18
6	18.000	20.882	-2.882	19
2	37.000	40.045	-3.045	20

Figure 5.4

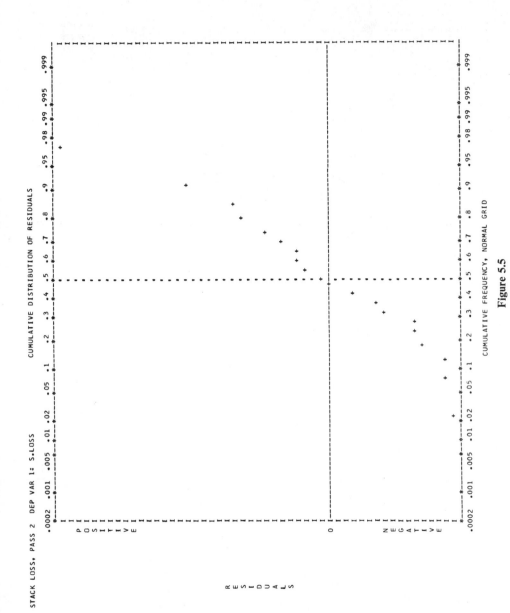

STACK LOSS, PASS 2 DEP VAR 1: S.LOSS

CUMULATIVE DISTRIBUTION OF RESIDUALS

CUMULATIVE FREQUENCY, NORMAL GRID

Figure 5.5

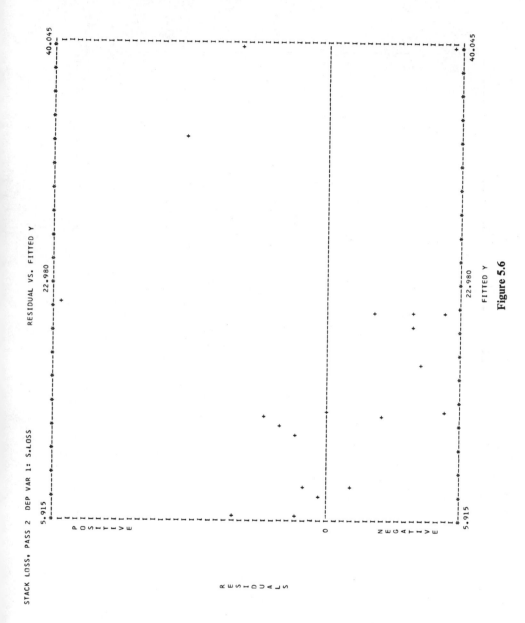

STACK LOSS, PASS 2   DEP VAR 1: S.LOSS   RESIDUAL VS. FITTED Y

Figure 5.6

LINEAR LEAST-SQUARES CURVE FITTING PROGRAM

STACK LOSS, PASS 3   DEP VAR 1: LOG Y   MIN Y = 8.451D-01   MAX Y = 1.623D 00   RANGE Y = 7.782D-01

```
LOG Y = B(0) + B(1)X1 + B(2)X2 + B(3)X3
Y = STACK LOSS
X1 = AIR FLOW
X2 = COOLING WATER TEMPERATURE
X3 = NITRIC ACID CONCENTRATION
ALL DATA USED
```

IND.VAR(I)	NAME	COEF.B(I)	S.E. COEF.	T-VALUE	R(I)SQRD	MIN X(I)	MAX X(I)	RANGE X(I)	REL.INF.X(I)
0		-4.12028D-01							
1	A.FLOW	1.50113D-02	3.19D-03	4.7	0.6559	5.000D 01	8.000D 01	3.000D 01	0.58
2	W.TEMP	2.75624D-02	8.70D-03	3.2	0.6113	1.700D 01	2.700D 01	1.000D 01	0.35
3	A.CONC	1.24388D-03	3.70D-03	0.3	0.2501	7.200D 01	9.300D 01	2.100D 01	0.03

```
NO. OF OBSERVATIONS          21
NO. OF IND. VARIABLES         3
RESIDUAL DEGREES OF FREEDOM  17
F-VALUE                      52.9
RESIDUAL ROOT MEAN SQUARE    0.07669506
RESIDUAL MEAN SQUARE         0.00588213
RESIDUAL SUM OF SQUARES      0.09999625
TOTAL SUM OF SQUARES         1.03390965
MULT. CORREL. COEF. SQUARED  .9033
```

ORDERED BY COMPUTER INPUT

IDENT.	OBSV.	WS DISTANCE	OBS. Y	FITTED Y	RESIDUAL
S LOSS	1	19.	1.623	1.644	-0.021
S LOSS	2	19.	1.568	1.643	-0.074
S LOSS	3	10.	1.568	1.515	0.053
S LOSS	4	0.	1.447	1.288	0.159
S LOSS	5	0.	1.255	1.233	0.022
S LOSS	6	1.	1.255	1.261	-0.006
S LOSS	7	1.	1.279	1.296	-0.017
S LOSS	8	1.	1.301	1.296	0.005
S LOSS	9	1.	1.176	1.201	-0.025
S LOSS	10	1.	1.146	1.054	0.092
S LOSS	11	1.	1.146	1.065	0.081
S LOSS	12	2.	1.114	1.037	0.077
S LOSS	13	1.	1.041	1.057	-0.015
S LOSS	14	1.	1.079	1.098	-0.019
S LOSS	15	5.	0.903	0.945	-0.042
S LOSS	16	5.	0.845	0.942	-0.097
S LOSS	17	5.	0.903	0.952	-0.049
S LOSS	18	5.	0.903	0.960	-0.057
S LOSS	19	4.	0.954	0.989	-0.035
S LOSS	20	4.	1.176	1.082	0.094
S LOSS	21	4.	1.176	1.303	-0.127

ORDERED BY RESIDUALS

OBSV.	OBS. Y	FITTED Y	ORDERED RESID.	SEQ
4	1.447	1.288	0.159	1
20	1.176	1.082	0.094	2
10	1.146	1.054	0.092	3
11	1.146	1.065	0.081	4
12	1.114	1.037	0.077	5
3	1.568	1.515	0.053	6
5	1.255	1.233	0.022	7
8	1.301	1.296	0.005	8
6	1.255	1.261	-0.006	9
13	1.041	1.057	-0.015	10
7	1.279	1.296	-0.017	11
14	1.079	1.098	-0.019	12
1	1.623	1.644	-0.021	13
9	1.176	1.201	-0.025	14
19	0.954	0.989	-0.035	15
15	0.903	0.945	-0.042	16
17	0.903	0.952	-0.049	17
18	0.903	0.960	-0.057	18
2	1.568	1.643	-0.074	19
16	0.845	0.942	-0.097	20
21	1.176	1.303	-0.127	21

Figure 5.7

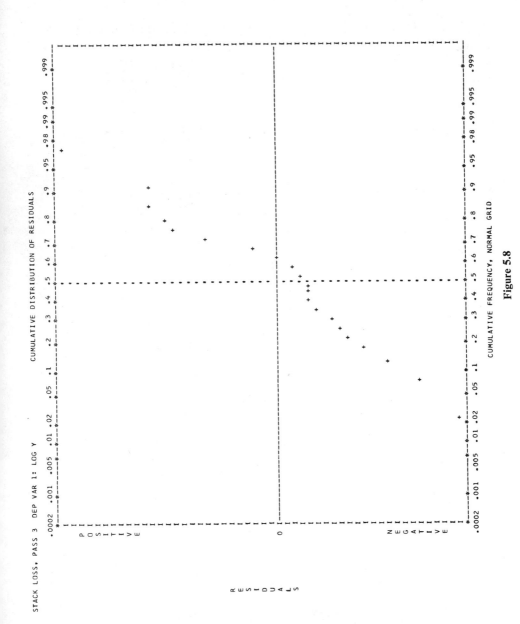

Figure 5.8

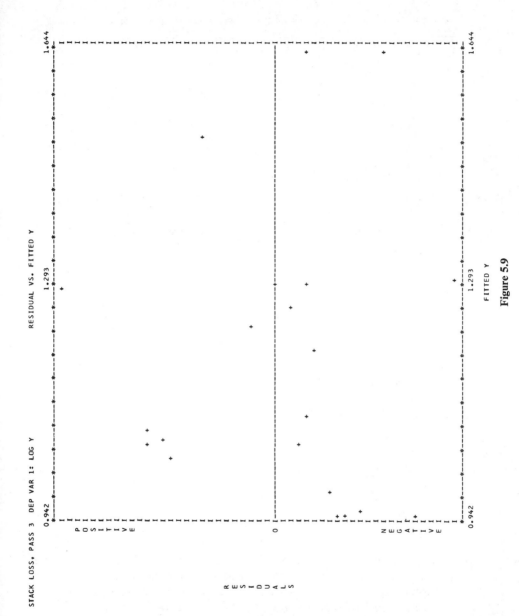

Figure 5.9

71

TABLE 5.2

EXAMINATION OF ALL ALTERNATIVES FOR SEVERAL INDEPENDENT REVISIONS

		Action			Result				
Pass	Factorial Symbol	Obs. 21 (A)	Response (B)	x_3 (C)	F-Value	R_y^2	RMS	Normal Plot	Residual vs. Fitted Y Plot
1	(1)	in	y	in	60	0.91	10.5	21 low	1, 2, 3, 4 outside points
2	a	out	y	in	99	0.95	6.6	Curved	1, 2, 3, 4 outside points
3	b	in	log y	in	55	0.90	0.0059	O.K.	O.K.
4	c	in	y	out	90	0.91	10.5	21 low	1, 2, 3, 4 outside points
5	ac	out	y	out	150	0.95	6.5	Curved	1, 2, 3, 4 outside points
6	bc	in	log y	out	83	0.90	0.0056	21 high	O.K.
7	ab	out	log y	in	66	0.92	0.0048	4 high	
8	abc	out	log y	out	102	0.92	0.0046	4 high	
			Observations 1, 3, 4, and 21 Omitted						
9	(1)	out	y	in	169	0.975	1.6	20 low	Curvature?
10	a	out	y	out	250	0.973	1.6	O.K.	Curvature?
11	b	out	log y	in	53	0.92	0.0032	20 high	Curvature?
12	ab	out	log y	out	83	0.92	0.0031	20 high 16, 2 low	Curvature?

through the estimation of average factorial effects on F, R_y^2, etc. It is clear that ac, which means "use y as response and drop *21* and x_3," gives highest F and R_y^2. Its nearest competitor is abc, which directs us to use log y as response, while dropping observation *21* and x_3.

5.5   Random Error Estimated from Near Neighbors

The results of Passes 1–8 supply clear directives as to factor x_3 and point *21*. Omitting x_3 gives no worse fit; dropping *21* produces a notable improvement. But the evidence has not sufficed to indicate clearly a choice between y and log y.

If log y is the right response, we expect the error to increase with y; in fact, the "per cent error in y" should be constant. If y is the correct response for our form of equation in the x_i, then the average error should be the same at all y. Now that x_3 has been dropped, we see that there are three sets of near replicates in the data as shown at the top of page 73.

If we dare assume that these three sets of consecutive duplicates are measuring random error among days, then two facts strike the eye. In the first

Observations	Residual SS	Degrees of Freedom	Mean Squares	Std. Dev.	Fitted Y-Value	Per Cent Error, s/Y
5, 6, 7, 8	2.8	3	0.93	1.0	21	5
10, 11, 12, 13, 14	6.8	4	1.70	1.3	12	11
15, 16, 17, 18, 19	2.0	4	0.50	0.7	8	9
Pooled	11.6	11	1.05	1.02		

place there is no evidence of trend in magnitude of error with Y. Second, the random error is much smaller than the value $s = 2.6$ of Pass 5 had led us to suppose. The trend in the estimated standard deviation, s (and hence in s^2), is not firmly ruled out, since we have only 11 degrees of freedom in all. But the new mean square of 1.05 gives definite information about lack of fit of our equation to the 20 data points.

$$\text{Mean square for lack of fit} = \frac{\text{RSS (Pass 5)} - \text{RSS (new replicates)}}{\text{df (Pass 5)} - \text{df (new replicates)}}$$

$$= (110.5 - 11.6)/(17 - 11) = 16.48,$$

$$F_{11}^{6} \text{ (lack of fit)} = \frac{(\text{Mean square for lack of fit})}{(\text{Mean square error})}$$

$$= 16.48/1.05$$

$$= 15.7,$$

$$F_{11}^{6} (0.0005) = 10.6.$$

We do not know yet whether this lack of fit is due to our having chosen a wrong equation form or to the presence of some "bad" data points.

5.6   Independence of Observations

Examining the second possibility first, we take a longer look at the data in the order of appearance of the points. It is distressing to see y drop from 42 to 37 with no change in x_1 or x_2 (points *1* and *2*). It is also upsetting to see y hold at 37 (point *3*) while x_1 drops from 80 to 75. Some light begins to appear on scanning the next five points, since x_1 is steady and y drops (in two days) to 18 and then holds fairly steady for four days. *There appears to be a lag of about one day in the response to a change in* x_1. Figure 5.10 shows all 21 points plotted in time order versus y, with x_1 and x_2 labeled for each point.

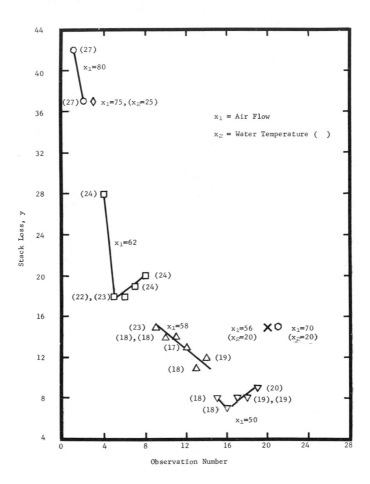

Figure 5.10 Sequence of observations.

We conclude that when x_1 was above 60 the plant required about a day to come to equilibrium ("line-out" is the term used by plant operators). We decide, therefore, to drop points *1*, *3*, *4*, and *21* permanently, since they correspond to transient states that cannot possibly be represented by the same equation as the "lined-out" points. We now return to the fray to see whether we can find an acceptable equation with satisfactory residual plots.

5.7   Systematic Examination of Alternatives Discovered Sequentially

Although we usually get our notions one at a time, it does not follow that we will find the best equation(s) by accepting or rejecting each notion after examining it *once*. It will often be possible to broaden our experience by looking at each new idea in the full context of all its predecessors.

Now that we have decided to drop four points because they are not representative, we must reconsider our equation for the remaining seventeen points. In strict analogy with factorial experimentation, we examine all the contingencies that we can *in a block* instead of one at a time. Thus, if we want to see the effects of logging y *and* of dropping x_3, we must make all four equation fits.

The lower panel of Table 5.2 gives the key statistics for each of the four equations under study. Now we see, as we did not before, that there is some hint of curvature in the residual versus fitted Y plots.

The term $b_{11}x_1^2$ is the natural one to choose, since x_1 is by far the more influential factor. But the curvature might be better represented by $b_{22}x_2^2$, by $b_{12}x_1x_2$, or by some combination of all three quadratic terms. Furthermore, we do not really know whether x_3 is uninfluential or has just been lying doggo beneath these quadratic terms. Finally, the dilemma of y versus $\log y$ is still with us.

We propose then 2^5 trials, each with a different equation, to test all combinations of x_1^2, x_2^2, x_1x_2, x_3, and $\log y$, keeping x_1 and x_2 in all equations. We will call these alternatives A, B, C, D, and E, respectively. Table 5.3 shows the residual sums of squares (RSS) in standard order. Those for $\log y$ obviously cannot be compared directly with those for y, but each set of sixteen can be internally compared. A more objective way of comparing large sets of RSS will be offered in Chapter 6; entirely by luck it is not needed here.

We note three equations—(1), b, and d—which give visibly poorer fits than the remainder, so we drop them forthwith. But there is something odd about the remaining thirteen equations in each set of sixteen: they all give very closely the same residual sums of squares. We would expect a drop of about 1.05 (the mean-square value from duplicates) when fitting an extra constant even to random responses, but nothing of the kind occurs. All thirteen values for y lie between 16.20 and 16.59. The values for $\log y$ are similarly restricted.

In the interests of simplicity we choose, as our best equation, a, that is,

$$Y = b_0 + b_1x_1 + b_2x_2 + b_{11}x_1^2,$$

TABLE 5.3

RESIDUAL SUMS OF SQUARES FOR 32 EQUATIONS

Observations 1, 3, 4, and 21 omitted, basic equation $= b_0 + b_1 x_1 + b_2 x_2$

$a = x_1^2$ in			$d = x_3$ in	
$b = x_2^2$ in			$e = \log y$ in place of y	
$c = x_1 x_2$ in			[x_1 and x_2 are always in]	

Pass	Factorial Symbol	RSS	Pass	Factorial Symbol	RSS
10	(1)	22.26	28	e	0.0434
13	a	16.44	29	ae	0.0211
14	b	18.34	30	be	0.0272
15	ab	16.43	31	abe	0.0209
16	c	16.59	32	ce	0.0201
17	ac	16.44	33	ace	0.0200
18	bc	16.59	34	bce	0.0201
19	abc	16.42	35	$abce$	0.0200
20	d	20.40	36	de	0.0420
21	ad	16.29	37	ade	0.0204
22	bd	16.29	38	bdc	0.0204
23	abd	16.20	39	$abde$	0.0204
24	cd	16.21	40	cde	0.0199
25	acd	16.21	41	$acde$	0.0199
26	bcd	16.18	42	$bcde$	0.0199
27	$abcd$	16.17	43	$abcde$	0.0199

since it gives us as good a fit as any of its competitors and has only one extra term. It should be noted that equation c (adding only the term $b_{12} x_1 x_2$) is almost as good.

Although equation a fits the present data far better than our old equation fitted the complete data, it is not a very nice fit anyway, as Figure 5.11 shows. A *residual* is attached to each point as it is located in x_1-x_2 space. We see immediately that the five largest residuals (both positive and negative) are close together in the middle of this plot. The largest residuals remain in these positions for all the other equations of the set tested. We conclude that the equation form we are using cannot accommodate this sort of wrinkle.

Nor do we think it worthwhile to press further. The data have been handled rather more rudely than is permissible without consultation with those who produced them. But the plant is long since shut down, and the process surely obsolete. Our methodological points have been made.

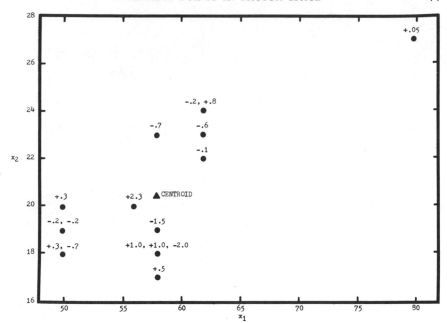

Figure 5.11   Stack loss residuals from $Y = b_0 + b_1 x_1 + b_2 x_2 + b_{11} x_1{}^2$. Observations 1, 3, 4, and 21 removed.

5.8   Remote Points in Factor Space

Figure 5.11 shows that point *2* is remote from the rest of the data. This remains true however x_1 and x_2 are scaled. (As already indicated in Chapter 4, it will be necessary in more complicated cases to weight each x_i by its importance, but this step is not needed here.)

A remote observation need not have a large *y*-residual. On the contrary, it will more often have a very small one, as in the present case. But point *2* was the first of the seventeen observations and may be suspected of controlling our estimates, b_1, b_2, and b_{11}. The only way to see if this is happening is to drop the point and repeat the whole computation. We have done this and find that b_1 and b_2 are only slightly changed while b_{11} drops to negligibility. The variances of the two b_i increase greatly. The reader is spared the documentation, but our conclusion is that point *2* is a valid one and should be retained.

5.9   Equation Using "Lined-Out" Data

We give the usual computer summary for our final equation in Figures 5.12–5.14 and a final summary in Table 5.4. Although we see nothing

LINEAR LEAST-SQUARES CURVE FITTING PROGRAM

STACK LOSS PASS 13   DEP VAR 1: S.LOSS   MIN Y = 7.000D 00   MAX Y = 3.700D 01   RANGE Y = 3.000D 01

Y = B(0) + B(1)X1 + B(2)X2 + B(3)X1SQRD
Y = STACK LOSS
X1 = AIR FLOW - MEAN VALUE OF AIR FLOW(57.7)
X2 = COOLING WATER TEMPERATURE - MEAN COOLING WATER TEMPERATURE(20.4)
1ST DAY OBSERVATIONS WITH FLOW GREATER THAN 60 OMITTED(1,3,4,21)

IND.VAR(I)	NAME	COEF.B(I)	S.E. COEF.	T-VALUE	R(II)SQRD	MIN X(I)	MAX X(I)	RANGE X(I)	REL.INF.X(I)
0		1.40685D 01							
1	A.FLOW	7.17691D-01	6.41D-02	11.2	0.6469	-7.700D 00	2.230D 00	3.000D 01	0.72
2	W.TEMP	5.27804D-01	1.50D-01	3.5	0.5750	-3.400D 00	6.600D 00	1.000D 01	0.18
3	FLOWSQ	6.81831D-03	3.18D-03	2.1	0.4351	9.000D-02	4.973D 02	4.972D 02	0.11

NO. OF OBSERVATIONS   17
NO. OF IND. VARIABLES   3
RESIDUAL DEGREES OF FREEDOM   13
F-VALUE   210.8
RESIDUAL ROOT MEAN SQUARE   1.12455407
RESIDUAL MEAN SQUARE   1.26462187
RESIDUAL SUM OF SQUARES   16.44008427
TOTAL SUM OF SQUARES   816.23529412
MULT. CORREL. COEF. SQUARED   .9799

-------ORDERED BY COMPUTER INPUT-------

IDENT.	OBSV.	WS DISTANCE	OBS. Y	FITTED Y	RESIDUAL
S LOSS	2	218.	37.000	36.947	-0.053
S LOSS	5	8.	18.000	18.125	-0.125
S LOSS	6	9.	18.000	18.653	-0.653
S LOSS	7	10.	19.000	19.181	-0.181
S LOSS	8	10.	20.000	19.181	0.819
S LOSS	9	2.	15.000	15.657	-0.657
S LOSS	10	1.	14.000	13.018	0.982
S LOSS	11	1.	14.000	13.018	0.982
S LOSS	12	3.	13.000	12.490	0.510
S LOSS	13	1.	11.000	13.018	-2.018
S LOSS	14	1.	12.000	13.546	-1.546
S LOSS	15	26.	8.000	7.680	0.320
S LOSS	16	26.	7.000	7.680	-0.680
S LOSS	17	25.	8.000	8.208	-0.208
S LOSS	18	25.	8.000	8.208	-0.208
S LOSS	19	25.	9.000	8.735	0.265
S LOSS	20	1.	15.000	12.657	2.343

-------ORDERED BY RESIDUALS-------

OBSV.	OBS. Y	FITTED Y	ORDERED RESID.	SEQ
20	15.000	12.657	2.343	1
10	14.000	13.018	0.982	2
11	14.000	13.018	0.982	3
8	20.000	19.181	0.819	4
12	13.000	12.490	0.510	5
15	8.000	7.680	0.320	6
19	9.000	8.735	0.265	7
2	37.000	36.947	-0.053	8
5	18.000	18.125	-0.125	9
7	19.000	19.181	-0.181	10
17	8.000	8.208	-0.208	11
18	8.000	8.208	-0.208	12
6	18.000	18.653	-0.653	13
9	15.000	15.657	-0.657	14
16	7.000	7.680	-0.680	15
14	12.000	13.546	-1.546	16
13	11.000	13.018	-2.018	17

Figure 5.12

STACK LOSS PASS 13   DEP VAR 1: S.LOSS   CUMULATIVE DISTRIBUTION OF RESIDUALS

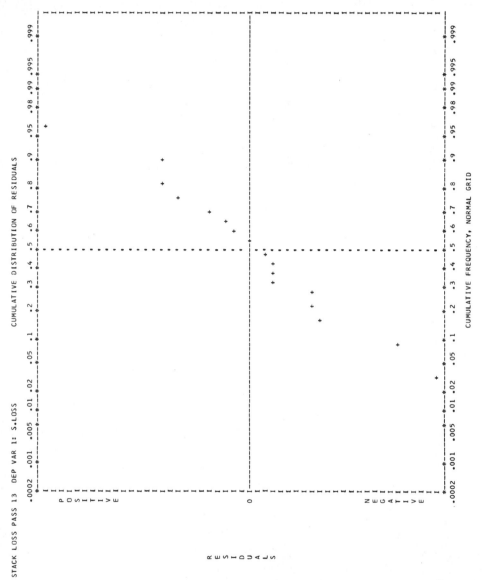

Figure 5.13

79

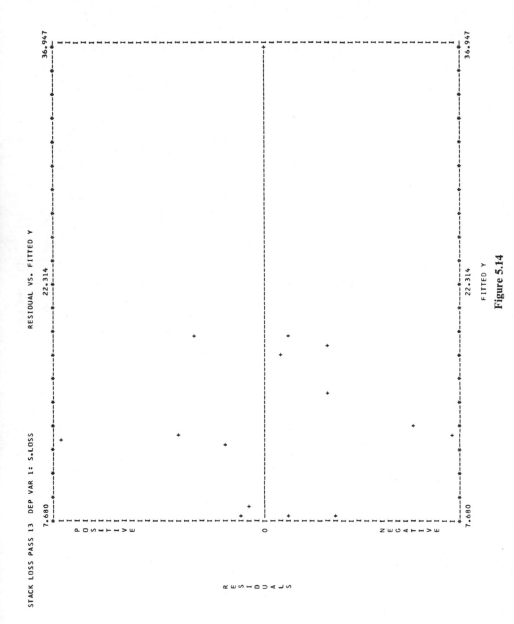

Figure 5.14

drastically wrong in any of the tables or plots, there is still some uncomfortable suggestion of lack of fit at the center of the data. We are now so near the ultimate error mean square of 1.05, however, that there is little hope of producing a notably better fit. (Our duplicates produce an expected residual sum of squares of 13 × 1.05 or 13.65. The current total RSS is 16.44; only 2.78, then, remains to be accounted for.)

TABLE 5.4

SUMMARY OF FINAL EQUATION OF STACK LOSS PROBLEM

$$Y = -15.6 - 0.06x_1 + 0.53x_2 + 0.0068x_1^2$$

Number of observations	17
Number of variables	3
F-value	211
Multiple correlation coefficient squared	0.98
Residual mean square	1.26
Normal plot	O.K.
Residual versus fitted Y plot	O.K.

i	Term	t_i	R_i^2	RI
1	x_1'	11.2	0.65	0.72
2	x_2'	3.5	0.57	0.18
3	$x_1'^2$	2.1	0.43	0.11

5.10   Conclusions on Stack Loss Problem

We conclude:

1. Of the twenty-one successive points given, four (*1, 3, 4, 21*) represent transitional states.

2. The remaining seventeen points are well fitted by either of the equations

$$Y_1 = 14.1 + 0.72x_1' + 0.53x_2' + 0.0068x_1'^2$$

$$= -15.4 - 0.07x_1 + 0.53x_2 + 0.0068x_1^2$$

or

$$Y_2 = 14.1 + 0.71x_1' + 0.51x_2' + 0.0254x_1'x_2',$$

where $x_1' = x_1 - 57.7$ and $x_2' = x_2 - 20.4$.

3. The residual error has a root-mean-square value of 1.1, with 14 degrees of freedom, inside the region of x_1-x_2 space spanned by the data (shown in Figure 5.11).

4. There is still some lack of fit in the five neighboring points *10, 11, 13, 14, 20*.

We return to these data in Chapter 7.

5. Retrospectively, we could have come to our final answer immediately after Passes 1 and 2 ($N = 21$, $K = 3, 2$)! The largest residuals (from the wrong equation, look you) are for points *1, 3, 4*, and *21*. If we had only had the perspicacity to ask then and there, "What is identifiable about these particular points?" we might have found our answer. They are at the highest Y-values and were taken immediately after changes in plant conditions. We have preferred to give a more nearly historical account of our trials and attendant fumblings in order to show the steps involved.

5.11 General Conclusions

1. Clues found in sequence should be tested all together. This often means that 2^L computer passes—or some balanced fraction thereof—should be made for L clues.

2. A better bound on random error can sometimes be obtained from near replicates (especially if these are not consecutive in time) than is given by the usual residual mean square from an empirical fitting equation.

3. Thorough search need not end with a single "best" equation.

4. The conclusions just given would be more persuasive if a larger set of data had been used in the example. In such larger sets, only drastic improvements should be accepted as proven.

CHAPTER 6

Selection of Independent Variables

6.1 Introduction

Assumptions

In the preceding chapters we have dealt with independent variables known to be influential. More strictly, we have been assuming that:

    a. All independent variables are known and are used in the fitting equation.*

    b. The form of the equation is known.

    c. The fitted equation is linear in its $(K + 1)$ unknown constants.

    d. The fitted equation is to be used only on a system that operates like the one producing the data, especially with regard to its correlation structure.

We now consider a modification of assumption *a*. We have a list of variables that quite surely include all the important factors, but we are not sure that every factor listed is really needed in the equation representing the data. We rely as before on others for assurance that the factors and their function forms [$x_1 = 1/T$, $x_2 = \log$ (space velocity), $x_3 = x_1 x_2$, etc.] are correct.

Obvious imperfections

Obvious imperfections in the data should of course be rectified before an elaborate search is made for the best equation or equations. But it is not possible to give a standard order of procedure that can be followed in all cases. Some defects will be visible at the very beginning; others will appear only after lengthy analysis. We mention here three sorts of defects, one familiar, two less well known.

---

* Some obviously uninfluential variables may have been dropped.

Extremely wild points are easy to spot. Often they are seen to represent conditions that are physically impossible. Many computer-oriented editing programs designed to spot such bad values are available. We do not recommend any "statistical" criterion for data deletion. The points that we are thinking of are those that would be removed as "impossible" by 99% of the technical experts in the field under study.

Since start-up difficulties are not rare in plants, initial runs or even early sets of data points are often not representable by the same equation as data collected subsequently. A plant that takes a long time to come to equilibrium after any large change in some factor (or factors) is really afflicted with repeated start-up difficulties. Chapter 5 has shown an example of this situation, and the reader has seen how tedious the spotting and elimination of such points may be.

Another common type of defect in plant data is "nesting." Some conditions (factor levels) are changed rarely; others, more frequently. It is unlikely that data taken under these conditions have uncorrelated errors. We will discuss a clear case of this defect in Chapter 8. Here it is only necessary to point out that *sorting* of the data table by each x_i will often suffice to spot nesting or to rule it out. The general method of equation choice proposed below may yield entirely wrong answers if the data are nested and if this fact is ignored.

2^K possible equations from K candidate variables

When there are K potential independent variables, there are 2^K possible equations. There will be one equation with none of the independent variables in it ($Y = \bar{y}$). There will be K equations each using one variable ($Y = b_0 + b_i x_i$). There will be $K(K - 1)/2$ equations using two variables, etc. Often there are several equations that represent the data equally well. Sometimes the simplest of these will be preferred. In other cases an equation with more factors will be chosen in the hope that it will avoid or decrease bias in the estimates of some constants.

On stepwise regression

One class of strategies for searching the 2^K possible equations is called by the general name "stepwise regression." There are two basic versions, forward and backward. In the forward version (Ralston and Wilf, Section 17*) independent variables are introduced one at a time. At each stage the variable which produces the greatest reduction in the residual sum of squares and whose $[b_i/s(b_i)]^2$ exceeds a preselected value is brought into the equation. The backward version begins with the full regression equation and eliminates the least important variables one at a time, using similar criteria.

* By M. A. Efroymson

Stepwise regression can lead to confusing results, however, when the independent variables are highly correlated (see the publications by Hamaker and by Oosterhoff). Moreover, it is based on the implicit assumption that there is *one* best equation and that stepwise regression will find it. In our experience there are often better equations with different sets of independent variables that are overlooked by this procedure.

F-test

The *F*-test is widely used to judge the need for adding to the equation a *set* of higher-order terms in the basic variables [see Brownlee, Section 13.8, and Davies (1958, Section 11B.32)]. The same procedure might be used to determine whether adding or deleting basic variables substantially affects the fit. But, as with stepwise regression, this search for essential variables is risky when the independent variables are correlated, because the result is not unique. Adding or deleting associated variables in different orders can lead to different equations.

All 2^K equations and fractions

The most comprehensive approach is to fit all 2^K equations and then to compare them. This is not a new idea, and even before the widespread use of electronic computers several authors published such results computed by hand (e.g., Hald, Section 20.3, and Moriceau). Now efficient programming permits the computation of 2^{12} residual sums of squares (equivalent to fitting 4096 equations) in less than ten seconds of IBM 360-65 computer time. (Space in the computer core, not time, is the limiting factor.)

There is, then, little point in looking for more efficient search methods when $K \leqslant 12$. By brute-strength application of our routines, it is possible to extend K to $(12 + Q)$ factors by doing 2^Q sets of size 2^{12} to exhaust all possibilities for small values of Q.

However, for searches with a large number of variables, two alternatives are recommended:

1. A $t_{K,i}$-directed search.
2. Fractional replication of the 2^K possible equations.

Neither is guaranteed to find all of the better-fitting equations, but both have worked well on all cases tried so far. When it is necessary to guarantee that all of the best equations have been found, the method described by Hocking and Leslie can be used for checking whether or not the list is complete.

6.2   Total Squared Error as a Criterion for Goodness of Fit—C_p

Definition

When a large number of alternative equations are being considered, it is imperative that some simple criterion of goodness (or badness) of fit be chosen to characterize each equation. We recommend a measure of "total squared error" given by Mallows. This statistic, called C_p, measures the sum of the *squared biases* plus the *squared random errors* in Y at all N data points. It is a simple function of the residual sum of squares from each fitting equation.

Derivation of the C_p statistic

The total squared error (bias plus random) for N data points, using a fitted equation with p terms (including b_0), is

$$\sum_{j=1}^{N} (\nu_j - \eta_j)^2 + \sum_{j=1}^{N} \text{Var.} \left\{ Y_{pj} \right\}$$

where $\nu_j = \nu(x_{1j}, x_{2j}, \ldots)$, expected value from *true* equation,

$\eta_j = \beta_0 + \sum_{i=1}^{k} \beta_i x_{ij}.$ expected value from the fitting equation being used,

$(\nu_j - \eta_j) =$ bias at the jth data point,
$p = k + 1$ when β_0 is present,
$\quad = k$ when β_0 is absent.

For convenience, let SSB_p stand for $\sum_{j=1}^{N} (\nu_j - \eta_j)^2$ and define a quantity, Γ_p, the standardized total squared error, by

(6.1) $$\Gamma_p = \frac{\text{SSB}_p}{\sigma^2} + \frac{1}{\sigma^2} \sum_{j=1}^{N} \text{Var.} (Y_{pj}).$$

It can be shown that:

(6.2) $$\sum_{j=1}^{N} \text{Var.} (Y_{pj}) = p\sigma^2.$$

Combining these equations gives:

(6.3) $$\Gamma_p = \frac{\text{SSB}_p}{\sigma^2} + p.$$

Now the residual sum of squares, RSS_p, from a p-term equation has the expectation:

$$E\{\text{RSS}_p\} = \text{SSB}_p + (N - p)\sigma^2.$$

Solving for SSB_p gives:

$$\text{SSB}_p = E\{\text{RSS}_p\} - (N - p)\sigma^2;$$

and so

$$\Gamma_p = \frac{E\{\text{RSS}_p\}}{\sigma^2} - (N - p) + p$$

$$= \frac{E\{\text{RSS}_p\}}{\sigma^2} - (N - 2p).$$

With a good estimate of σ^2 (call it s^2), the statistic C_p is an estimate of Γ_p.

$$(6.4) \qquad\qquad C_p = \frac{\text{RSS}_p}{s^2} - (N - 2p).$$

When a p-term equation has negligible bias, $\text{SSB}_p \approx 0$, $E\{\text{RSS}_p\} = (N - p)\sigma^2$, and

$$E\{C_p \mid \text{SSB}_p = 0\} = \frac{(N - p)\sigma^2}{\sigma^2} - (N - 2p) = p.$$

Mallows' graphical method of comparing fitted equations

If the C_p's for a number of equations are plotted against p, those for equations with small bias will tend to cluster about the line $C_p = p$ (Figure 6.1, point A), while those for equations with substantial bias will fall above the line (Figure 6.1, point B).* Although point B in Figure 6.1 is above the line $C_p = p$, it is still below point A and consequently represents an equation with slightly lower total error. Thus, adding terms may reduce bias, but at the expense of increasing the total variance of prediction for the N points and, consequently, the average variance per point. If an equation is needed for interpolation in the region of factor space spanned by the data points, it may pay to drop a few terms in order to obtain a simpler equation. In other words, it may pay to accept some bias in order to get a lower average error of prediction.

We need an unbiased estimate of σ^2 to calculate C_p. Often the mean square from the complete equation will serve this purpose. This forces C_p to be p_{max} or $(K + 1)$ for the complete equation. Substituting this estimate for σ^2 in the C_p calculation assumes that the complete equation has been carefully chosen to give negligible bias. A step is made in Chapter 7 toward finding a less biased estimator of σ^2 than s^2 from the complete equation.

* We see from the form of equation 6.4 that C_p may decrease by as much as 2 units as p decreases by 1 because of dropping a term from an equation. (The RSS cannot decrease on dropping a term; s^2 and N are fixed.)

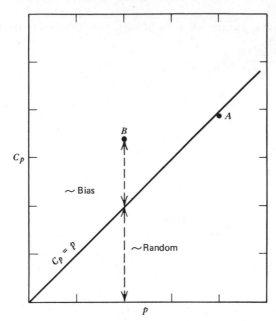

Figure 6.1 C_p plot.

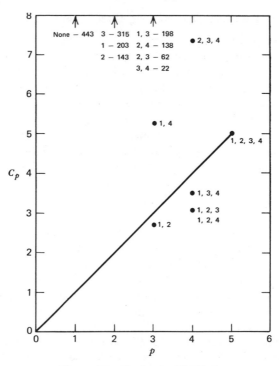

Figure 6.2 C_p plot for Hald's data.

88

6.3 A Four-Variable Example

Hald's book (Section 20.3) contains an interesting example involving four independent variables. The data, given in Appendix 6A.1, concern the heat evolved during the hardening of certain cements, as it depends on the percentage of four constituents. The equation is

$$Y = b_0 + b_1 x_1 + b_2 x_2 + b_3 x_3 + b_4 x_4.$$

Statistics for the complete equation and C_p's for all sixteen equations, using the four variables in all combinations, are given in Table 6.1.

It is easy to see from the listing (and from Figure 6.2) which are the

TABLE 6.1

STATISTICS FOR HALD'S DATA

$N = 13$

(b_0 included in all equations)

Term	Complete Equation				Simple Correlation Coefficients, r_{ii}'					
	b_i	$s(b_i)$	t_i	R_i^2		x_1	x_2	x_3	x_4	y
Constant	62.40									
x_1	1.55	0.74	2.08	0.97	x_1	1.0				
x_2	0.51	0.72	0.70	0.99	x_2	0.23	1.0			
x_3	0.10	0.75	0.14	0.98	x_3	−0.82	−0.14	1.0		
x_4	−0.14	0.71	0.20	0.99	x_4	−0.24	−0.97	0.03	1.0	
F-value	111.4				y	0.73	0.82	−0.53	−0.82	1.0
R_y^2	0.98									
RSS	47.9									
$s^2(y)$	5.98									

C_p's for All Equations

Variables in Equation	p	C_p	Variables in Equation	p	C_p
None	1	443.2	x_4	2	138.8
x_1	2	202.7	x_1, x_4	3	5.5
x_2	2	142.6	x_2, x_4	3	138.3
x_1, x_2	3	2.7	x_1, x_2, x_4	4	3.0
x_3	2	315.3	x_3, x_4	3	22.4
x_1, x_3	3	198.2	x_1, x_3, x_4	4	3.5
x_2, x_3	3	62.5	x_2, x_3, x_4	4	7.3
x_1, x_2, x_3	4	3.0	x_1, x_2, x_3, x_4	5	5.0

best-fitting equations and that should end our story. But we can hardly help noticing that, when factor x_1 is omitted, C_p is always greater than 7. It seems that we have computed at least half of our C_p's only to discard them. Table 6.1 shows that t_1 is 2.08, the largest of the four. If we had decided to retain factor x_1 in all equations, we would have computed only eight C_p's:

Variables in Equation		C_p
1	(0)	202.7
1	2	2.7*
1	3	198.2
1	23	3.0*
1	4	5.5
1	24	3.0*
1	34	3.5*
1	234	5.0

In this case, t_i is an excellent indicator; all four best C_p's are spotted in seven trials. We note that, if we had added the next largest t_i (only 0.7!), we would have gone directly to equation 12 and its neighbors 123 and 124, finding the winners immediately.

Even though the sixteen C_p-values constitute a 2^4 arrangement, there is little point in putting them through the usual Yates computation (see Davies, 1960, Sections 7.45 and 7.61) to find "effects of entering factors" and their interactions. Inspection of the first four values in Table 6.1 shows immediately that there is no hope of getting a simple factorial representation in terms of average effects of entering factors. Similarly, a short look at the eight C_p's including x_1—just given—tells us that factors 2, 3, and 4 do not have simple effects on C_p. We have all the information we need in the eight values as they stand.

Here is our conclusion, to be documented further, about interpreting sets of 2^K C_p-values and their fractions 2^{K-Q}. We do full factorial sets of C_p when we can, for the sake of completeness, not hoping to discover factors whose entry always produces the same decrease in C_p. When we must do some fraction of the full set, we do not use the usual fractions because these are bound to omit influential factors in half of their runs. The useful fractions are always in corners of the full experiment-space. *Clues to the location of these corners are always given by the factors that have largest t_i-values in the full equation.*

Hamaker used two standard "stepwise regression" methods on these data but found that the forward and backward procedures selected different variables. These data, which are only a part of a much larger set, are discussed again in Chapter 9.

6.4 A Six-Variable Example

a. Search of all 2^6 equations

The data that we shall consider [previously discussed by Gorman and Toman (1966)] are measurements of the "rate of rutting" of 31 experimental asphalt pavements, each prepared under different conditions specified by the values of five of the independent variables. An indicator variable is used to separate the results for 16 pavements tested in one set of runs from those for the 15 tested in the second set. The following equation was first used to fit the data:

$$Y = b_0 + b_1 x_1 + b_2 x_2 + b_3 x_3 + b_4 x_4 + b_5 x_5 + b_6 x_6$$

where $Y = $ log (change of rut depth in inches/million wheel passes),

$x_1 = $ log (viscosity of asphalt),

$x_2 = $ percentage of asphalt in surface course,

$x_3 = $ percentage of asphalt in base course,

$x_4 = $ indicator variable to separate two sets of runs,

$x_5 = $ percentage of fines in surface course, and

$x_6 = $ percentage of voids in surface course.

The original data are given in Figure 6B.1.

The objectives of the analysis were (1) to estimate the effects of the six individual factors, thus providing information for the design of an experiment, and (2) to find a simple interpolation formula for interim use. We will use the smallest equation that shows no serious lack of fit for the first objective, but will find that there is no "simple" equation which will satisfy the second objective.

The familiar statistics b_i, $s(b_i)$, t_i, R_i^2 and $r_{ii'}$ are given in Table 6.2, along with the complete set of sixty-four C_p-values for all subsets of the six factors. Using the usual 2^K factorial nomenclature, we indicate the presence of a factor by lower-case letters a, b, c, d, e, and f for $x_1, \ldots, x_6$, respectively, and so identify each equation by a "word" of six letters or less. Figure 6.3 shows all C_p less than 100 plotted versus p. It is obvious that a and b are present in all equations with C_p less than 18.

The list of C_p's contains a set of eight that are smallest (starred in the table). Once found, they are seen to have a simple structure: each contains abf plus a combination of c, d, and e. Figure 6.4, which shows these eight plotted against their respective p-values on a larger scale, makes it clear that $abdf$ is the best-fitting equation. Inclusion of d improves the fit over abf for all combinations of c and e. We see further that inclusion of c or e or both only adds random error and reduces bias negligibly.

TABLE 6.2
STATISTICS FOR SIX-VARIABLE EXAMPLE
$N = 31$
(b_0 included in all equations)

Complete Equation

Term	b_i	$s(b_i)$	t_i	R_i^2
Constant	−2.64	—	—	—
x_a	−0.51	0.073	7.0	0.91
x_b	0.50	0.12	4.3	0.32
x_c	0.10	0.14	0.7	0.32
x_d	−0.13	0.064	2.1	0.90
x_e	0.02	0.034	0.6	0.31
x_f	0.14	0.047	2.9	0.44

F-value 140
R_y^2 0.972
RSS 0.3070
$s^2(y)$ 0.01279

Simple Correlation Coefficients, r_{ii}'

	a	b	c	d	e	f	y
a	1.0						
b	−0.06	1.0					
c	−0.22	−0.26	1.0				
d	0.93	0.01	−0.12	1.0			
e	−0.30	−0.14	0.44	−0.23	1.0		
f	0.46	−0.43	−0.03	0.40	0.12	1.0	
y	−0.97	0.15	0.19	−0.93	0.32	−0.41	1.0

C_p's for All Equations

Variables in Equation	p	C_p	Variables in Equation	p	C_p	Variables in Equation	p	C_p	Variables in Equation	p	C_p
(1)—none	1	835.6	f	2	748.8	e	2	692	ef	3	575.5
a	2	20.43	af	3	21.68	ae	3	20.00	aef	4	21.87
b	2	817.0	bf	3	716.1	be	3	693.7	bef	4	577.4
ab	3	13.98	abf	4	14.29	abe	4	5.67*	abef	5	7.45*
c	2	806.7	cf	3	748.3	ce	3	666.8	cef	4	577.3
ac	3	21.90	acf	4	22.53	ace	4	21.37	acef	5	22.84
bc	3	770.0	bcf	4	709.0	bce	4	668.0	bcef	5	579.1
abc	4	15.95	abcf	5	16.17	abce	5	7.56*	abcef	6	9.43*
d	2	92.0	df	3	84.2	de	3	92.8	def	4	82.7
ad	3	20.53	adf	4	21.46	ade	4	20.61	adef	5	22.24
bd	3	70.23	bdf	4	57.56	bde	4	70.21	bdef	5	59.09
abd	4	12.22	abdf	5	11.32	abde	5	4.26*	abdef	6	5.51*
cd	3	88.6	cdf	4	85.1	cde	4	89.36	cdef	5	83.9
acd	4	22.42	acdf	5	22.86	acde	5	22.36	acdef	6	23.65
bcd	4	58.38	bcdf	5	53.92	bcde	5	56.83	bcdef	6	54.38
abcd	5	13.51	abcdf	6	13.24	abcde	6	5.31*	abcdef	7	7.00*

It is well to compare the coefficients of all the better equations in order to judge the consequences of dropping other factors. Table 6.3 shows all b_i for the better-fitting equations. We see that these effects are about the same for the four equations that contain a, b, d, and f (i.e., x_1, x_2, x_4, and x_6). But when d is dropped, b_1 changes considerably. The factor x_1 is then measuring $(x_1 + 1.007x_4)$, since x_1 and x_4 have nearly the same spread and also high correlation ($[44]/[11] = 1.14$; $r_{14} = 0.94$).

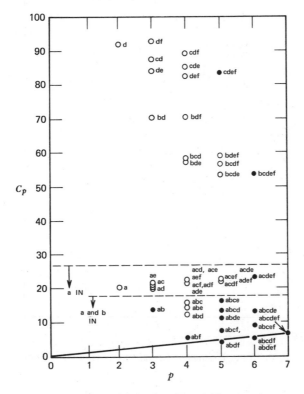

Figure 6.3 C_p plot for six-variable example.

Since equation *abdf* fits best, the harried data analyst might simply advise the experimenters to use it for prediction or extrapolation. Unfortunately, the equation has a serious shortcoming: x_4 is not reproducible. As you will recall, x_4 is the indicator variable that separates the two sets of runs. We do not know what caused the two runs to respond differently. Hence, we cannot set factor x_4 at either of its former levels. The data do show that there was a

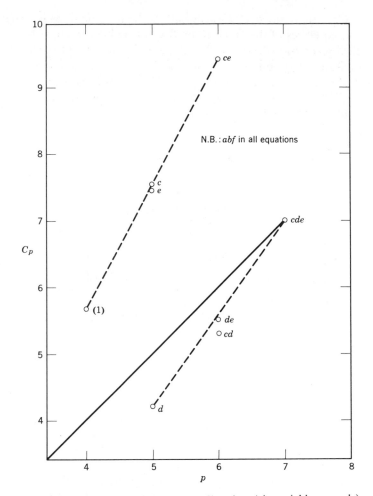

Figure 6.4   Plot of eight smallest C_p-values (six-variable example).

considerable difference between the log rutting rates in the two sets of data that is not accounted for by the change in asphalt viscosity. It is not possible to judge whether this discrepancy is a result of some difference between batches of asphalt, a change in tires used on the test machine, a change in some other raw material, or a combination of these. It will not do, either, to christen this variation a random one. All we can do at this time is to record the difference and study the data to see whether it is a consistent one.

In order to challenge the form of equation being used, we added the five cross-product terms—$x_i x_4$ with $i = 1, 2, 3, 5, 6$—in a separate computer

TABLE 6.3
COMPARISON OF EQUATIONS FOR SIX-VARIABLE EXAMPLE

	Equation									
	abf		*abdf*		*abcdf*		*abdef*		*abcdef*	
Term	b_i	(t_i)	b_i	(t_i)	b_i	(t_i)	b_i	(t_i)	b_i	(t_i)
Constant	−1.57	(—)	−1.85	(—)	−2.69	(—)	−2.05	(—)	−2.64	(—)
$x_1 = a$	−0.66	(25.5)	−0.55	(8.4)	−0.52	(7.4)	−0.52	(7.5)	−0.51	(7.0)
$x_2 = b$	0.43	(3.9)	0.46	(4.3)	0.50	(4.4)	0.47	(4.4)	0.50	(4.3)
$x_3 = c$	—	(—)	—	(—)	0.12	(1.0)	—	(—)	0.10	(0.7)
$x_4 = d$	—	(—)	−0.11	(1.9)	−0.13	(2.0)	−0.12	(2.0)	−0.13	(2.1)
$x_5 = e$	—	(—)	—	(—)	—	(—)	0.03	(0.9)	0.02	(0.6)
$x_6 = f$	0.15	(3.1)	0.14	(3.2)	0.14	(3.2)	0.13	(2.8)	0.14	(2.9)
s	0.112		0.111		0.111		0.112		0.113	
C_p	5.67		4.26		5.31		5.51		7.00	

pass, not shown. We found all t-values for the five new terms to be less than 1.0, while all the linear terms maintained their former importance. We also made separate least-squares fits for each of the two sets of data. Again we found no large differences in effects or in residual mean squares. Thus, we are left with *abdf* as our best equation for summarizing these data.

As estimates of the true effects of x_1, x_2, and x_6, we would recommend those given by the equation *abdf*. The usual statistics are all shown in the computer printout reproduced in Figure 6B.5. But the experimenter cannot safely use this equation for prediction, since he cannot set the level of x_4 in future data.

In regard to factors x_3 and x_5, which have been dropped, we must also hedge a little. If these factors had been varied over realistic ranges in these tests, we could boast that our estimated values, b_1, b_2, and b_6, have some degree of validity, since they hold over quite a range of experience. (In actual fact, wider variations of x_3 and x_5 were possible and, for x_3, profitable.)

The indefensible blocking of the present data serves only to provide us with a cautionary tale. It is not only post hoc that we would recommend *proper* blocking of any future runs. Since the test machine holds only sixteen pavement segments, it is obvious that future blocks will be at most of that size. Every factor must be varied in each block. The rules for doing this are not given here but are well known to design statisticians.

b. $t_{6,i}$-Directed search

Ordering the six (in general K) factors by their decreasing t-values (called $t_{K,i}$ in order to distinguish them from the t-values found in subsidiary equations, we have the results shown in the table. Computing directly the six C_p-values for equations 1, 12, 126, 1246, 12346, and 123456, we obtain the values given to the right in the same table. This method immediately gives us, at least in this example, the winner and two competitors (indicated by **).

i	Added Factor	$t_{k,i}$	Cumulative p	C_p
1	a	7.0	2	20.4
2	b	4.3	3	14.0
6	f	2.9	4	5.7**
4	d	2.1	5	4.3**
3	c	0.7	6	5.3**
5	e	0.6	7	7.0

Since $t_i^2 s^2$ measures the increase in the *fitting* sum of squares due to x_i, and since the *residual* sum of squares determines C_p, there is a simple identity relating the $t_{K,i}$ for factor x_i in the K-constant equation to the C_p (called now $C_{K,-i}$), measuring the total squared error made by the equation *with* x_i *removed*:

$$C_{K,-i} = t_{K,i}^2 + K - 1,$$

since $C_{K+1} = K + 1$.

Thus, if a $t_{K,i}$ as large as 3 appears, the $C_{K,-i}$ found on dropping the corresponding factor will be $(K + 8)$. The C_p's related to this one by further deletions of factors cannot be more than 2 units below $(K + 8)$ for each factor dropped. It will only be possible to arrive at a C_p roughly equal to p along this route by dropping at least 8 factors. If the uninfluential factors being dropped produce their expected decrease in the residual sum of squares of s^2 (rather than the minimum possible decrease of zero), then 16 factors will have to be dropped to negate the effect on C_p of dropping a factor with a $t_{K,i}$ of 3. We have not seen this happen, partly no doubt because we have not run many cases with 16 uninfluential factors.

By similar reasoning, a factor with a $t_{K,i}$ of 2 can be omitted from a well-fitting equation only if from three to six other factors are dropped along with it to bring C_p down to p or less. We will see several cases (not chosen for this reason) in which factors with $t_{K,i}$-values as small as 1 have remained in all acceptable equations.

c. Fractional replicates

The fractions of 2^K needed here are not the usual ones, since our aim is so different from that of the factorial experimenter. We are looking for subsets of small C_p-values. We are not looking for the factors whose presence uniformly reduces C_p by a constant amount over the whole system. There are, usually, no such factors. Indeed every correlation between two factors appears in the 2^K set of C_p as a two-factor interaction. It may well be that once we have included a controlling factor another set of less influential factors operates nearly additively. This subset will not usually be discovered by a balanced fraction, however, since even in the simplest case (one dominant factor) only half of the fraction gives information about the operation of the other factors in the favorable context. More commonly two or more factors are present in *all* favorable equations. We are then painted into a quadrant, an octant, or an *n*-tant of the full set of 2^K possibilities and correspondingly into a *fraction* of the balanced fraction usually chosen. It is only in this favorable "corner" that a fractional replicate may show which secondary factors are helpful, and which are not.

In the present case factors *a* and *b* (x_1 and x_2) are visibly important ($t_{K,i} = 7.0, 4.3$). As a further check, the C_p's for the equations dropping *a*, *b*, or both are calculated and compared with the value for the full equation:

	p	C_p
cdef	5	83.9
acdef	6	23.6
bcdef	6	54.4
abcdef	7	7.0

From these results it is plausible to conclude that only equations containing both *a* and *b* will be satisfactory.

Consider equation *acdef* with $C_p = 23.6$. The only way in which an equation with a lower C_p and p can be obtained is by eliminating variables that contribute little to the fit. Such eliminations will on the average reduce C_p by one unit (and cannot possibly reduce it by more than two units) per variable dropped. If all five variables are dropped from equation *acdef*, the value of C_p cannot be reduced by more than ten units, or to 13.6, which still exceeds 7.0, the value for the full equation. Hence, no equation containing *a* but not *b* can have a C_p less than 7.0.

There are 2^4 or 16 equations involving *c*, *d*, *e*, and *f* in all combinations, with *a* and *b* always included. For illustrative purposes we select a half-replicate of these, using the identifying letters of the equations as factorial treatment combinations. We choose the defining contrast I + *CDEF*.

		p	C_p
ab		3	13.98
ab	*cf*	5	7.56
ab	*df*	5	4.26
ab	*cd*	5	13.51
ab	*ef*	5	7.45
ab	*ce*	5	16.17
ab	*de*	5	11.32
ab	*cdef*	7	7.0

The usual Yates calculation (Davies, 1960, Section 7.45) is performed on the C_p's, as shown in the table. Effects C, D, E, etc., show how the presence of the corresponding variable affects the value of C_p. [The pattern, which shows clearly in the eight ordered C_p's in the table, is repeated in the

		p	C_p	I	II	III	III/4	Effects
ab		3	13.98	21.54	39.31	81.25	—	I + CDEF
ab	cf	5	7.56	17.77	41.94	7.23	1.8	C
ab	df	5	4.26	23.62	2.83	−9.07	−2.3	D
ab	cd	5	13.51	18.32	4.40	2.63	0.7	CD + EF
ab	ef	5	7.45	−6.42	−3.77	2.63	0.7	E
ab	ce	5	16.17	9.25	−5.30	1.57	0.4	CE + DF
ab	de	5	11.32	8.72	15.67	−1.53	−0.4	DE + CF
ab	cdef	7	7.00	−4.32	−13.04	−28.71	−7.2	F

half-replicate not used. The effects found also agree closely. This is a rather stringent test of our "factorial assumption," since the half used first contains the smallest C_p (4.26), whereas the other half does not.]

Effect F predominates and on the average reduces C_p by 7.2 units. One more check on the importance of F is made by calculating C_p for *abcde*; we get 13.2 (Table 6.3).

The average effect of inserting d (x_4) into equations containing *ab* is to reduce C_p by 2.3 units. If d were without effect in reducing the residual sum of squares at all, it would *increase* C_p by 1 unit. We have little doubt, then, that d is also a useful term. This assumption is tested in both directions by first dropping d from our best equation, *abdf*, to get *abf* ($C_p = 5.7$), and then by looking at the set of four equations that contain *abdf* along with c and/or e:

		C_p	Effect	
abdf	(1)	4.26		
	c	4.31	+1.27	C
	e	5.51	+1.47	E
	ce	7.00	+0.22	CE

Summarizing, we have found that *abdf* must be in any well-fitting equation, and that c and e then only add variance (the expected increase in C_p for a factor having no effect on bias is 1.0).

The purpose of factorial exploration is to locate variables common to all the better equations. The $t_{K,i}$-values for the coefficients in the complete equation are good first indicators, but variables picked in this way must be checked by wider sampling to make sure that they are essential. The remaining useful factors are found by using full factorials or fractional sampling plans. This approach has been used successfully on problems with as many as twenty-five variables.

In this discussion, we have used very small fractional replicates for expository purposes. In real life fractional replicates will of course always be much larger and involve much larger sets of factors. Appendix 6C gives three sets of fractions that may be found useful. It is perhaps needless (and certainly redundant) to remind the user of this appendix that these fractions are recommended for use on the set of *less* influential factors, with the more surely influential ones firmly fixed in all equations to be tested.

APPENDIX 6A

COMPUTER PRINTOUTS OF FOUR-VARIABLE EXAMPLE (page 100)

Pass	Figures
1	6A.1 and 6A.2
2	6A.3

APPENDIX 6B

COMPUTER PRINTOUTS OF SIX-VARIABLE EXAMPLE (page 103)

Pass	Figures
1	6B.1–6B.4
2	6B.5–6B.7

APPENDIX 6C

FRACTIONAL REPLICATION FOR 2^K EQUATIONS

Three sets of fractions of 2^K are given below. The first set is of "Resolution V." This means that all simple nonadditivities (usually called two-factor interactions and abbreviated 2fi) are estimable. Thus, if factors A and B (or x_1 and x_2) have small effects in reducing the total squared error, C_p, the data analyst may want to know whether their simultaneous presence in the fitting equation has a large effect. If this is the case, it is called an "AB interaction." It can be detected only by fitting all four equations:

Factors in Equation	Literal Designation	Equation
None	(1)	$Y = \bar{y}$
x_1	a	$Y = \bar{y} + b_1(x_1 - \bar{x}_1)$
x_2	b	$Y = \bar{y} + b_2(x_2 - \bar{x}_2)$
x_1, x_2	ab	$Y = \bar{y} + b_1(x_1 - \bar{x}_1) + b_2(x_2 - \bar{x}_2)$

(*continued on page 112*)

LINEAR LEAST-SQUARES CURVE FITTING PROGRAM

4 VARIABLES PASS 1

EQUATION 1 OF A MULTI-EQUATION PROBLEM

DATA READ WITH SPECIAL FORMAT

FORMAT CARD 1 (A6,I4,I2, 4F6.3, 30X, F6.3)

BZERO = CALCULATED VALUE

DATA INPUT   4 INDEPENDENT VARIABLES   1 DEPENDENT VARIABLES

OBSV.	SEQ.	1-11-21	2-12-22	3-13-23	4-14-24	5-15-25	6-16-26	7-17-27	8-18-28	9-19-29	10-20-30
1	1	7.000	26.000	6.000	60.000	78.500					
2	1	1.000	29.000	15.000	52.000	74.300					
3	1	11.000	56.000	8.000	20.000	104.300					
4	1	11.000	31.000	8.000	47.000	87.600					
5	1	7.000	52.000	6.000	33.000	95.900					
6	1	11.000	55.000	9.000	22.000	109.200					
7	1	3.000	71.000	17.000	6.000	102.700					
8	1	1.000	31.000	22.000	44.000	72.500					
9	1	2.000	54.000	18.000	22.000	93.100					
10	1	21.000	47.000	4.000	26.000	115.900					
11	1	1.000	40.000	23.000	34.000	83.800					
12	1	11.000	66.000	9.000	12.000	113.300					
13	1	10.000	68.000	8.000	12.000	109.400					

MEANS OF VARIABLES
7.46154D 00 4.81538D 01 1.17692D 01 3.00000D 01 9.54231D 01

ROOT MEAN SQUARES OF VARIABLES
5.88239D 00 1.55609D 01 6.40513D 00 1.67382D 01 1.50437D 01

SIMPLE CORRELATION COEFFICIENTS, R(I,I PRIME)

1	1.000	0.229	-0.824	-0.245	0.731
2	0.229	1.000	-0.139	-0.973	0.816
3	-0.824	-0.139	1.000	0.030	-0.535
4	-0.245	-0.973	0.030	1.000	-0.821
5	0.731	0.816	-0.535	-0.821	1.000

100

Figure 6A.1

LINEAR LEAST-SQUARES CURVE FITTING PROGRAM

4 VARIABLES PASS 1   DEP VAR 1: HEAT   MIN Y = 7.250D 01   MAX Y = 1.159D 02   RANGE Y = 4.340D 01

Y = B(0) + B(1)X1 + B(2)X2 + B(3)X3 + B(4)X4
Y = CUMULATIVE HEAT OF HARDENING AFTER 180 DAYS, CALORIES/GRAM OF CEMENT
X1 = PCT. TRICALCIUM ALUMINATE, C3A
X2 = PCT. TRICALCIUM SILICATE, C3S
X3 = PCT. CALCIUM ALUMIUM FERRATE, CAF
X4 = PCT. DICALCIUM SILICATE, C2S
NOTE: CLINKER COMPOSITION CALCULATED FROM ANALYSIS OF OXIDES.

IND.VAR(I)	NAME	COEF.B(I)	S.E. COEF.	T-VALUE	R(I)SQRD	MIN X(I)	MAX X(I)	RANGE X(I)	REL.INF.X(I)
0		6.24054D 01							
1	C3A	1.55110D 00	7.45D-01	2.1	0.9740	1.000D 00	2.100D 01	2.000D 01	0.71
2	C3S	5.10168D 00	7.24D-01	0.7	0.9961	2.600D 01	7.100D 01	4.500D 01	0.53
3	CAF	1.01909D-01	7.55D-01	0.1	0.9787	4.000D 00	2.300D 01	1.900D 01	0.04
4	C2S	-1.44061D-01	7.09D-01	0.2	0.9965	6.000D 00	6.000D 01	5.400D 01	0.18

NO. OF OBSERVATIONS	13
NO. OF IND. VARIABLES	4
RESIDUAL DEGREES OF FREEDOM	8
F-VALUE	111.5
RESIDUAL ROOT MEAN SQUARE	2.44600796
RESIDUAL MEAN SQUARE	5.98295492
RESIDUAL SUM OF SQUARES	47.86363935
TOTAL SUM OF SQUARES	2715.76307692
MULT. CORREL. COEF. SQUARED	.9824

-------ORDERED BY COMPUTER INPUT-------

IDENT.	OBSV.	WS DISTANCE	OBS. Y	FITTED Y	RESIDUAL
C 22	1	25.	78.500	78.495	0.005
E 23	2	34.	74.300	72.789	1.511
M 92	3	8.	104.300	105.971	-1.671
E 88	4	19.	87.600	89.327	-1.727
N 96	5	1.	95.900	95.649	0.251
T 85	6	7.	109.200	105.275	3.925
94	7	33.	102.700	104.149	-1.449
S 24	8	30.	72.500	75.675	-3.175
A 89	9	14.	93.100	91.722	1.378
M 90	10	74.	115.900	115.618	0.282
P 25	11	20.	83.800	81.809	1.991
L 95	12	20.	113.300	112.327	0.973
E 91	13	21.	109.400	111.694	-2.294

--------ORDERED BY RESIDUALS--------

OBSV.	OBS. Y	FITTED Y	ORDERED RESID.	SEQ
6	109.200	105.275	3.925	1
11	83.800	81.809	1.991	2
2	74.300	72.789	1.511	3
9	93.100	91.722	1.378	4
12	113.300	112.327	0.973	5
10	115.900	115.618	0.282	6
5	95.900	95.649	0.251	7
1	78.500	78.495	0.005	8
7	102.700	104.149	-1.449	9
3	104.300	105.971	-1.671	10
4	87.600	89.327	-1.727	11
13	109.400	111.694	-2.294	12
8	72.500	75.675	-3.175	13

Figure 6A.2

LINEAR LEAST-SQUARES CURVE FITTING PROGRAM

4 VARIABLES PASS 2 DEP VAR 1: HEAT   MIN Y = 7.250D 01 MAX Y = 1.159D 02 RANGE Y = 4.340D 01

Y = B(0) + B(1)X1 + B(2)X2
Y = CUMULATIVE HEAT OF HARDENING AFTER 180 DAYS, CALORIES/GRAM OF CEMENT
X1 = PCT. TRICALCIUM ALUMINATE, C3A
X2 = PCT. TRICALCIUM SILICATE, C3S
NOTE: CLINKER COMPOSITION CALCULATED FROM ANALYSIS OF OXIDES.

IND.VAR(I)	NAME	COEF.B(I)	S.E. COEF.	T-VALUE	R(I)SQRD	MIN X(I)	MAX X(I)	RANGE X(I)	REL.INF.X(I)
0		5.25773D 01							
1	C3A	1.46831D 00	1.21D-01	12.1	0.0522	1.000D 00	2.100D 01	2.000D 01	0.68
2	C3S	6.62250D-01	4.59D-02	14.4	0.0522	2.600D 01	7.100D 01	4.500D 01	0.69

NO. OF OBSERVATIONS   13
NO. OF IND. VARIABLES   2
RESIDUAL DEGREES OF FREEDOM   10
F-VALUE   229.5
RESIDUAL ROOT MEAN SQUARE   2.40633504
RESIDUAL MEAN SQUARE   5.79044832
RESIDUAL SUM OF SQUARES   57.90448318
TOTAL SUM OF SQUARES   2715.76307692
MULT. CORREL. COEF. SQUARED   .9787

		--------ORDERED BY COMPUTER INPUT--------			
IDENT.	OBSV.	WS DISTANCE	OBS. Y	FITTED Y	RESIDUAL
C 22	1	37.	78.500	80.074	-1.574
E 23	2	43.	74.300	73.251	1.049
M 92	3	9.	104.300	105.815	-1.515
E 88	4	27.	87.600	89.258	-1.658
N 96	5	1.	95.900	97.293	-1.393
T 85	6	8.	109.200	105.152	4.048
94	7	47.	102.700	104.002	-1.302
S 24	8	38.	72.500	74.575	-2.075
A 89	9	14.	93.100	91.275	1.825
M 90	10	68.	115.900	114.538	1.362
P 25	11	21.	83.800	80.536	3.264
L 95	12	29.	113.300	112.437	0.863
E 91	13	32.	109.400	112.293	-2.893

--------ORDERED BY RESIDUALS--------				
OBSV.	OBS. Y	FITTED Y	ORDERED RESID.	SEQ
6	109.200	105.152	4.048	1
11	83.800	80.536	3.264	2
9	93.100	91.275	1.825	3
10	115.900	114.538	1.362	4
2	74.300	73.251	1.049	5
12	113.300	112.437	0.863	6
7	102.700	104.002	-1.302	7
5	95.900	97.293	-1.393	8
3	104.300	105.815	-1.515	9
1	78.500	80.074	-1.574	10
4	87.600	89.258	-1.658	11
8	72.500	74.575	-2.075	12
13	109.400	112.293	-2.893	13

Figure 6A.3

LINEAR LEAST-SQUARES CURVE FITTING PROGRAM

6 VARIABLES PASS 1

EQUATION 1 OF A MULTI-EQUATION PROBLEM

DATA READ WITH STANDARD FORMAT (A6, I4, I2, 10F6.3)

3ZERO = CALCULATED VALUE

DATA INPUT   6 INDEPENDENT VARIABLES   1 DEPENDENT VARIABLES

OBSV.	SEQ.	1-11-21	2-12-22	3-13-23	4-14-24	5-15-25	6-16-26	7-17-27	8-18-28	9-19-29	10-20-30
1	1	2.800	4.680	4.870	-1.000	8.400	4.916	6.750			
2	1	1.400	5.190	4.500	-1.000	6.500	4.563	13.000			
3	1	1.400	4.820	4.730	-1.000	7.900	5.321	14.750			
4	1	3.300	4.850	4.760	-1.000	8.300	4.865	12.600			
5	1	1.700	4.860	4.950	-1.000	8.400	3.776	8.250			
6	1	2.900	5.160	4.450	-1.000	7.400	4.397	10.670			
7	1	3.700	4.820	5.050	-1.000	6.800	4.867	7.280			
8	1	1.700	4.860	4.700	-1.000	8.600	4.828	12.670			
9	1	0.920	4.780	4.840	-1.000	6.700	4.865	12.580			
10	1	0.680	5.160	4.760	-1.000	7.700	4.034	20.600			
11	1	6.000	4.570	4.820	-1.000	7.400	5.450	3.580			
12	1	4.300	4.610	4.650	-1.000	6.700	4.853	7.000			
13	1	0.600	5.070	5.100	-1.000	7.500	4.257	26.200			
14	1	1.800	4.660	5.090	-1.000	8.200	5.144	11.670			
15	1	6.000	5.420	4.410	-1.000	5.800	3.718	7.670			
16	1	4.400	5.010	4.740	-1.000	7.100	4.715	12.250			
17	1	88.000	4.970	4.660	1.000	6.500	4.625	0.760			
18	1	62.000	5.010	4.720	1.000	8.000	4.977	1.350			
19	1	50.000	4.960	4.900	1.000	6.800	4.322	1.440			
20	1	58.000	5.200	4.700	1.000	8.200	5.087	1.600			
21	1	90.000	4.800	4.600	1.000	6.600	5.971	1.100			
22	1	66.000	4.980	4.690	1.000	6.400	4.647	0.850			
23	1	140.000	5.350	4.760	1.000	7.300	5.115	1.200			
24	1	240.000	5.040	4.800	1.000	7.800	5.939	0.560			
25	1	420.000	4.800	4.800	1.000	7.400	5.916	0.720			
26	1	500.000	4.830	4.600	1.000	6.700	5.471	0.470			
27	1	180.000	4.660	4.720	1.000	7.200	4.602	0.330			
28	1	270.000	4.670	4.500	1.000	6.300	5.043	0.260			
29	1	170.000	4.720	4.700	1.000	6.800	5.075	0.760			
30	1	98.000	5.000	5.070	1.000	7.200	4.334	0.800			
31	1	35.000	4.700	4.800	1.000	7.700	5.705	2.000			

Figure 6B.1

103

DATA TRANSFORMATIONS

POSITION	CODE	OPERATION
1		COMMON LOG
2	2	COMMON LOG
3		NONE
4		NONE
5		NONE
6		NONE
7	2	COMMON LOG

CONSTANT	LOCATION	OMIT	VARIABLE
	1	0	1
	2	0	2
	3	0	3
	4		4
	5		5
	6	0	6
	7	0	7

DATA AFTER TRANSFORMATIONS   THE FITTED EQUATION HAS   6 INDEPENDENT VARIABLES,   1 DEPENDENT VARIABLES

	A	B	C	D	E	F	LOG Y			
OBSV.	1-11-21	2-12-22	3-13-23	4-14-24	5-15-25	6-16-26	7-17-27	8-18-28	9-19-29	10-20-30
1	4.47158D-01	4.68000D 00	4.87000D 00	-1.00000D 00	8.40000D 00	4.91600D 00	8.29304D-01			
2	1.46128D-01	5.19000D 00	4.50000D 00	-1.00000D 00	6.50000D 00	4.56280D 00	1.11394D 00			
3	1.46128D-01	4.82000D 00	4.73000D 00	-1.00000D 00	7.90000D 00	5.32080D 00	1.16879D 00			
4	5.18514D-01	4.85000D 00	4.76000D 00	-1.00000D 00	8.30000D 00	4.86550D 00	1.10037D 00			
5	2.30449D-01	4.86000D 00	4.95000D 00	-1.00000D 00	8.40000D 00	3.77560D 00	9.16454D-01			
6	4.62398D-01	5.16000D 00	4.45000D 00	-1.00000D 00	7.40000D 00	4.39720D 00	1.02816D 00			
7	5.68202D-01	4.82000D 00	5.05000D 00	-1.00000D 00	6.80000D 00	4.86700D 00	8.62131D-01			
8	2.30449D-01	4.86000D 00	4.70000D 00	-1.00000D 00	8.60000D 00	4.82840D 00	1.10278D 00			
9	-3.62122D-02	4.78000D 00	4.84000D 00	-1.00000D 00	6.70000D 00	4.86540D 00	1.09968D 00			
10	-1.67491D-01	5.16000D 00	4.76000D 00	-1.00000D 00	7.70000D 00	4.03430D 00	1.31387D 00			
11	7.78151D-01	4.57000D 00	4.82000D 00	-1.00000D 00	7.40000D 00	5.45050D 00	5.53383D-01			
12	6.33468D-01	4.61000D 00	4.65000D 00	-1.00000D 00	7.50000D 00	4.85260D 00	8.45098D-01			
13	-2.21849D-01	5.07000D 00	5.10000D 00	-1.00000D 00	5.80000D 00	4.25740D 00	1.41830D 00			
14	2.55273D-01	4.66000D 00	5.09000D 00	1.00000D 00	8.20000D 00	5.14380D 00	1.06707D 00			
15	7.78151D-01	5.42000D 00	4.41000D 00	1.00000D 00	7.10000D 00	3.71800D 00	8.84795D-01			
16	6.43453D-01	5.01000D 00	4.74000D 00	1.00000D 00	6.50000D 00	4.71460D 00	1.08814D 00			
17	1.94448D 00	4.97000D 00	4.66000D 00	1.00000D 00	8.00000D 00	4.62550D 00	1.19186D-01			
18	1.79239D 00	5.01000D 00	4.72000D 00	1.00000D 00	8.00000D 00	4.97690D 00	1.30334D-01			
19	1.69897D 00	4.96000D 00	4.90000D 00	1.00000D 00	6.80000D 00	4.32220D 00	1.58362D-01			
20	1.76343D 00	5.20000D 00	4.70000D 00	1.00000D 00	8.20000D 00	5.08740D 00	2.04120D-01			
21	1.95424D 00	4.98000D 00	4.60000D 00	1.00000D 00	6.60000D 00	5.97080D 00	4.13927D-02			
22	1.81954D 00	5.35000D 00	4.76000D 00	1.00000D 00	6.40000D 00	4.64660D 00	-7.05811D-02			
23	2.14613D 00	5.04000D 00	4.80000D 00	1.00000D 00	7.30000D 00	5.11540D 00	7.91812D-02			
24	2.38021D 00	5.04000D 00	4.80000D 00	1.00000D 00	7.80000D 00	5.93940D 00	-2.51812D-01			
25	2.62325D 00	4.83000D 00	4.60000D 00	1.00000D 00	7.40000D 00	5.91630D 00	-1.42668D-01			
26	2.69897D 00	4.66000D 00	4.72000D 00	1.00000D 00	6.70000D 00	5.47090D 00	-3.27902D-01			
27	2.55270D 00	4.72000D 00	4.50000D 00	1.00000D 00	7.20000D 00	5.04320D 00	-5.85027D-01			
28	2.43136D 00	4.67000D 00	4.70000D 00	1.00000D 00	6.30000D 00	5.07480D 00	-1.19186D-01			
29	2.23045D 00	4.72000D 00	4.70000D 00	1.00000D 00	6.80000D 00	4.33360D 00	-9.69100D-02			
30	1.99123D 00	5.00000D 00	5.07000D 00	1.00000D 00	7.20000D 00	4.33360D 00	3.01030D-01			
31	1.54407D 00	4.70000D 00	4.80000D 00	1.00000D 00	7.70000D 00	5.70510D 00				

MEANS OF VARIABLES
1.18343D 00 4.91000D 00 4.75613D 00 -3.22581D-02 7.30000D 00 4.88386D 00 4.87498D-01

ROOT MEAN SQUARES OF VARIABLES
9.44940D-01 2.17286D-01 1.76459D-01 1.01600D 00 7.28926D-01 5.78012D-01 6.07145D-01

SIMPLE CORRELATION COEFFICIENTS, R(I,I PRIME)

1	1.000	-0.058	-0.217	0.939	-0.303	0.469	-0.972
2	-0.058	1.000	-0.263	0.012	-0.138	-0.432	0.154
3	-0.217	-0.263	1.000	-0.120	0.437	-0.027	0.189
4	0.939	0.012	-0.120	1.000	-0.234	0.405	-0.929
5	-0.303	-0.138	0.437	-0.234	1.000	0.116	0.320
6	0.469	-0.432	-0.027	0.405	0.116	1.000	-0.410
7	-0.972	0.154	0.189	-0.929	0.320	-0.410	1.000

Figure 6B.1 (*continued*)

104

LINEAR LEAST-SQUARES CURVE FITTING PROGRAM

6 VARIABLES PASS 1   DEP VAR 1: LOG Y   MIN Y = -5.850D-01   MAX Y = 1.418D 00   RANGE Y = 2.003D 00

$Y = B(0) + B(1)X1 + B(2)X2 + B(3)X3 + B(4)X4 + B(5)X5 + B(6)X6$
Y = LOG(RATE OF CHANGE OF RUT DEPTH IN INCHES / MILLION WHEEL PASSES)
X1 = A, LOG(VISCOSITY OF ASPHALT)
X2 = B, PER CENT ASPHALT IN SURFACE COURSE
X3 = C, PER CENT ASPHALT IN BASE COURSE
X4 = D, INDICATOR VARIABLE TO SEPARATE TWO SETS OF RUNS
X5 = E, PER CENT FINES IN SURFACE COURSE
X6 = F, PER CENT VOIDS IN SURFACE COURSE

IND.VAR(I)	NAME	COEF.B(II)	S.E. COEF.	T-VALUE	R(II)SQRD	MIN X(I)	MAX X(I)	RANGE X(I)	REL.INF.X(I)
0		-2.64431D 00							
1	A	-5.13307D-01	7.31D-02	7.0	0.9105	-2.218D-01	2.699D 00	2.921D 00	0.75
2	B	4.97988D-01	1.15D-01	4.3	0.3210	4.570D 00	5.420D 00	8.500D-01	0.21
3	C	1.01094D-01	1.42D-01	0.7	0.3191	4.410D 00	5.100D 00	6.900D-01	0.03
4	D	-1.34394D-01	6.39D-02	2.1	0.8989	-1.000D 00	1.000D 00	2.0000D 00	0.13
5	E	1.88562D-02	3.42D-02	0.6	0.3150	5.800D 00	8.600D 00	2.800D 00	0.03
6	F	1.37463D-01	4.79D-02	2.9	0.4439	3.718D 00	5.971D 00	2.253D 00	0.15

NO. OF OBSERVATIONS   31
NO. OF IND. VARIABLES   6
RESIDUAL DEGREES OF FREEDOM   24
F-VALUE   140.1
RESIDUAL ROOT MEAN SQUARE   0.11312210
RESIDUAL MEAN SQUARE   0.01279208
RESIDUAL SUM OF SQUARES   0.30701002
TOTAL SUM OF SQUARES   11.05875356
MULT. CORREL. COEF. SQUARED   .9722

		ORDERED BY COMPUTER INPUT					ORDERED BY RESIDUALS			
IDENT.	OBSV.	WS DISTANCE	OBS. Y	FITTED Y	RESIDUAL	OBSV.	OBS. Y	FITTED Y	ORDERED RESID.	SEQ
108	1	14.	0.829	0.918	-0.088	16	1.088	0.916	0.172	1
109	2	25.	1.114	1.204	-0.090	4	1.100	0.946	0.155	2
110	3	24.	1.169	1.174	-0.005	25	-0.143	-0.297	0.154	3
111	4	11.	1.100	0.946	0.155	29	-0.119	-0.272	0.153	4
112	5	22.	0.916	0.970	-0.053	19	0.158	0.037	0.121	5
113	6	14.	1.028	1.016	0.012	12	0.845	0.724	0.121	6
114	7	9.	0.862	0.907	-0.044	31	0.301	0.184	0.117	7
115	8	20.	1.103	1.093	0.010	26	-0.328	-0.415	0.088	8
116	9	32.	1.100	1.173	-0.074	13	1.418	1.371	0.047	9
117	10	41.	1.314	1.327	-0.013	15	0.885	0.856	0.029	10
118	11	7.	0.554	0.743	-0.189	21	0.041	0.019	0.023	11
119	12	9.	0.845	0.724	0.121	18	0.130	0.108	0.022	12
120	13	43.	1.418	1.371	0.047	6	1.028	1.016	0.012	13
121	14	20.	1.067	1.056	0.011	14	1.067	1.056	0.011	14
122	15	12.	0.885	0.856	0.029	8	1.103	1.093	0.010	15
123	16	8.	1.088	0.916	0.172	3	1.169	1.174	-0.005	16
124	17	14.	-0.119	-0.072	-0.047	10	1.314	1.327	-0.013	17
125	18	9.	0.130	0.108	0.022	23	0.079	0.106	-0.027	18
126	19	8.	0.158	0.037	0.121	30	-0.097	-0.067	-0.030	19
127	20	10.	0.204	0.235	-0.031	20	0.204	0.235	-0.031	20
128	21	16.	0.041	0.019	0.023	7	0.862	0.907	-0.044	21
129	22	10.	-0.071	0.001	-0.071	17	-0.119	-0.072	-0.047	22
130	23	24.	0.079	0.106	-0.027	5	0.916	0.970	-0.053	23
131	24	33.	-0.252	-0.042	-0.210	22	-0.071	0.001	-0.071	24
132	25	24.	-0.143	-0.297	0.154	9	1.100	1.173	-0.074	25
133	26	33.	-0.328	-0.415	0.088	1	0.829	0.918	-0.088	26
134	27	46.	-0.585	-0.434	-0.151	2	1.114	1.204	-0.090	27
135	28	49.	-0.481	-0.370	-0.111	28	-0.481	-0.370	-0.111	28
136	29	26.	-0.119	-0.272	0.153	27	-0.585	-0.434	-0.151	29
137	30	35.	-0.097	-0.067	-0.030	11	0.554	0.743	-0.189	30
138	31	25.	0.301	0.184	0.117	24	-0.252	-0.042	-0.210	31

Figure 6B.2

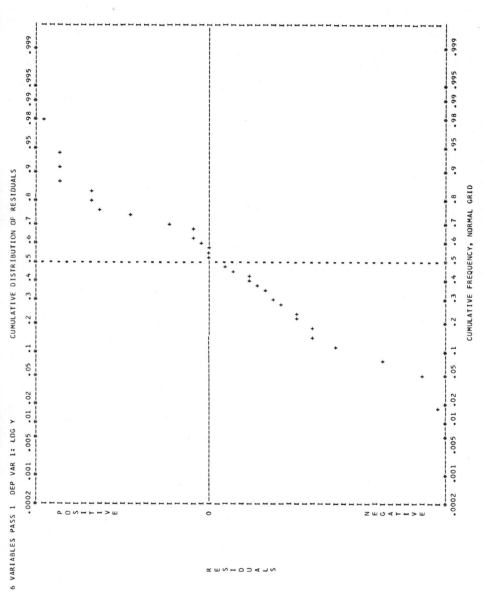

Figure 6B.3

106

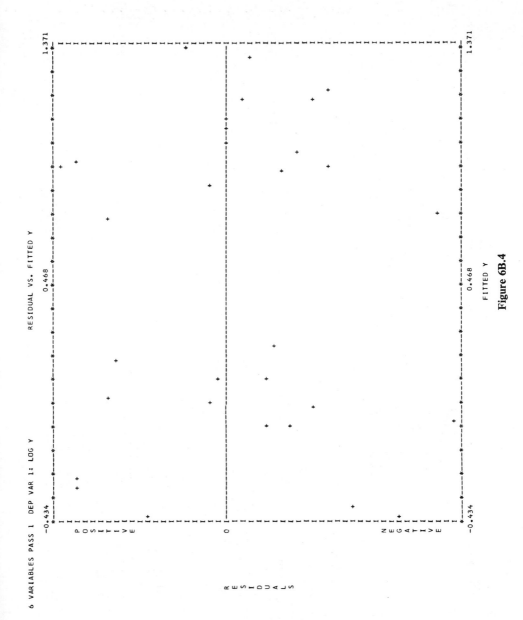

Figure 6B.4

107

LINEAR LEAST-SQUARES CURVE FITTING PROGRAM

6 VARIABLES PASS 2 DEP VAR 1: LOG Y   MIN Y = -5.850D-01   MAX Y = 1.418D 00   RANGE Y = 2.003D 00

$Y = B(0) + B(1)X1 + B(2)X2 + B(3)X4 + B(4)X6$
Y = LOG(RATE OF CHANGE OF RUT DEPTH IN INCHES / MILLION WHEEL PASSES)
X1 = A, LOG(VISCOSITY OF ASPHALT)
X2 = B, PER CENT ASPHALT IN SURFACE COURSE
X4 = D, INDICATOR VARIABLE TO SEPARATE TWO SETS OF RUNS
X6 = F, PER CENT VOIDS IN SURFACE COURSE

IND.VAR(I)	NAME	COEF.B(II)	S.E. COEF.	T-VALUE	R(II)SQRD	MIN X(I)	MAX X(I)	RANGE X(I)	REL.INF.X(I)
1	A	-1.85226D 00	6.55D-02	8.4	0.8918	-2.218D-01	2.699D 00	2.921D 00	0.80
2	B	-5.47451D-01	1.07D-01	4.3	0.2331	4.576D 00	5.420D 00	8.500D-01	0.20
3	D	-1.11184D-01	5.94D-02	1.9	0.8863	-1.000D 00	1.000D 00	2.000D 00	0.11
4	F	1.43649D-01	4.49D-02	3.2	0.3847	3.718D 00	5.971D 00	2.253D 00	0.16

```
NO. OF OBSERVATIONS              31
NO. OF IND. VARIABLES            4
RESIDUAL DEGREES OF FREEDOM      26
F-VALUE                          216.0
RESIDUAL ROOT MEAN SQUARE        0.11148083
RESIDUAL MEAN SQUARE             0.01242798
TOTAL SUM OF SQUARES             0.32312739
MULT. CORREL. COEF. SQUARED      .9708
```

COMPONENT EFFECT OF EACH VARIABLE ON EACH OBSERVATION (IN UNITS OF Y)
(VARIABLES ORDERED BY THEIR RELATIVE INFLUENCE --- OBSERVATIONS ORDERED BY INFLUENCE OF MOST INFLUENTIAL VARIABLE)

SEQ.	OBSV.	1 A	2 B	4 F	3 D
1	13	0.77	0.07	-0.09	0.11
2	10	0.74	0.12	-0.12	0.11
3	9	0.67	-0.06	-0.00	0.11
4	2	0.57	0.13	-0.05	0.11
5	3	0.57	-0.04	0.06	0.11
6	5	0.52	-0.02	-0.16	0.11
7	8	0.52	-0.02	-0.01	0.11
8	14	0.51	-0.12	0.04	0.11
9	1	0.40	-0.11	0.00	0.11
10	6	0.39	0.12	-0.07	0.11
11	4	0.36	-0.03	-0.00	0.11
12	7	0.34	-0.04	-0.00	0.11
13	12	0.30	-0.14	-0.00	0.11
14	16	0.30	0.05	-0.02	0.11
15	11	0.22	-0.16	0.08	0.11
16	15	0.22	0.24	-0.17	0.11
17	31	-0.20	-0.10	-0.12	-0.11
18	19	-0.28	0.02	-0.08	-0.11
19	20	-0.32	0.13	0.03	-0.11
20	18	-0.33	0.05	0.01	-0.11
21	22	-0.35	0.03	-0.03	-0.11
22	17	-0.42	0.03	-0.04	-0.11
23	21	-0.42	-0.05	0.16	-0.11
24	30	-0.44	0.04	-0.08	-0.11
25	23	-0.53	-0.20	0.03	-0.11
26	29	-0.57	-0.09	0.03	-0.11
27	27	-0.59	-0.12	-0.04	-0.11
28	24	-0.66	0.06	0.15	-0.11
29	28	-0.68	-0.11	0.02	-0.11
30	25	-0.79	-0.05	0.15	-0.11
31	26	-0.83	-0.04	0.08	-0.11

Figure 6B.5

LINEAR LEAST-SQUARES CURVE FITTING PROGRAM

6 VARIABLES   PASS 2   DEP VAR 1: LOG Y

ORDERED BY INPUT — ORDERED BY WS DISTANCE (COMPUTER)

IDENT.	OBSV.	WS DISTANCE	OBS. Y	FITTED Y	RESIDUAL
108	1	15.	0.829	0.896	-0.067
109	2	28.	1.114	1.247	-0.133
110	3	27.	1.169	1.184	-0.015
111	4	12.	1.100	0.929	0.172
112	5	25.	0.916	0.934	-0.018
113	6	15.	1.028	1.036	-0.008
114	7	10.	0.862	0.888	-0.026
115	8	23.	1.103	1.086	0.017
116	9	37.	1.100	1.200	-0.100
117	10	47.	1.314	1.329	-0.015
118	11	7.	0.554	0.740	-0.186
119	12	10.	0.845	0.752	0.093
120	13	50.	1.418	1.349	0.069
121	14	23.	1.067	1.024	0.043
122	15	12.	0.885	0.887	-0.002
123	16	8.	1.088	0.913	0.175
124	17	15.	-0.119	-0.053	-0.066
125	18	10.	0.130	0.099	0.031
126	19	8.	0.158	0.033	0.125
127	20	11.	0.204	0.219	-0.015
128	21	18.	0.041	0.056	-0.014
129	22	11.	-0.071	0.023	-0.094
130	23	27.	0.079	0.083	-0.004
131	24	38.	-0.252	-0.070	-0.181
132	25	53.	-0.143	-0.318	0.176
133	26	57.	-0.328	-0.410	0.082
134	27	30.	-0.481	-0.371	-0.111
135	28	40.	-0.585	-0.399	-0.186
136	29	28.	-0.119	-0.261	0.142
137	30	17.	-0.097	-0.107	0.010
138	31	6.	0.301	0.196	0.105

ORDERED BY RESIDUALS

OBSV.	OBS. Y	FITTED Y	ORDERED RESID.	SEQ
25	-0.143	-0.318	0.176	1
16	1.088	0.913	0.175	2
4	1.100	0.929	0.172	3
29	-0.119	-0.261	0.142	4
19	0.158	0.033	0.125	5
31	0.301	0.196	0.105	6
12	0.845	0.752	0.093	7
26	-0.328	-0.410	0.082	8
13	1.418	1.349	0.069	9
14	1.067	1.024	0.043	10
18	0.130	0.099	0.031	11
8	1.103	1.086	0.017	12
30	-0.097	-0.107	0.010	13
15	0.885	0.887	-0.002	14
23	0.079	0.083	-0.004	15
6	1.028	1.036	-0.008	16
21	0.041	0.056	-0.014	17
10	1.314	1.329	-0.015	18
3	1.169	1.184	-0.015	19
20	0.204	0.219	-0.015	20
5	0.916	0.934	-0.018	21
7	0.862	0.888	-0.026	22
17	-0.119	-0.053	-0.066	23
1	0.829	0.896	-0.067	24
22	-0.071	0.023	-0.094	25
9	1.100	1.200	-0.100	26
27	-0.481	-0.371	-0.111	27
2	1.114	1.247	-0.133	28
24	-0.252	-0.070	-0.181	29
28	-0.585	-0.399	-0.186	30
11	0.554	0.740	-0.186	31

Figure 6B.5 (continued)

109

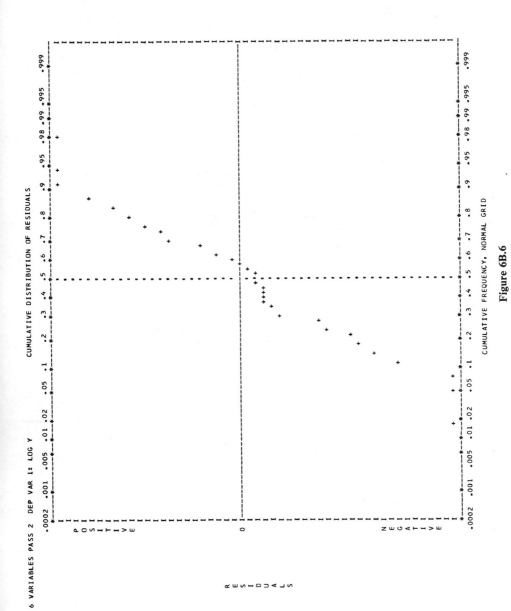

Figure 6B.6

110

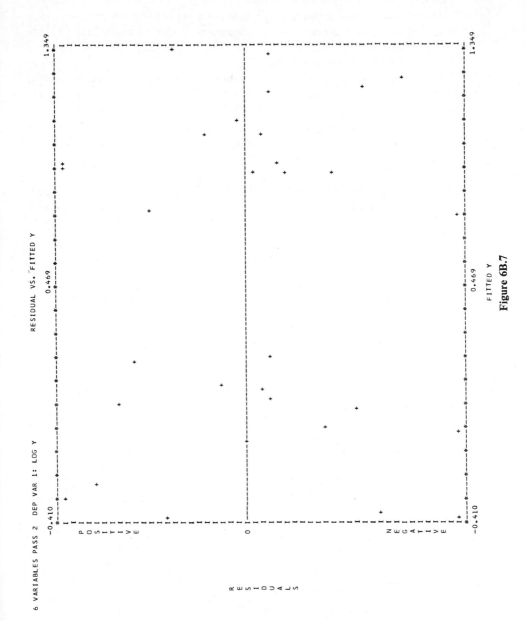

Figure 6B.7

Of course we do not want to double the number of equations for each new factor. The device of "Resolution V" fractional replication permits estimation of all 2fi *provided* there are no combinations of *three* factors that produce unexpectedly large reductions in C_p. If 3fi are expected, we will require a "Resolution VII" plan. See Table 6C.3.

The plans given in Table 6C.1 are adapted from the National Bureau of Standards booklet, "Fractional Factorial Experimental Designs for Factors at Two Levels," *Applied Mathematics Series* 48 (reprinted with corrections in 1962). Some effort has been made to select the simplest set of generators to ease transcription and multiplication. All relevant information is given for identifying each 2fi.

The resulting 2^{K-Q} values of C_p must be put through the usual Yate's algorithm for estimation of the "effects and interactions" of the various factors on the goodness of fit (see Davies, 1960, Sections 7.45 and 7.61). Thus,

TABLE 6C.1

"Resolution V" Plans for 2^K Equations (K Factors)

N.B.S. page	K	Q	2^{K-Q} =N	Equation Generators	Key Interactions in Alias Subgroup
5	5	1	16	ae,be,ce,de	-ABCD(E)
7	6	1	32	af,bf,cf,df,ef	+ABCDE(F)
8	7	1	64	ag,bg,cg,dg,eg,fg	-ABCDEF(G) *
22	8	2	64	agh,bgh,cg,dh,eg,fh	-ABCE(G), -ABDF(H)
24	9	2	128	aj,bj,chj,dh,ehj,fh,ghj	-ABCDG(J),CDEFG(H) **
36	10	3	128	ahk, bjk, ck, dhjk, eh,fhj,gj	-ABCD(K), -ADEF(H), -BDFG(J) ***
46	11	4	128	ahkl,bhj,chk,dhjl, ehl, fhjk, ghjkl	-ADEG(L),-ACFG(K), -BDFG(J), ABCDEFG(H) -BCFH(L), ACEH(J)
57	12	5	128	ahkl,bhjl,chjm,dhkm, ehklm,fhjlm,ghjk	-ABEF(L),-CDEF(M), -BCFG(J),-ADEG(K), ABCDEFG(H),-CDEF(M), ABGH(M), CDGH(L)

* actually of Resolution VII

** actually of Resolution VI

*** GH+JK, GJ+HK, GK+HJ estimable

for the 2^{5-1} the standard order, bringing in one live letter (underlined in the table) at a time, is given at the left below. At the right the relevant aliases of each effect are listed.

	Equation (1)	Aliases
gen.	*ae*	A
gen.	$\overline{be}$	B
	$\overline{ab}$	$AB - CDE$
gen.	*ce*	C
	$\overline{ac}$	$AC - BDE$
	bc	$BC - ADE$
	abce	$ABC - DE$
gen.	*de*	D
	$\overline{ad}$	$AD - BCE$
	bd	$BD - ACE$
	abde	$ABD - CE$
	cd	$CE - ABE$
	acde	$ACD - BE$
	bcde	$BCD - AE$
	abcd	$ABCD - E$

As a further example, for 12 factors in 128 equations, we generate the 128 C_p-values to be computed from the 7 generators of lower-case letters given at the bottom of Table 6C.1. These are then put through Yate's algorithm to obtain the "effects on C_p." All the largest effects, *when negative* (since we want to consider the equations with the smallest C_p), are inspected for their usefulness. A "half-normal plot" [Daniel, 1959] of the full set of effects and interactions may be useful in choosing the factors with influential separate or combined effects.

It will be necessary to know the two-factor aliases of each effect, since these are the most likely interpretations of any large interaction contrasts. These aliases are determined from the entries in the column of capitalized letters at the right of Table 6C.1. The reader will see, for example, that if the contrast labeled $ABCDEFG$ in terms of the live letters $A, \ldots, G$ came out large, this would mean that H has a large effect. Similarly, if CDE came out large, this would probably mean FM, its two-factor interaction alias; thus factors F and M do not operate additively on C_p. The analyst would then have to put the contrasts measuring F, $M (= CDEF)$ and $FM (= CDE)$ through the *reverse* Yates algorithm [Yates] to determine what combination of f and m gives the lowest C_p. We see that negative interactions are favorable, and that positive interactions indicate that the two factors together do not produce as much decrease in C_p as their individual effects led us to expect.

When it appears that these sets of 2^{K-Q} equations are too large, then, with greater risk, we can try a fractional replicate of "Resolution IV," which inevitably measures *sums* of two or more 2fi. A list of these is given in Table 6.C2, with sympathy for those who must use them. The detailed listing of 2fi aliases for the 2^{16-11} is also given.

TABLE 6C.2
"Resolution IV" Plans for 2^K Equations (K Factors)

K	Q	2^{K-Q} = N	Equation Generators	Key Interactions in Alias Subgroup
4	1	8	ad,bd,cd	ABCD
5	1	16	ae,be,ce,de	-ABCD(E)*
6	2	16	adf,bdf,cd,ef	ABC(D),ABE(F)
7	3	16	adfg,bdf,cdg,efg	ABC(D),ABE(F),ACE(G)
8	4	16	adfg,bdfh,cdgh,efgh	ABC(D),ABE(F),ACE(G), ABG(H)
9-15		32	Drop letters from end of generators below	Drop generators containing letters not used.
16	11	32	adfgklnq, bdfhkmoq, cdghlmpq, efghnopq, jklmnopq	See next page for aliasing of 2fi.
32	26	64	adfg klnq ruwx b'c'e'h' ** bdfh kmoq suwy b'd'f'h' cdgh lmpq tuxy c'd'g'h' efgh nopq vwxy e'f'g'h' jklm nopq a'b'c'd' e'f'g'h' rstu vwxy a'b'c'd' e'f'g'h'	

* actually of Resolution V
** letters r to h' are in correspondence with letters a to q:
      ABCD EFGH JKLM NOPQ
      RSTU VWXY A'B'C'D' E'F'G'H'

THE 15 STRINGS OF $2fi$ IN THE SYMMETRICAL 2^{16-11}

AB	CD	EF	GH	JK	LM	NO	PQ
AC	BD	EG	FH	JL	KM	NP	OQ
AD	BC	EH	FG	JM	KL	NQ	OP
AE	BF	CG	DH	JN	KO	LP	MQ
AF	BE	CH	DG	JO	KN	LQ	MP
AG	BH	CE	DF	JP	KQ	LN	MO
AH	BG	CF	DE	JQ	KP	LO	MN
AJ	BK	CL	DM	EN	FO	GP	HQ
AK	BJ	CM	DL	EO	FN	GQ	HP
AL	BM	CJ	DK	EP	FQ	GN	HO
AM	BL	CK	DJ	EQ	FP	GO	HN
AN	BO	CP	DQ	EJ	FK	GL	HM
AO	BN	CQ	DP	EK	FJ	GM	HL
AP	BQ	CN	DO	EL	FM	GJ	HK
AQ	BP	CO	DN	EM	FL	GK	HJ

TABLE 6C.3
"RESOLUTION VII" PLANS FOR 2^K EQUATIONS (K FACTORS)

K	Q	2^{K-Q} = N	Equation Generators	Key Interactions in Alias Subgroup
8	1	128	ah,bh,ch,dh,eh,fh,gh	ABCD EFG(H)
9	1	256	aj,bj,cj,dj,ej,fj, gj,hj	-ABCD EFGH(J)
11	2	512	akl,bkl,ckl,dkl, el,fl,gk,hk,jk	-ABCDEF(L), ABCD GHJ(K)
13	3	1024	almn,blm,clm,dln,eln, fl, gmn, hmn, jm,kn	-ABCDEF(L), -ABCGHJ(M), -ADEGHK(N)

Some Consequences of the Disposition of the Data Points

7.1 Introduction

A small set of pilot-plant data ($N = 36$, $K = 10$) is first studied, both to extend the earlier work to a ten-factor case and to make three new suggestions. These have to do with the *arrangement* of the data points, that is, with their relative positions in factor space. Later, the new suggestions are applied to some data sets used in earlier chapters.

Points taken under extreme conditions have great influence on the coefficients found and on their apparent correlations. On the other hand, runs made under nearly the same conditions may be used to estimate random error relatively free from function bias. Means are given for spotting remote and neighboring points. In the case examined first, the outlying points show that we have curvature in one direction. For the same case we find twenty-four pairs of near neighbors that give a value just matching the residual root mean square from the fitting equation. There is, then, no evidence of systematic lack of fit of the equation proposed.

When an equation has been chosen, it is often informative to the research worker or engineer to see in detail the effect of each variable on each observation. In this way the operation of trade-offs and reinforcements can be seen and studied. A table of component effects is given, which shows in an orderly way how each variable contributes to the fitted value of each observation.

7.2 Description of Example with Ten Independent Variables

The data shown in Figure 7A.1, which were first discussed by Gorman and Toman (1966), were used to derive an interpolation formula for describing the operation of a petroleum refining unit. The points are given in

116

chronological order. The month and day are listed in the identification column, and each response is the result of a week's operation under the conditions indicated. There are two 2-week gaps (points *7–9* and *15–16*) and two 3-week gaps (*17–19* and *35–36*). Points numbered *8* and *18* are, for unknown reasons, missing.

First steps

Technical knowledge of the process indicated that the ten independent variables should be entered linearly without any square or cross-product terms, and that the dependent variable should be entered as a common log function. The results of fitting this full equation are shown in Figures 7A.1–7A.6 and in summary in the first line of Table 7.1.

The plots and standard statistics appear acceptable. However, we note that the largest residual is caused by the first day's observation.

TABLE 7.1

SUMMARY OF COMPUTER PASSES: TEN-VARIABLE EXAMPLE

		Conditions			Results	
		Variables		Observations		C_p, Drop
Pass	Figures	Out	New In	Out	RMS	Variables
1	7A.1–7A.6				0.0122	4, 9
2	7A.7	4, 9			0.0125	
3	7A.8			19, 20	0.0104	
4	7A.9	4, 9		19, 20	0.0117	
5	7A.10			1	0.0101	
6	7A.11	4, 9		1	0.0100	
7	7A.12–7A.14		x_3^2		0.0105	7
8	7A.15–7A.19	7	x_3^2		0.0105	
9	7A.20–7A.22		x_3^2	1	0.0084	4
10	7A.23–7A.27	4	x_3^2	1	0.0084	

Search of all 2^K possible equations

From the C_p search on all 2^{10} potential equations, we see in Figure 7A.5 that variables 1, 2, 3, 5, 6, 7, and 8 are in all four of the equations with the smallest C_p's. The smallest C_p is obtained by omitting variables 4 and 9. However, from the p versus C_p plot (Figure 7A.6) we note that there is possibly less bias if 10 is excluded and variables 4 and 9 are retained.

$t_{K,i}$-**directed search**

Although a $t_{K,i}$-directed search (described in Chapter 6) is not essential with $K = 10$, we observe that it would have provided us with the same set of candidate equations as the full search, but with less effort. Table 7.2 shows the variables rearranged in the order of their decreasing t-values in the full equation. The ellipsis "Cumulative C_p" means that the equation used to obtain each C_p-value includes all preceding variables as well as the new one added on each line.

TABLE 7.2
CUMULATIVE p- AND C_p-VALUES

(Variables Ordered by $t_{K,i}$)

| Variables | $t_{K,i}$ | Cumulative | |
		p	C_p
6	4.1	2	
8	3.8	3	
3	3.6	4	
5	2.9	5	28
7	2.6	6	22
2	2.4	7	12
1	1.6	8	9.7*
9	1.3	9	10.2
4	1.2	10	10.3
10	1.1	11	11.0

* Minimum value.

In this case (but not always) the turning point in the column of C_p-values tells us correctly which variables should be in the "basic set." A search is then carried out on all combinations of the remaining variables. In less clear-cut cases, we have found that we must include in our basic set one *less* variable or, more rarely, two less, To be conservative, the computer search subroutine has been programmed to include in the basic set two less variables than the minimum C_p requires.

The clue to whether this caution is necessary is given by the relation of the minimal C_p to p. If the smallest "cumulative" C_p is equal to or less than p, we can safely include all variables down to the turning point. If, on the contrary, our smallest C_p were, say, 8 with $K = 10$ and $p = 4$, we would take as the "basic set" all the variables down to two less than the turning point. The C_p-search option of the Linear Least-Squares Curve Fitting Program has been written so that it will do this automatically.

In this example, with variables 3, 5, 6, 7, and 8 in the "basic set" of variables (two less than the minimum C_p set), the search of all combinations of the remaining variables (1, 2, 4, 9, and 10) led to the same candidates as did the search of all 10 variables (1024 equations). Thus, with the $t_{K.i}$-directed search the C_p-values of only 32 equations—plus the original 8—had to be calculated.

In Pass 2 we remove variables 4 and 9. The printout in Figure 7A.7 shows no evidence of anything wrong in the standard statistics or in the usual plots (not shown). But being of suspicious nature, we look further.

7.3   Interior Analysis I. Effects of Outermost Points

Most of the operations proposed so far are global in that they describe the whole set of data points without dissection. The statistics t_i, F, R_y^2, C_p, s^2, and even the b_i are global in this sense. The residuals, d_j, are interior statistics, but they concern Y and do not provide any insight into the conditions, x, under which the data were taken. We need some way of distinguishing data taken under extreme conditions from the rest. This capability will be useful for spotting points that are especially influential, points that are in error, and perhaps points that indicate curvature.

It is obvious that we should not combine x_i of different dimensions and units directly, in order to find extreme points in factor space. Minimal scaling will require dimensionless transforms of the x-coordinates. The simplest suggestion might be to put

$$X_{ij} = \frac{x_{ij} - \bar{x}_i}{s_i}$$

where $\bar{x}_i$ is the average of the x_{ij}, and s_i is $([ii]/N)^{1/2}$.

We could then compute a "standardized squared distance from mean" for each point, D_j^2:

$$D_j^2 = \sum_{i=1}^{K} X_{ij}^2$$

The principal disadvantage of this measure of distance from the X centroid is its equal weighting of coordinates that may differ widely in their importance. Indeed, if several factors have no effect whatever on the response, their contribution to this D_j^2 may mislead us into believing a point to be outermost when it is not extreme in any important respect.

Wishing to gauge each x_i-coordinate by its apparent effect in moving responses from $\bar{y}$, we might use

$$X_{ij} = b_i(x_{ij} - \bar{x}_i).$$

Since this has the dimensions of y, the corresponding squared distance will need some normalizing factor to make it unit-free. We suggest s_y^2.

Our "weighted squared standardized distance" (WSSD) for point j is now:

$$\text{WSSD}_j = D_j{}^2 = \sum_{i=1}^{K} \left[\frac{b_i(x_{ij} - \bar{x}_i)}{s_y} \right]^2 = \frac{1}{s_y{}^2} \sum_{i=1}^{K} C_{ij}{}^2$$

This quantity, although it has the disadvantage of its uncertainty (due to the presence of the b_i), does weight the x-coordinates as fairly as we can at each stage.

The WSSD can be used to spot points which may be controlling the global statistics, including the b_i and the $R_i{}^2$. To take the simplest pathological case, suppose that $K = 2$, and that *two* points have much larger WSSD's than all the rest, These two points will control the b_i and their large correlation. None of the standard global statistics will detect this, nor will the residuals, which will be *small* for the two outlying points.

If, again, we have about $(K + 1)$ remote points, they may well control all b_i ($i = 1, 2, \ldots, K$) with small correlations. Looking at the usual statistics will not reveal anything wrong. Indeed, there may not be anything wrong, but if a small minority of points is controlling all the estimated b_i, the data analyst will surely want to know this. It may happen that the data taken on "bad" days, under extreme conditions, do not provide the best equation for predicting more nearly normal operation.

Once the outlying points are spotted, it is easy to carry through one or two additional passes, to see whether the interior data are telling a story compatible with that told by the outer points.

To return to the data under study, points *19* and *20* are clearly quite far out. In difficult cases we would want to see all K components, C_{ij}, that added together give $(Y_j - \bar{Y})$ and whose squares, summed and divided by s_y^2, give $D_j{}^2$. These are all given in Figure 7A.16:

$$C_{ij} = b_i(x_{ij} - \bar{x}_i).$$

The present situation, however, is simple enough so that inspection suffices. We note the location of each coordinate value with respect to its maximum and minimum values. Points *19* and *20* are at the low ends of the ranges of x_5 and x_6 and at the high end of the range of x_3. They represent the first two weeks of operation after a three-week shutdown (or at least omission of data). To see whether these points are decisive for any of our estimates, we delete them in Pass 3 with all ten variables present and again in Pass 4 with variables 4 and 9 omitted.

In Table 7.3 the ten b_i for the two cases with and without variables 4 and 9 are compared with those for the corresponding two cases with points *19* and *20* removed. It appears that b_3 and b_7 are both changed by nearly 50%

TABLE 7.3

COMPARISON OF b_i IN PRESENCE AND ABSENCE OF OBSERVATIONS 19 AND 20

(Variables in Decreasing Order of Relative Influence)

Pass:	1	2	3	4
Variables 4, 9:	In	Out	In	Out
Observations *19*, *20*:	In	In	Out	Out
Variables				
6	0.54	0.56	0.51	0.54
5	−4.0	−4.7	−4.4	−5.4
3	−1.0	−0.98	−0.47	−0.51
8	0.048	0.042	0.046	0.040
7	0.028	0.028	0.015	0.016
1	−0.028	−0.034	−0.035	−0.041
2	−0.011	−0.009	−0.014	−0.011
10	0.008	0.009	0.009	0.011

when points *19* and *20* are dropped. This might well mean curvature in the x_3 and x_7 directions, induced by *19* and *20* being well off the plane through the other 34 points. A computer pass, not shown, including terms in x_3^2 and x_7^2 gave *t*-values of 2.1 and 0.1. Later we discuss Pass 7, in which x_3^2 only is included.

Although it is not really needed in this problem, we mention here a simple transformation for avoiding high correlation between x and x^2. We use $(x - \bar{x})$ and $(x - d)^2$ as independent variables, where

$$d = \frac{\sum_{j=1}^{N} x_j^2 (x_j - \bar{x})}{2 \sum_{j=1}^{N} (x_j - \bar{x})^2} .$$

This formula is found by requiring the covariation between $(x_j - \bar{x})$ and $(x_j - d)^2$ to be zero. Thus $\sum_{j=1}^{N} (x_j - \bar{x})(x_j - d)^2 = 0$. Solving directly for d gives the expression shown above. We thank O. Dykstra for pointing this out to us. For the present case, $d = 0.59$, and so we use $(x_3 - 0.59)^2$ as our quadratic term.

A digression on point 1

We note that point 1 has given the largest (negative) residual in all passes made so far. We see from the WSSD value given in each printout that this is not an outlying point in "influence space." Its only uniqueness seems to be its chronological position. We may suspect some sort of failure to reach equilibrium, as in the data studied in Chapter 5, but, unlike that case, there is no supporting evidence here.

Passes 5 and 6, Table 7.1, show the results of dropping point *1* with $K = 10$ as in Pass 1, and with factors 4 and 9 removed as in Pass 2. The new residual mean square is 0.0101, as compared with 0.0122 earlier. This reduction is not excessive, considering that we have deliberately chosen this point because of its large contribution to the RMS. Table XX (Fisher and Yates) shows that the expected value of the largest of 25 (not 36) standard normal deviates is about 1.97, and our point 1 has a residual of -0.218, which is almost exactly 1.97 times our Pass 1 RRMS value of 0.110. So we retain point *1* for the nonce.

Study of basic equation with x_3^2 added

Earlier we uncovered evidence of curvature in the x_3 direction. Pass 7 ($N = 36$, $K = 11$), Figure 7A.12, shows a *t*-value for the new variable, x_3^2, of 2.2, a reduction of the RMS by nearly 20% (to 0.0105), an increase in the *t*-values of seven variables, *and* point *1* still low. The C_p search now suggests that only variable 7 be dropped, in contrast with Pass 1, which indicated that 4, 9, and possibly 10 should be omitted!

We drop variable 7 in Pass 8, Figure 7A.15. As a result, we find everything acceptable except the large but not impossible residual for point *1*.

Since we are still doubtful about point *1*, Pass 9, Figure 7A.20, is made ($N = 35$, $K = 11$, x_3^2 included). The usual C_p search, Figures 7A.21 and 7A.22, suggests that only x_4 is droppable. Pass 10, Figure 7A.23, makes this change ($N = 35$, $K = 10$). No dramatic improvements are visible, so once again we decide to retain point *1*.

As a parting gesture, we try putting in all square and cross-product terms for the four most influential variables: x_3, x_5, x_6, x_8. We find no new *t*-values as large as 1.0, and the C_p search admits no new terms.

Our conclusion from all this boat-rocking is that the conditions of Pass 8 ($N = 36$, $K = 10$, x_7 dropped, x_3^2 added) give a satisfactory representation of the whole set of data.

7.4   "Component Effects" Table. Partially Ordered Multifactor Data*

From the very beginning our objective has been to summarize a mass of multifactor data in an equation or in a few alternative equations. We have concentrated on scrutinizing all aspects of lack of fit of the equation to the data points.

But suppose now that we have an equation that appears to satisfy the

---

* Component-plus-residual plots, the component effect plus the residual of each observation plotted against each x_i, are useful as an aid (1) in choosing the appropriate form of equation, (2) in visualizing the distribution of the observations over the range of each independent variable, and (3) in judging the influence of each observation on each component of the equation. (See 1973 Wood reference for examples.)

requirements we have put down. We want to *expand* the equation, to see in detail what its various additive parts are doing. One way to gain this insight is to display and study the numerical value of each summand of the current fitting equation:

$$Y_j - \bar{y} = \sum_{i=1}^{K} b_i(x_{ij} - \bar{x}_i) = \sum_{i=1}^{K} C_{ij},$$

as it appears at each of the data points.

The individual terms, $C_{ij} = b_i(x_{ij} - \bar{x}_i)$, may be dubbed the *component effects* of x_i on Y_j. In Figure 7A.16 they are ordered by their decreasing relative influences in columns and by decreasing effects of the most influential variable in rows.

This ordering is chosen so that the largest effects will be easily visible. The table may be expected to have smaller values in the later columns and in the middle rows. Simple correlations among the x_i become apparent in a more informative way. Thus, in the present example, x_5 is clearly correlated with x_6, and this correlation is important since the effect of x_6 is often nearly as large as that of x_5. On the other hand, the correlation of x_6 with x_1, although formally even larger, is of less importance, as the column of C_{1j} shows. Only in observations *4, 9, 19,* and *20* does variable x_1 have effects greater than 0.1.

7.5   Interior Analysis II. Error Estimation from Near Neighbors

Up to this point we have continually depended on some measure of lack of fit for our estimate of random error. We have in fact always assumed that some "full" equation gave us an *over*fit, so that the RMS estimated a variance measuring only random error. But how can we gain some assurance that this is the case? We have just been working on a problem in which a new term, $x_3{}^2$, turned up, which reduced our estimate by about 20%. It is disturbing to reflect that one or more large terms may be lying hidden in the residual sum of squares. The classical requirement for safe inference—a good estimate of random error from randomized replicates—can hardly ever be met by historical data. Must we then admit that no safe inferences can be drawn from any such data? We do not think so.

Some data, although containing no exact replicates, will have some "near replicates," that is, pairs of observations taken far apart in time but under nearly the same x-conditions. It is of little importance if two points differ widely in their levels of an independent variable that has negligible influence. But it is important, if points differ widely in their values for important x_i—those which make a large difference in Y.

As a measure of (squared) distance in "effect space" between any two points j and j', we will use:

$$D_{jj'}^2 = \sum_{i=1}^{K} \left[\frac{b_i(x_{ij} - x_{ij'})}{s_y} \right]^2 = \frac{1}{s_y{}^2} \sum_i (C_{ij} - C_{ij'})^2 \equiv \text{WSSD}_{jj'}.$$

The purpose of the divisor s_y^2 is not to "Studentize" the statistic $\text{WSSD}_{jj'}$ (since the s^2 is often an upward-biased estimate of the true variance), but rather to render the distances dimensionless and small. The numerator terms are all correlated random variables and are also somewhat biased when the equation is, as we now suspect, incomplete. It will not be advisable to flood the D^2 with many uninfluential terms, since these bring with them more random variability and so tend to inflate the sum. For these two partly compensating reasons, we will not set a critical value, say 1, for $D_{jj'}^2$ in the hope that lesser values might locate pairs that would yield an estimate of σ^2 of bounded bias.

We will look for pairs of points with small $D_{jj'}^2$ in order to correct their (presumably small) y-differences, $(y_j - y_{j'})$, by their corresponding equation-differences, $(Y_j - Y_{j'})$, to get a single local corrected estimate of random error. Now

$$(y_j - y_{j'}) - (Y_j - Y_{j'}) = (y_j - Y_j) - (y_{j'} - Y_{j'})$$

$$= d_j - d_{j'},$$

where d_j is obviously the residual for point j. For ease in computation we will use the absolute values of these differences in residuals as if they were *ranges* of pairs of observations:

$$|d_j - d_{j'}| \equiv \Delta d_{jj'} \equiv \Delta d.$$

Since the expected value of the range for pairs of independent observations from a normal distribution is 1.128σ, we will average n of our Δd's and multiply by $0.886 (= 1/1.128)$ to get a running estimate, s_n, of σ.

Since the number of distances for N points is $\binom{N}{2}$, which increases roughly as N^2, a *double* sorting is proposed to first screen out points that cannot be neighbors, that is, those with widely separated Y-values. Points with nearly the same Y-values *may* be neighbors, or they may be on or near single contours of constant Y but far apart in some x_i.

Having ordered the data by increasing Y, we now compute the squared distances for the $(N - 1)$ pairs with adjacent Y-values. We then compute the same statistic for pairs separated by one, by two, and by three intervening Y-values. It is for these $(4N - 10)$ points, then, that we compute D^2.

We now print the $\Delta_n d$ ordered by their increasing $D_{jj'}^2$ and cumulate these by the formula:

$$s_n = 0.886 \left(\sum_n \Delta_n d \right) / n,$$

where n is a new index going from 1 for the closest pair to $(4N - 10)$ for the pair with the largest D^2 in our set. The details of this double sorting are

TABLE 7.4
DETAILS OF NEAR-NEIGHBOR CALCULATIONS
(10 Variables, Pass 8)

Obser-vation	Ordered Fitted Y	Residual	Delta Residuals and the Weighted Standardized Squared Distances of Near Neighbors							
			Adjacent		1 Apart		2 Apart		3 Apart	
			Del.	WSSD	Del.	WSSD	Del.	WSSD	Del.	WSSD
20	1.41	−0.01	0.04	33.36	0.09	38.30	0.00	6.04	0.16	28.03
9	1.54	0.03	0.13	2.75	0.04	30.85	0.11	3.04	0.11	20.72
4	1.64	−0.10	0.09	39.58	0.24	4.60	0.24	15.62	0.06	4.27
19	1.73	−0.01	0.16	26.57	0.16	77.47	0.03	35.59	0.08	41.54
16	1.74	0.15	0.00	16.40	0.19	1.31(7)	0.24	3.91	0.04	4.40
38	1.86	0.15	0.19	9.73	0.24	6.90	0.04	6.75	0.01	4.03
14	1.87	−0.04	0.05	0.91(3)	0.15	1.30(6)	0.18	10.68	0.15	3.42
12	1.94	−0.09	0.20	0.46(1)	0.23	7.23	0.10	2.15	0.01	2.02
13	1.98	0.11	0.02	8.49	0.30	2.71	0.21	1.65(11)	0.10	6.60
11	1.98	0.14	0.33	4.70	0.23	6.87	0.12	2.11	0.19	2.64
1	2.01	−0.19	0.10	1.91(15)	0.21	3.03	0.14	12.54	0.13	4.73
10	2.02	−0.10	0.11	4.80	0.04	15.79	0.03	7.03	0.03	16.54
2	2.06	0.02	0.07	5.37	0.08	3.46	0.08	6.51	0.09	0.79(2)
7	2.07	−0.05	0.01	4.33	0.02	2.65	0.16	3.49	0.09	13.67
21	2.07	−0.06	0.00	4.59	0.17	2.99	0.10	7.27	0.04	3.52
36	2.08	−0.07	0.18	5.90	0.10	15.75	0.04	4.51	0.02	15.98
5	2.10	0.11	0.07	4.42	0.22	1.89(14)	0.15	8.92	0.01	5.69
15	2.11	0.04	0.14	10.38	0.08	4.00	0.09	5.96	0.06	7.86
6	2.13	−0.11	0.06	11.94	0.23	6.40	0.08	3.69	0.23	2.83
17	2.14	−0.05	0.17	3.62	0.02	6.99	0.17	6.50	0.06	16.17
23	2.16	0.12	0.15	4.06	0.01	4.58	0.11	9.49	0.13	18.39
34	2.17	−0.03	0.15	1.40(8)	0.04	5.99	0.02	9.50	0.10	18.10
35	2.18	0.12	0.11	9.50	0.13	11.17	0.25	16.78	0.14	6.65
33	2.20	0.01	0.02	14.48	0.14	19.91	0.03	7.04	0.07	14.83
37	2.21	−0.01	0.12	6.01	0.01	14.88	0.05	4.16	0.06	7.15
27	2.24	−0.13	0.11	20.27	0.07	3.15	0.18	2.00	0.07	43.03
32	2.30	−0.02	0.04	12.01	0.07	11.91	0.04	12.52	0.02	12.41
29	2.33	−0.06	0.11	2.18	0.00	30.18	0.06	3.74	0.20	6.42
31	2.36	0.05	0.11	33.73	0.05	2.64	0.09	3.68	0.06	21.85
22	2.37	−0.06	0.06	24.71	0.20	30.44	0.05	33.08	0.00	29.34
24	2.43	0.00	0.14	1.83(13)	0.01	12.68	0.06	4.13	0.00	2.15
25	2.45	0.14	0.15	10.38	0.20	1.47(9)	0.14	1.16(5)	0.09	1.12(4)
3	2.47	0.00	0.05	11.01	0.00	12.60	0.06	12.17		
28	2.49	−0.06	0.06	1.73(12)	0.11	3.69				
30	2.49	0.00	0.06	1.48(10)						
26	2.61	0.06								

shown in Table 7.4, where the 15 closest pairs are indicated in parentheses, adjacent to their WSSD.

The computer printout, Figure 7A.17, shows the successive estimates of the standard deviation, s_n, at the left. The third column gives the WSSD between the two points identified in the fourth and fifth columns. We see s_n stabilizing at about 0.10 which agrees with s_y found from the best-fitting equation. We conclude that there is no evidence of lack of fit.

7.6   Other Examples of Error Estimation from Near Neighbors

Six-variable example of Chapter 6

Since variables A, B, D, F (x_1, x_2, x_4, x_6) have clear effects, whereas C and E (x_3 and x_5) have small ones (relative influence = 0.03 for each), it makes sense to drop the latter two variables and to look for near neighbors in the resulting four-dimensional effect space.

As in the preceding section, the data are first ordered by their Y-values (see Figure 7B.1 at *right*), and the D^2 are computed for all pairs within four steps of each other. The second ordering starts with the smallest of these ($4N - 10$) distances. The corresponding cumulative standard deviations, s_n, are listed in the first column of this figure. Again we see that the s_n-values go as low as 0.10, which is quite close to the RRMS of 0.11, and then increase steadily as D^2 becomes much greater than 1.

Our conclusion is that the four-variable equation fits well. There is no evidence to the contrary.

Stack loss example of Chapter 5

Inspection of the ordering by fitted Y (the second and third right-hand columns in Figure 7C.2) shows that points *15* to *19* are in a single clump. Points *10* to *14* are in another clump, and *5* to *8* in another. In the third to fifth left-hand columns we see that many of the exact replicates (WSSD = 0.0) were taken on successive days: points *7* and *8*, *10* and *11*, *15* and *16*, and *17* and *18*. Our s_n, then, is reflecting the "autocorrelation" of operations on successive days under unchanged conditions. These pairs all meet our near-neighbor requirement but they are obviously too close together in time to be considered independent.

The only way to look at the actual effects of real changes in operating conditions is to average these adjacent points and to fit the resulting six pieces of data! Figures 7C.3–7C.5 summarize this little exercise. The soundest conclusion is that only x_1 affects y, with coefficient 0.97 (standard error 0.07, based on 4 degrees of freedom), and that the system standard deviation was about 1.5. The thinness of the evidence needs no emphasis, but it should be pointed out that this is the fault of the data. Even though the original table appeared to give 21 three-factor data points, we find that a linear equation in one factor amply fits the *six* different stable conditions listed.

The conclusions that we reported at the end of Chapter 5 are somewhat modified:

1. After removal of the four transitional points we have but *six* really different conditions, with points *5–8*, *10–14*, and *15–19* judged to be local nests with their own error, s_w, of 1.0 (11 degrees of freedom).

2. The six points remaining are well fitted by:

$$Y = -41.2 + 0.97x_1.$$

3. The residual root mean square, s, from this equation is 1.5 (with, alas, four degrees of freedom).

We have overdone this study, hoping that readers will understand our purposes and that they will regard only charitably our final reduction of the problem almost to absurdity.

7.7 Summary

1. A ten-variable example is put through the routines recommended earlier. All 2^{10} C_p-values are computed, and all less than 10 are shown and plotted. Variables x_4 and x_9 are found to be unhelpful.

2. The use of the $t_{K,i}$ from the full equation to spot likely subsets of good equations is demonstrated in detail.

3. A measure of the *relevant* distance of data points from their centroid is introduced. It allows for the apparent differences in the influences of each x_i on Y.

4. The weighted standardized squared distance (WSSD) of item 3 is used to spot two outlying points (*19* and *20*). It is shown that these points are off the plane in the x_3 direction and hence that a term in $x_3{}^2$ improves the fit of the whole set.

5. The equation with $x_3{}^2$ is again put through the C_p-search routine. Now x_7 is dropped and x_4 and x_9 are retained.

6. Two passes, not shown, were made to see whether any other squared or cross-product terms in x_3, x_5, x_6, and x_8 were justified. None of the new terms was accepted.

7. A table showing the effect of each variable on each observation—a component effects table—is presented. Its columns list the variables in order of their decreasing relative influences. Its rows are ordered by decreasing values of the effect of the most influential variable. This table makes explicit the many cases of combined effects and of compensating effects present in all sets of historical data.

8. A measure of "distance between points in Y-space" is used first to find near replicates and then to find less biased estimates of the error standard deviation. When it is applied to the set of data of item 1 above, we find no evidence of lack of fit.

9. Acceptably close neighbors are found in the six-variable data of Chapter 6. They give the same standard deviation as the fitting equation, which is therefore judged to fit well.

10. The stack loss data of Chapter 5 are next studied. A smaller standard deviation is found from near neighbors, but this is due to adjacency in time of the neighboring points. We conclude that there are really only six different stable conditions in these data, which a single variable fits admirably.

APPENDIX 7A

COMPUTER PRINTOUTS OF TEN-VARIABLE EXAMPLE(page 129)

See Table 7.1 for descriptions.

Pass	Figures
1	7A.1–7A.6
2	7A.7
3	7A.8
4	7A.9
5	7A.10
6	7A.11
7	7A.12–7A.14
8	7A.15–7A.19
9	7A.20–7A.22
10	7A.23–7A.27

APPENDIX 7B

COMPUTER PRINTOUT OF SIX-VARIABLE EXAMPLE(page 157)

Pass 2   Figure 7B.1

APPENDIX 7C

COMPUTER PRINTOUTS OF STACK LOSS EXAMPLE(page 158)

Pass	Figures
13	7C.1 and 7C.2
28	7C.3–7C.6
30	7C.7

LINEAR LEAST-SQUARES CURVE FITTING PROGRAM

10VARIABLES PASS 1

EQUATION 1 OF A MULTI-EQUATION PROBLEM

DATA READ WITH SPECIAL FORMAT

FORMAT CARD 1 (A6,I4,I2,6X,6F6.3,6X,2F6.3/12X,3F6.3//18X,F6.3)

BZERO = CALCULATED VALUE

DATA INPUT   11 INDEPENDENT VARIABLES   1 DEPENDENT VARIABLES

OBSV.	SEQ.	1-11-21	2-12-22	3-13-23	4-14-24	5-15-25	6-16-26	7-17-27	8-18-28	9-19-29	10-20-30
1	1	7.890	79.000	0.570	2.500	0.190	1.422	91.700	5.000	26.000	4.100
	2	0.0	66.000								
2	1	8.680	83.000	0.500	1.520	0.190	1.642	87.800	4.490	7.000	2.700
	2	0.0	120.000								
3	1	5.610	82.000	0.680	1.690	0.220	2.310	87.000	8.460	26.000	1.600
	2	0.0	293.000								
4	1	14.750	78.000	0.510	1.210	0.180	1.150	87.000	1.770	6.000	4.300
	2	0.0	35.000								
5	1	7.260	86.000	0.510	1.450	0.200	1.661	88.400	5.530	13.000	2.600
	2	0.0	160.000								
6	1	6.620	82.000	0.490	1.220	0.200	1.799	87.600	3.610	3.000	0.100
	2	0.0	106.000								
7	1	6.250	88.000	0.500	1.200	0.230	1.892	88.900	4.160	6.000	3.700
	2	0.0	104.000								
9	1	14.550	77.000	0.550	2.910	0.170	1.031	90.200	2.860	3.000	3.500
	2	0.0	37.000								
10	1	9.440	74.000	0.500	1.840	0.200	1.348	88.800	5.130	10.000	5.300
	2	0.0	84.000								
11	1	6.540	84.000	0.530	1.430	0.220	1.638	91.800	3.490	10.000	3.900
	2	0.0	132.000								
12	1	8.390	75.000	0.630	1.450	0.190	1.305	92.200	3.450	11.000	4.400
	2	0.0	71.000								
13	1	9.680	72.000	0.600	1.590	0.190	1.335	90.200	3.410	12.000	4.100
	2	0.0	123.000								
14	1	10.090	76.000	0.620	1.470	0.190	1.180	90.500	4.140	8.000	6.300
	2	0.0	67.000								
15	1	8.600	82.000	0.540	1.640	0.200	1.370	89.900	7.800	26.000	6.600
	2	0.0	141.000								
16	1	11.480	77.000	0.650	1.930	0.180	1.031	91.400	4.650	28.000	6.600
	2	0.0	77.000								
17	1	8.840	73.000	0.590	1.580	0.200	1.347	91.400	5.030	13.000	11.900
	2	0.0	125.000								
19	1	12.060	72.000	0.780	1.970	0.090	0.641	83.200	9.310	4.000	0.200
	2	0.0	52.000								
20	1	12.210	69.000	0.840	1.310	0.120	0.618	85.300	6.690	35.000	0.600
	2	0.0	25.000								
21	1	6.720	80.000	0.510	2.870	0.220	1.731	87.500	5.120	10.000	5.800
	2	0.0	102.000								
22	1	8.650	79.000	0.540	1.460	0.160	1.381	89.800	7.970	75.000	3.800
	2	0.0	206.000								
23	1	8.010	76.000	0.510	1.920	0.170	1.466	88.800	3.650	19.000	5.400
	2	0.0	190.000								
24	1	5.960	70.000	0.480	1.650	0.230	1.994	90.200	4.170	16.000	7.600
	2	0.0	270.000								
25	1	5.860	71.000	0.430	1.960	0.230	2.131	89.100	5.490	6.000	6.000
	2	0.0	390.000								

Figure 7A.1

SIMPLE CORRELATION / DATA OUTPUT TABLE

#											
26	1	5.630	71.000	0.430	1.510	0.220	2.204	88.800	4.290	30.000	10.400
	2	0.0	458.000								
27	1	6.190	71.000	0.440	1.620	0.260	2.101	90.300	1.530	17.000	10.700
	2	0.0	129.000								
28	1	5.900	72.000	0.420	2.050	0.240	2.063	87.600	7.660	12.000	9.400
	2	0.0	268.000								
29	1	6.570	74.000	0.440	1.040	0.240	1.909	88.500	2.560	9.000	17.900
	2	0.0	188.000								
30	1	5.840	73.000	0.420	1.780	0.220	2.017	87.200	5.480	33.000	10.700
	2	0.0	310.000								
31	1	5.840	69.000	0.420	1.900	0.230	2.011	86.400	2.060	22.000	12.900
	2	0.0	260.000								
32	1	8.230	72.000	0.460	2.020	0.170	1.589	89.400	1.830	50.000	12.500
	2	0.0	190.000								
33	1	8.120	82.000	0.470	1.760	0.190	1.706	90.100	0.650	185.000	6.500
	2	0.0	164.000								
34	1	9.250	83.000	0.500	1.310	0.190	1.674	94.900	3.120	28.000	11.700
	2	0.0	138.000								
35	1	8.310	82.000	0.470	1.340	0.200	1.667	90.300	5.130	4.000	9.600
	2	0.0	202.000								
36	1	7.800	85.000	0.630	1.860	0.220	1.962	92.200	3.160	24.000	8.500
	2	0.0	102.000								
37	1	6.580	77.000	0.620	1.510	0.240	1.959	95.000	1.330	15.000	16.600
	2	0.0	160.000								
38	1	8.970	76.000	0.620	1.920	0.240	1.592	95.500	2.130	22.000	8.800
	2	0.0	101.000								

MEANS OF VARIABLES
8.26028D 00 7.70000D 01 5.38889D-01 1.70528D 00 2.00833D-01 1.60769D 00 8.95806D 01 4.34194D 00 2.28889D 01 6.86944D 00
2.10952D 00

ROOT MEAN SQUARES OF VARIABLES
2.37164D 00 5.26986D 00 9.87863D-02 4.18306D-01 3.39222D-02 4.13473D-01 2.58853D 00 2.10839D 00 3.11385D 01 4.35384D 00
2.89905D-01

SIMPLE CORRELATION COEFFICIENTS, R(I,I PRIME)

I	1	2	3	4	5	6	7	8	9	10	11
1	1.000	-0.041	0.511	0.118	-0.708	-0.868	-0.069	-0.000	-0.044	-0.355	-0.812
2	-0.041	1.000	-0.003	-0.157	0.064	0.095	0.241	0.006	0.091	-0.297	-0.096
3	0.511	-0.003	1.000	0.003	-0.590	-0.649	-0.018	0.338	-0.078	-0.437	-0.630
4	0.118	-0.157	0.003	1.000	-0.067	-0.089	-0.028	0.080	0.019	-0.093	-0.102
5	-0.708	0.064	-0.590	-0.067	1.000	0.841	0.378	-0.365	-0.135	0.539	0.560
6	-0.868	0.095	-0.649	-0.089	0.841	1.000	0.134	-0.198	0.035	0.450	0.814
7	-0.069	0.241	-0.018	-0.028	0.378	0.134	1.000	-0.476	0.074	0.399	0.040
8	-0.000	0.006	0.338	0.080	-0.365	-0.198	-0.476	1.000	-0.184	-0.458	0.056
9	-0.044	0.091	-0.078	0.019	-0.135	0.035	0.074	-0.184	1.000	0.046	0.160
10	-0.355	-0.297	-0.437	-0.093	0.539	0.450	0.399	-0.458	0.046	1.000	
11	-0.812	-0.096	-0.630	-0.102	0.560	0.814	0.040	0.056	0.160		1.000

Figure 7A.1 (*continued*)

130

LINEAR LEAST-SQUARES CURVE FITTING PROGRAM

10VARIABLES   PASS 1   DEP VAR 1: LOG Y   MIN Y = 1.398D 00   MAX Y = 2.661D 00   RANGE Y = 1.263D 00

ALL 10 VARIABLES ARE PRESENT.
NO OBSERVATIONS ARE OMITTED.

IND.VAR(I)	NAME	COEF.B(I)	S.E. COEF.	T-VALUE	R(I)SQRD	MIN X(I)	MAX X(I)	RANGE X(I)	REL.INF.X(I)
0		9.61476D-01	1.75D-02	1.6	0.7971	5.61CD 00	1.475D 01	9.140D 00	0.20
1	X1	-2.81065D-02	4.450D-03	2.4	0.3830	5.900D-01	8.800D-01	1.200D-01	0.16
2	X2	-1.09610D-02	2.780D-01	3.6	0.5404	4.200D-01	8.400D-01	4.200D-01	0.13
3	X3	-9.94835D-01	2.690D-02	1.2	0.0988	4.200D-01	2.600D-01	1.870D-01	0.08
4	X4	-5.46406D-02	1.35D 00	2.9	0.8346	9.000D-02	2.600D-01	1.700D-01	0.53
5	X5	-3.95960D 00	1.080D-01	4.1	0.8862	9.000D-02	2.600D-01	1.692D 00	0.73
6	X6	5.44901D-01	1.250D-02	2.8	0.5523	6.100D-02	9.550D-01	1.230D 01	0.27
7	X7	2.78181D-02	6.850D-04	3.8	0.5005	8.320D-01	9.310D 00	8.660D 00	0.33
8	X8	4.80904D-02	6.56D-03	1.3	0.2451	3.000D 00	1.850D 02	1.820D 02	0.13
9	X9	8.69075D-04		1.2	0.5740	1.000D-01	1.790D 01	1.780D 01	0.11
10	X10	7.57204D-03							

NO. OF OBSERVATIONS   36
NO. OF IND. VARIABLES   10
RESIDUAL DEGREES OF FREEDOM   25
F-VALUE   21.7
RESIDUAL ROOT MEAN SQUARE   0.11029818
RESIDUAL MEAN SQUARE   0.01216569
RESIDUAL SUM OF SQUARES   0.30414222
TOTAL SUM OF SQUARES   2.94156556
MULT. CORREL. COEF. SQUARED   .8966

----ORDERED BY COMPUTER INPUT----

IDENT.	OBSV.	WS DISTANCE	OBS. Y	FITTED Y	RESIDUAL
7 14	1	2.	1.820	2.038	-0.218
7 21	2	2.	2.079	2.055	0.025
7 28	3	19.	2.467	2.386	0.081
8 4	4	10.	1.544	1.575	-0.031
8 11	5	1.	2.097	2.204	-0.107
8 18	6	2.	2.204	2.125	-0.099
8 25	7	5.	2.025	2.085	-0.068
9 8	8	13.	2.017	2.017	-0.001
9 15	9	2.	1.568	1.567	-0.071
9 22	10	2.	1.924	1.995	-0.088
9 29	11	4.	2.121	1.930	-0.079
10 6	12	3.	1.851	1.906	-0.184
10 13	13	5.	2.090	1.810	0.016
10 20	14	4.	2.149	2.097	0.052
11 10	15	11.	1.826	1.732	0.154
11 17	16	2.	1.886	2.068	0.029
12 1	17	52.	2.097	1.710	0.006
12 8	18	44.	1.398	1.546	-0.148
12 15	19	2.	1.716	2.037	-0.028
12 22	20	2.	1.546	2.330	-0.016
12 29	21	9.	2.009	2.120	0.159
1 5	22	6.	2.279	2.416	0.015
1 12	23	10.	2.431	2.527	0.064
1 19	24	11.	2.591	2.626	0.035
1 26	25	13.	2.661	2.280	-0.169
2 2	26	7.	2.111	2.537	-0.109
2 9	27	9.	2.428	2.289	-0.015
2 16	28	3.	2.537	2.509	-0.018
2 23	29	3.	2.274	2.324	0.091
3 1	30	1.	2.491	2.279	0.000
3 8	31	3.	2.415	2.196	0.019
3 15	32	8.	2.279	2.285	-0.146
3 22	33	1.	2.140	2.240	-0.066
4 12	34	6.	2.305	2.082	-0.073
4 21	35	6.	2.009	2.046	0.008
4 28	36	6.	2.204	1.917	0.088
			2.004		

----ORDERED BY RESIDUALS----

OBSV.	OBS. Y	FITTED Y	ORDERED RESID.	SEQ
13	2.090	1.906	0.184	1
23	2.279	2.120	0.159	2
16	1.886	1.732	0.154	3
5	2.204	2.097	0.107	4
31	2.415	2.324	0.091	5
11	2.121	2.033	0.088	6
38	2.004	1.917	0.088	7
35	2.467	2.386	0.081	8
3	2.305	2.240	0.066	9
25	2.591	2.527	0.064	10
15	2.149	2.097	0.052	11
26	2.661	2.626	0.035	12
17	2.097	2.068	0.029	13
2	2.079	2.055	0.025	14
33	2.215	2.196	0.019	15
14	1.826	1.810	0.016	16
37	2.204	2.196	0.015	17
19	1.716	1.710	0.006	18
9	1.568	1.567	0.001	19
32	2.279	2.279	0.000	20
29	2.274	2.289	-0.015	21
22	2.314	2.330	-0.016	22
30	2.491	2.509	-0.018	23
21	2.009	2.037	-0.028	24
4	1.544	1.575	-0.031	25
7	2.017	2.085	-0.068	26
10	1.995	2.063	-0.071	27
36	1.851	1.930	-0.074	28
12	2.009	2.082	-0.079	29
6	2.025	2.125	-0.099	30
28	2.428	2.537	-0.109	31
34	2.140	2.285	-0.146	32
20	1.398	1.546	-0.148	33
27	2.111	2.280	-0.169	34
1	1.820	2.038	-0.218	35
			-0.218	36

Figure 7A.2

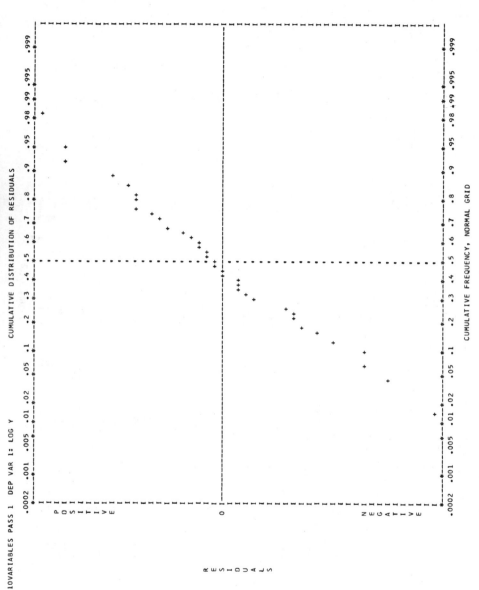

Figure 7A.3

132

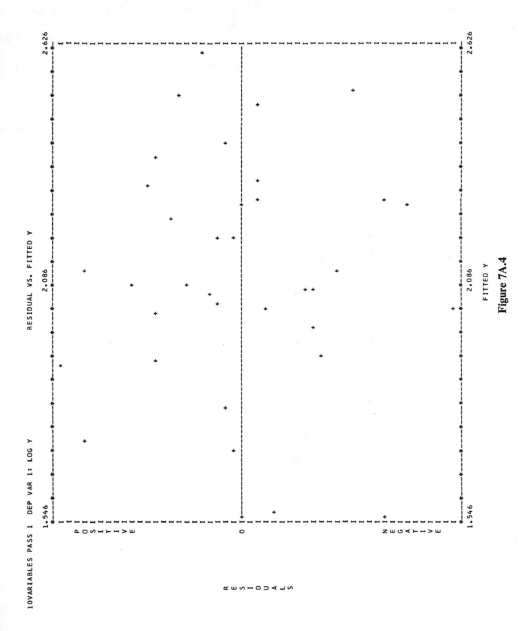

Figure 7A.4

CP VALUES FOR THE SELECTION OF VARIABLES

10VARIABLES PASS 1

```
NUMBER OF OBSERVATIONS                     36
NUMBER OF VARIABLES IN FULL EQUATION       10
NUMBER OF VARIABLES IN BASIC EQUATION       0
REMAINDER OF VARIABLES TO BE CONSIDERED    10
```

P   CP   VARIABLES IN EQUATION

P	CP										
9	9.7	X1	X2	X3		X5	X6	X7	X8		X10
10	10.3	X1	X2	X3	X4	X5	X6	X7	X8	X9	
10	10.4	X1	X2	X3		X5	X6	X7	X8	X9	X10
11	11.0	X1	X2	X3	X4	X5	X6	X7	X8	X9	X10

Figure 7A.5

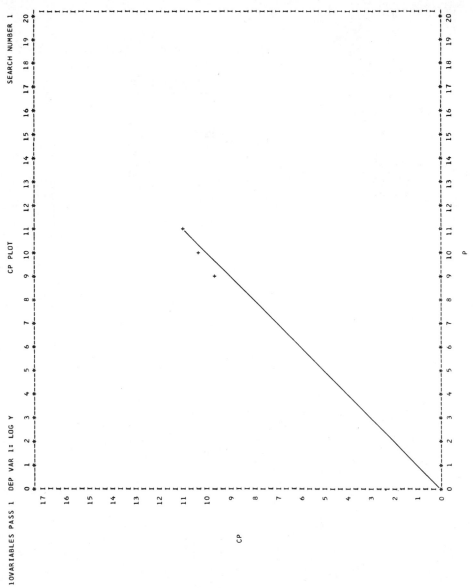

Figure 7A.6

LINEAR LEAST-SQUARES CURVE FITTING PROGRAM

IOVARIABLES PASS 2   DEP VAR 1: LOG Y   MIN Y = 1.398D 00   MAX Y = 2.661D 00   RANGE Y = 1.263D 00

ALL VARIABLES ARE PRESENT EXCEPT 4 AND 9.
NO OBSERVATIONS ARE OMITTED.

IND.VAR(I)	NAME	COEF.B(I)	S.E. COEF.	T-VALUE	R(I)SQRD	MIN X(I)	MAX X(I)	RANGE X(I)	REL.INF.X(I)
0		9.40004D-01							
1	X1	-3.37636D-02	1.730D-02	1.9	0.7890	5.610D 00	1.475D 01	9.140D 00	0.24
2	X2	-9.36202D-03	4.420D-03	2.1	0.3418	6.900D 01	8.000D 01	1.900D 01	0.14
3	X3	-9.78527D-01	2.790D-01	3.5	0.5302	4.200D 01	8.400D 01	4.200D 01	0.33
5	X5	-4.73700D 00	1.230D 00	3.9	0.7943	9.000D-02	2.600D-01	1.700D-01	0.64
6	X6	5.65135D-01	1.330D-01	4.2	0.8827	6.180D-01	2.310D 00	1.692D 00	0.76
7	X7	2.78364D-02	1.070D-02	2.6	0.5358	8.320D 01	9.500D 01	1.230D 01	0.27
8	X8	4.22291D-02	1.210D-02	3.5	0.4539	6.500D 01	9.310D 01	8.660D 00	0.29
	X10	9.08730D-03	6.550D-03	1.4	0.5605	1.000D-01	1.790D 01	1.780D 01	0.13

NO. OF OBSERVATIONS   36
NO. OF IND. VARIABLES   8
RESIDUAL DEGREES OF FREEDOM   27
F-VALUE   26.0
RESIDUAL ROOT MEAN SQUARE   0.11179569
RESIDUAL MEAN SQUARE   0.01249828
RESIDUAL SUM OF SQUARES   0.33745349
TOTAL SUM OF SQUARES   2.94156556
MULT. CORREL. COEF. SQUARED   .8853

ORDERED BY COMPUTER INPUT

IDENT.	OBSV.	WS DISTANCE	OBS. Y	FITTED Y	RESIDUAL
7 14	1	2.	1.820	2.081	-0.261
7 21	2	1.	2.079	2.067	0.012
7 28	3	19.	2.467	2.374	0.092
8 4	4	11.	1.544	1.546	-0.001
8 11	5	1.	2.204	2.100	0.104
8 18	6	2.	2.025	2.130	-0.105
8 25	7	5.	2.017	2.080	-0.063
9 8	8	14.	1.568	1.630	-0.062
9 15	9	2.	1.924	1.990	-0.066
9 22	10	2.	2.121	2.036	0.085
9 29	11	4.	1.851	1.928	-0.077
10 6	12	3.	2.090	1.899	0.191
10 13	13	6.	1.826	1.799	0.027
10 20	14	3.	2.149	2.072	0.077
11 3	15	11.	1.886	1.726	0.160
11 10	16	1.	2.097	2.059	0.037
12 1	17	58.	1.716	1.742	-0.026
12 8	18	48.	1.398	1.503	-0.105
12 15	19	20.	2.009	2.106	-0.097
12 22	20	1.	2.314	2.273	0.041
12 29	21	2.	2.279	2.157	0.121
1 2	22	6.	2.531	2.507	0.024
1 12	23	6.	2.561	2.538	0.023
1 26	24	10.	2.111	2.615	0.045
2 2	25	15.	2.428	2.532	-0.156
2 9	26	12.	2.274	2.267	-0.104
2 16	27	7.	2.491	2.271	0.003
2 23	28	9.	2.415	2.342	-0.011
2 30	29	3.	2.009	2.502	-0.031
3 1	30	3.	2.279	2.310	-0.073
3 8	31	3.	2.215	2.097	0.118
3 15	32	1.	2.140	2.244	-0.147
3 22	33	6.	2.305	2.287	0.061
4 12	34	10.	2.140	2.108	-0.100
4 21	35	6.	2.204	2.212	-0.008
4 28	36	6.	2.004	1.910	0.094

ORDERED BY RESIDUALS

OBSV.	OBS. Y	FITTED Y	ORDERED RESID.	SEQ
13	2.090	1.899	0.191	1
16	1.886	1.726	0.160	2
23	2.279	2.157	0.121	3
33	2.215	2.097	0.118	4
38	2.204	2.100	0.104	5
5	2.004	1.910	0.094	6
3	2.467	2.374	0.092	7
11	2.121	2.036	0.085	8
15	2.149	2.072	0.077	9
31	2.415	2.342	0.073	10
35	2.305	2.244	0.061	11
25	2.591	2.538	0.053	12
26	2.661	2.615	0.045	13
22	2.314	2.273	0.041	14
17	2.097	2.059	0.037	15
14	1.826	1.799	0.027	16
24	2.431	2.407	0.024	17
2	2.079	2.067	0.012	18
29	2.274	2.271	0.003	19
4	1.544	1.546	-0.001	20
37	2.204	2.212	-0.008	21
30	2.491	2.502	-0.011	22
19	1.716	1.742	-0.026	23
32	2.279	2.310	-0.031	24
9	1.568	1.630	-0.062	25
7	2.017	2.080	-0.063	26
10	1.924	1.990	-0.066	27
12	1.851	1.928	-0.077	28
21	1.990	2.106	-0.097	29
36	1.928	2.108	-0.100	30
28	2.106	2.532	-0.104	31
20	2.532	2.130	-0.105	32
6	1.503	2.287	-0.105	33
34	2.130	2.267	-0.147	34
27	2.111	2.212	-0.156	35
1	1.820	2.081	-0.261	36

Figure 7A.7

LINEAR LEAST-SQUARES CURVE FITTING PROGRAM

10VARIABLES PASS 3   DEP VAR 1: LOG Y   MIN Y = 1.544D 00   MAX Y = 2.661D 00   RANGE Y = 1.117D 00

ALL 10 VARIABLES ARE PRESENT.
OBSERVATIONS 19 AND 20 ARE OMITTED.

IND.VAR(I)	NAME	COEF.B(I)	S.E. COEF.	T-VALUE	R((I)SQRD	MIN X(I)	MAX X(I)	RANGE X(I)	REL.INF.X(I)
0		2.36860D 00							
1	X1	-1.46557D-02	1.650D-02	2.1	0.7672	5.610D 00	1.475D 01	9.140D 00	0.28
2	X2	-1.41048D-02	4.350D-03	3.2	0.3746	6.900D 00	8.800D 01	1.900D 01	0.24
3	X3	-4.73371D-01	3.400D-01	1.4	0.5105	4.200D-01	6.800D-01	2.600D-01	0.11
4	X4	-8.00876D-02	4.500D-02	1.8	0.1306	1.040D 00	2.910D 00	1.870D 00	0.13
5	X5	-4.39810D 00	1.350D 00	3.3	0.7206	1.600D-01	2.600D-01	1.000D-01	0.39
6	X6	5.08103D-01	1.250D-01	4.1	0.8310	1.030D 00	2.310D 00	1.290D 00	0.58
7	X7	1.45717D-02	1.160D-02	1.3	0.5498	8.640D 01	9.550D 01	9.100D 00	0.12
8	X8	4.58197D-02	1.160D-02	3.9	0.3769	6.500D-01	8.460D 01	7.810D 00	0.32
9	X9	1.03808D-03	6.500D-04	1.6	0.2739	3.000D 00	1.850D 02	1.820D 02	0.17
10	X10	9.18776D-03	6.260D-03	1.5	0.5382	1.000D-01	1.790D 01	1.780D 01	0.15

NO. OF OBSERVATIONS   34
NO. OF IND. VARIABLES   10
RESIDUAL DEGREES OF FREEDOM   23
F-VALUE   19.3
RESIDUAL ROOT MEAN SQUARE   0.10200231
RESIDUAL MEAN SQUARE   0.01040447
RESIDUAL SUM OF SQUARES   0.23930284
TOTAL SUM OF SQUARES   2.24445456
MULT. CORREL. COEF. SQUARED   .8934

----ORDERED BY COMPUTER INPUT----

IDENT.	OBSV.	WS DISTANCE	OBS. Y	FITTED Y	RESIDUAL
7	14	3.	1.820	2.028	-0.208
7	21	2.	2.079	2.055	0.025
7	28	16.	2.467	2.464	0.003
8	4	15.	1.544	1.606	-0.062
8	11	2.	2.204	2.090	0.114
8	18	2.	2.025	2.133	-0.108
8	25	2.	2.017	2.054	-0.037
9	8	19.	1.568	1.542	0.026
9	15	3.	1.924	2.007	-0.083
9	22	11.	2.121	2.000	0.120
9	29	4.	1.851	1.987	-0.135
10	6	4.	1.826	1.874	-0.048
10	13	7.	2.149	2.099	0.051
10	20	5.	2.097	2.193	-0.097
11	3	13.	1.886	1.789	0.097
11	10	5.	2.009	2.021	-0.013
12	15	17.	2.314	2.367	-0.053
12	22	10.	2.279	2.152	0.127
12	29	4.	2.431	2.409	0.022
1	5	5.	2.591	2.486	0.105
1	12	8.	2.661	2.617	0.044
1	19	9.	2.111	2.241	-0.130
1	26	13.	2.428	2.505	-0.076
2	2	10.	2.274	2.301	-0.027
2	9	6.	2.491	2.507	-0.015
2	16	5.	2.415	2.301	0.068
2	23	8.	2.347	2.347	-0.068
3	1	5.	2.279	2.213	-0.073
3	8	2.	2.215	2.505	-0.022
3	15	1.	2.140	2.213	-0.073
3	22	4.	2.305	2.205	0.100
4	12	8.	2.009	2.073	-0.064
4	21	4.	2.204	2.193	0.011
4	28	4.	2.004	1.885	0.120

----ORDERED BY RESIDUALS----

OBS. Y	FITTED Y	ORDERED RESID.	SEQ
2.279	2.152	0.127	1
2.121	2.000	0.120	2
2.090	1.970	0.120	3
1.885	1.885	0.120	4
2.204	2.090	0.114	5
2.591	2.486	0.105	6
2.305	2.205	0.100	7
1.789	1.789	0.097	8
2.415	2.347	0.068	9
2.149	2.099	0.051	10
2.661	2.617	0.044	11
1.568	1.542	0.026	12
2.079	2.055	0.025	13
2.431	2.409	0.022	14
2.204	2.193	0.022	15
2.204	2.193	0.011	16
2.467	2.464	0.003	17
2.009	2.021	-0.013	18
2.491	2.507	-0.015	19
2.097	2.117	-0.020	20
2.279	2.301	-0.023	21
2.017	2.054	-0.027	22
2.017	2.017	-0.037	23
1.826	1.874	-0.048	24
2.314	2.367	-0.053	25
1.544	1.606	-0.062	26
1.564	1.564	-0.064	27
2.140	2.213	-0.073	28
2.428	2.505	-0.076	29
1.924	2.007	-0.083	30
2.025	2.025	-0.100	31
2.025	2.133	-0.108	32
2.111	2.241	-0.130	33
1.851	1.987	-0.135	34
1.820	2.028	-0.208	

Figure 7A.8

137

LINEAR LEAST-SQUARES CURVE FITTING PROGRAM

10VARIABLES PASS 4   DEP VAR 1: LOG Y   MIN Y = 1.544D 00   MAX Y = 2.661D 00   RANGE Y = 1.117D 00

ALL VARIABLES ARE PRESENT EXCEPT 4 AND 9.
OBSERVATIONS 19 AND 20 ARE OMITTED.

IND.VAR(I)	NAME	COEF.B(II)	S.E. COEF.	T-VALUE	R(II)SQRD	MIN X(II)	MAX X(II)	RANGE X(II)	REL.INF.X(II)
0		2.19648D 00							
1	X1	-4.09886D-02	1.72D-02	2.4	0.7599	5.610D 00	1.475D 01	9.140D 00	0.34
2	X2	-1.13949D-02	4.40D-03	2.6	0.3118	6.900D 01	8.800D 01	1.900D 01	0.19
3	X3	-5.13574D-01	3.59D-01	1.4	0.5083	4.200D-01	6.800D-01	2.600D-01	0.12
4	X5	-5.39766D 00	1.27D 00	4.2	0.6466	1.600D-01	2.600D-01	1.000D-01	0.48
5	X6	5.39356D-01	1.30D-01	4.2	0.8249	1.031D 00	2.310D 00	1.279D 00	0.62
6	X7	1.55073D-02	1.22D-02	1.3	0.5437	8.640D 01	9.550D 01	9.100D 00	0.13
7	X8	3.97228D-02	1.19D-02	3.3	0.3328	6.500D-01	8.460D 00	7.810D 00	0.28
8	X10	1.15123D-02	6.53D-03	1.8	0.5232	1.000D-01	1.790D 01	1.780D 01	0.18

NO. OF OBSERVATIONS	34
NO. OF IND. VARIABLES	8
RESIDUAL DEGREES OF FREEDOM	25
F-VALUE	20.9
RESIDUAL ROOT MEAN SQUARE	0.10805287
RESIDUAL MEAN SQUARE	0.01167542
RESIDUAL SUM OF SQUARES	0.29188557
TOTAL SUM OF SQUARES	2.24445456
MULT. CORREL. COEF. SQUARED	.8700

—————ORDERED BY COMPUTER INPUT—————

IDENT.	OBSV.	WS DISTANCE	OBS. Y	FITTED Y	RESIDUAL
7 14	1	3.	1.820	2.089	-0.270
7 21	2	1.	2.079	2.069	0.010
7 28	3	16.	2.467	2.445	0.022
8 4	4	16.	1.544	1.559	-0.015
8 11	5	2.	2.204	2.094	0.110
8 18	6	5.	2.025	2.133	-0.108
8 25	7	20.	2.017	2.046	-0.029
9 8	9	3.	1.568	1.631	-0.063
9 15	10	2.	1.924	1.999	-0.075
9 22	11	5.	2.121	2.002	0.118
9 29	12	2.	1.851	1.970	-0.119
10 6	13	4.	2.090	1.947	0.143
10 13	14	7.	1.826	1.850	-0.024
10 20	15	4.	2.149	2.072	0.078
11 3	16	14.	1.886	1.777	0.109
11 10	17	3.	2.097	2.100	-0.004
11 17	18	1.	2.009	2.121	-0.112
12 15	21	10.	2.009	2.121	-0.111
12 22	22	4.	2.314	2.299	0.015
12 29	23	5.	2.279	2.198	0.081
1 5	24	8.	2.431	2.394	0.037
1 12	25	9.	2.591	2.503	0.088
1 19	26	14.	2.661	2.604	0.057
1 26	27	10.	2.009	2.121	-0.111
2 2	28	5.	2.428	2.507	-0.079
2 9	29	7.	2.274	2.272	0.002
2 16	30	3.	2.491	2.503	-0.012
2 23	31	2.	2.415	2.369	0.046
3 1	32	4.	2.279	2.345	-0.066
3 8	33	8.	2.215	2.080	0.135
3 15	34	5.	2.140	2.222	-0.082
3 22	35		2.305	2.214	0.091
4 12	36		2.009	2.109	-0.100
4 21	37		2.204	2.209	-0.005
4 28	38		2.004	1.874	0.130

—————ORDERED BY RESIDUALS—————

OBSV.	OBS. Y	FITTED Y	ORDERED RESID.	SEQ
13	2.090	1.947	0.143	1
33	2.215	2.080	0.135	2
38	2.004	1.874	0.130	3
11	2.121	2.002	0.118	4
5	2.204	2.094	0.110	5
16	1.886	1.777	0.109	6
35	2.305	2.214	0.091	7
25	2.591	2.503	0.088	8
23	2.279	2.198	0.081	9
15	2.149	2.072	0.078	10
26	2.661	2.604	0.057	11
31	2.415	2.369	0.046	12
24	2.431	2.394	0.037	13
3	2.467	2.445	0.022	14
22	2.314	2.299	0.015	15
2	2.079	2.069	0.010	16
29	2.274	2.272	0.002	17
17	2.097	2.100	-0.004	18
37	2.204	2.209	-0.005	19
30	2.491	2.503	-0.012	20
4	1.544	1.559	-0.015	21
14	1.826	1.850	-0.024	22
7	2.017	2.046	-0.029	23
9	1.568	1.631	-0.063	24
32	2.279	2.345	-0.066	25
10	1.924	1.999	-0.075	26
28	2.428	2.507	-0.079	27
34	2.140	2.222	-0.082	28
36	2.009	2.109	-0.100	29
6	2.025	2.133	-0.108	30
27	2.009	2.121	-0.111	31
21	2.009	2.121	-0.112	32
12	1.851	1.970	-0.119	33
1	1.820	2.089	-0.270	34

Figure 7A.9

138

LINEAR LEAST-SQUARES CURVE FITTING PROGRAM

IOVARIABLES PASS 5   DEP VAR 1: LOG Y

MIN Y = 1.398D 00   MAX Y = 2.661D 00   RANGE Y = 1.263D 00

ALL 10 VARIABLES ARE PRESENT.
OBSERVATION 1 IS OMITTED.

IND.VAR(I)	NAME	COEF.B(I)	S.E. COEF.	T-VALUE	MIN X(I)	MAX X(I)	RANGE X(I)	REL.INF.X(I)
0		6.82166D-01						
1	X1	-3.80476D-02	1.640-02	2.3	5.610D 00	1.475D 01	9.140D 00	0.28
2	X2	-1.05190D-02	4.11D-03	2.6	6.900D 00	8.800D 01	1.900D 01	0.16
3	X3	-1.03277D 00	2.540-01	4.1	4.200D-01	8.400D-01	4.200D-01	0.34
4	X4	-1.55029D-02	4.560-02	0.3	1.040D 00	2.910D 00	1.870D 00	0.02
5	X5	-4.16900D 00	1.230D 00	3.4	9.000D-02	2.600D-01	1.700D-01	0.56
6	X6	5.00553D-01	1.230D-01	4.1	6.180D-01	2.310D 00	1.692D 00	0.67
7	X7	3.24637D-02	9.980D-03	3.3	8.320D 01	9.550D 01	1.230D 01	0.32
8	X8	4.75094D-02	1.140-02	4.2	6.500D-01	9.310D 01	8.660D 01	0.33
9	X9	7.99321D-04	6.240-04	1.3	3.000D 00	1.850D 02	1.820D 02	0.12
10	X10	6.16035D-03	6.000-03	1.0	1.000D 00	1.790D 01	1.780D 01	0.09

NO. OF OBSERVATIONS   35
NO. OF IND. VARIABLES   10
RESIDUAL DEGREES OF FREEDOM   24
F-VALUE   25.9
RESIDUAL ROOT MEAN SQUARE   0.10044791
RESIDUAL MEAN SQUARE   0.01008978
RESIDUAL SUM OF SQUARES   0.24215478
TOTAL SUM OF SQUARES   2.8507933
MULT. CORREL. COEF. SQUARED   .9152

----ORDERED BY COMPUTER INPUT----

IDENT.	OBSV.	(ORDERED BY WS DISTANCE) OBSV.	WS DISTANCE	OBS. Y	FITTED Y	RESIDUAL
7 21	3	21.		2.079	2.055	0.025
7 28	2	14.		2.467	2.374	0.093
8 4	4	14.		1.544	1.520	0.024
8 11	5	2.		2.204	2.109	0.095
8 18	6	2.		2.025	2.128	-0.102
8 25	7	6.		2.017	2.083	-0.066
9 8	9	16.		1.568	1.601	-0.033
9 15	10	2.		1.924	2.008	-0.083
9 22	11	2.		2.121	2.061	0.060
9 29	12	4.		1.851	1.955	-0.104
10 6	13	3.		2.090	1.913	0.177
10 13	14	6.		1.826	1.814	0.012
10 20	15	4.		2.149	2.112	0.037
11 3	16	12.		1.886	1.751	0.136
11 10	17	3.		2.097	2.074	0.023
12 1	19	63.		1.716	1.723	-0.007
12 8	20	52.		1.398	1.532	-0.134
12 15	21	2.		2.009	2.091	-0.082
12 22	22	8.		2.314	2.343	-0.029
12 29	23	2.		2.279	2.151	0.128
1 5	24	7.		2.431	2.423	0.008
1 12	25	11.		2.591	2.541	0.050
1 19	26	12.		2.661	2.614	0.047
1 26	27	13.		2.428	2.542	-0.114
2 2	28	8.		2.274	2.250	0.024
2 9	29	8.		2.491	2.507	-0.015
2 16	30	11.		2.415	2.318	0.097
2 23	31	6.		2.279	2.298	-0.019
3 1	32	4.		2.215	2.204	0.011
3 8	33	6.		2.140	2.290	-0.150
3 15	34	1.		2.305	2.235	0.070
3 22	35	6.		2.009	2.092	-0.083
4 12	36	12.		2.204	2.200	0.004
4 21	37	8.		2.004	1.941	0.063
4 28	38					

----ORDERED BY RESIDUALS----

OBSV.	OBS. Y	FITTED Y	ORDERED RESID.	SEQ
13	2.090	1.913	0.177	1
16	1.886	1.751	0.136	2
23	2.279	2.151	0.128	3
31	2.415	2.318	0.097	4
5	2.204	2.109	0.095	5
3	2.467	2.374	0.093	6
35	2.305	2.235	0.070	7
38	2.004	1.941	0.063	8
11	2.121	2.061	0.060	9
25	2.591	2.541	0.050	10
26	2.661	2.614	0.047	11
15	2.149	2.112	0.037	12
2	2.079	2.055	0.025	13
4	1.544	1.520	0.024	14
29	2.274	2.250	0.024	15
17	2.097	2.074	0.023	16
14	1.826	1.814	0.012	17
33	2.215	2.204	0.011	18
24	2.431	2.423	0.008	19
37	2.204	2.200	0.004	20
21	1.716	1.723	-0.007	21
36	2.491	2.507	-0.015	22
32	2.279	2.298	-0.019	23
22	2.314	2.343	-0.029	24
9	1.568	1.601	-0.033	25
7	2.017	2.083	-0.066	26
21	2.009	2.091	-0.082	27
36	2.009	2.092	-0.083	28
10	1.924	2.008	-0.083	29
6	2.025	2.128	-0.102	30
12	1.851	1.955	-0.104	31
28	2.428	2.542	-0.114	32
20	1.398	1.532	-0.134	33
34	2.140	2.290	-0.150	34
27	2.111	2.272	-0.161	35

Figure 7A.10

139

LINEAR LEAST-SQUARES CURVE FITTING PROGRAM

10VARIABLES PASS 6   DEP VAR 1: LOG Y   MIN Y = 1.398D 00   MAX Y = 2.661D 00   RANGE Y = 1.263D 00

ALL VARIABLES ARE PRESENT EXCEPT 4 AND 9.
OBSERVATION 1 IS OMITTED.

IND.VAR(I)	NAME	COEF.B(I)	S.E. COEF.	T-VALUE	R(I)SQRD	MIN X(I)	MAX X(I)	RANGE X(I)	REL.INF.X(I)
0		6.54624D-01							
1	X1	-4.19392D-02	1.580-02	2.7	0.7961	5.61CD 00	1.475D 01	9.140D 00	0.30
2	X2	-9.86601D-03	3.950-03	2.5	0.3403	6.900D 01	8.800D 01	1.900D 01	0.15
3	X3	-1.04875D 00	2.500-01	4.2	0.5335	4.200D-01	8.400D-01	4.200D-01	0.35
4	X5	-4.87839D 00	1.10D 00	4.4	0.7941	9.000D-02	2.600D-01	1.700D-01	0.66
5	X6	5.22431D-01	1.200-01	4.3	0.8839	6.180D-01	2.310D 00	1.692D 00	0.70
6	X7	3.39452D-02	9.810-03	3.5	0.5498	8.320D 01	9.550D 01	1.230D 01	0.33
7	X8	4.33426D-02	1.080-02	4.0	0.4530	6.500D-01	9.310D 00	8.660D 00	0.30
8	X10	6.59323D-03	5.910-03	1.1	0.5653	1.000D-01	1.790D 01	1.780D 01	0.09

NO. OF OBSERVATIONS   35
NO. OF IND. VARIABLES   8
RESIDUAL DEGREES OF FREEDOM 26
F-VALUE   32.5
RESIDUAL ROOT MEAN SQUARE   0.09984454
RESIDUAL MEAN SQUARE   0.00996893
RESIDUAL SUM OF SQUARES   0.25919221
TOTAL SUM OF SQUARES   2.85507933
MULT. CORREL. COEF. SQUARED .9092

----ORDERED BY COMPUTER INPUT----

IDENT.	OBSV.	MS DISTANCE	OBS. Y	FITTED Y	RESIDUAL
7 21	2	1.	2.079	2.071	0.008
7 28	3	22.	2.467	2.361	0.106
8 4	4	16.	1.544	1.513	0.031
8 11	5	2.	2.204	2.116	0.088
8 18	6	3.	2.025	2.149	-0.124
8 25	7	6.	2.017	2.089	-0.072
9 8	9	19.	1.568	1.626	-0.058
9 15	10	3.	1.924	2.004	-0.080
9 22	11	3.	2.121	2.071	0.049
9 29	12	5.	1.851	1.965	-0.114
10 6	13	4.	2.090	1.916	0.174
10 13	14	7.	1.826	1.814	0.012
10 20	15	4.	2.149	2.092	0.057
11 3	16	14.	1.886	1.740	0.147
11 10	17	3.	2.097	2.072	-0.025
12 1	19	74.	1.716	1.745	-0.029
12 8	20	60.	1.398	1.508	-0.110
12 15	21	2.	2.009	2.110	-0.101
12 22	22	8.	2.314	2.306	0.008
1 5	23	3.	2.279	2.179	0.100
1 12	24	8.	2.431	2.423	0.008
1 19	25	12.	2.591	2.551	0.040
1 26	26	13.	2.661	2.614	0.047
2 2	27	19.	2.111	2.264	-0.154
2 9	28	14.	2.428	2.531	-0.103
2 16	29	9.	2.274	2.247	0.027
2 23	30	9.	2.491	2.498	-0.006
3 1	31	12.	2.415	2.324	0.091
3 8	32	5.	2.279	2.314	-0.035
3 15	33	4.	2.215	2.106	0.109
3 22	34	5.	2.140	2.305	-0.165
4 12	36	7.	2.009	2.103	-0.094
4 21	37	14.	2.204	2.213	-0.009
4 28	38	10.	2.004	1.931	0.073

----ORDERED BY RESIDUALS----

OBSV.	OBS. Y	FITTED Y	ORDERED RESID.	SEQ
13	2.090	1.916	0.174	1
16	1.886	1.740	0.147	2
33	2.215	2.106	0.109	3
3	2.467	2.361	0.106	4
23	2.279	2.179	0.100	5
31	2.415	2.324	0.091	6
5	2.204	2.116	0.088	7
38	2.004	1.931	0.073	8
15	2.149	2.092	0.057	9
35	2.305	2.250	0.055	10
11	2.121	2.071	0.049	11
26	2.661	2.614	0.047	12
25	2.591	2.551	0.040	13
4	1.544	1.513	0.031	14
29	2.274	2.247	0.027	15
14	1.826	1.814	0.012	16
2	2.079	2.071	0.008	17
22	2.314	2.306	0.008	18
24	2.431	2.423	0.008	19
30	2.491	2.498	-0.006	20
37	2.204	2.213	-0.009	21
17	2.097	2.072	-0.025	22
19	1.716	1.745	-0.029	23
32	2.279	2.314	-0.035	24
9	1.568	1.626	-0.058	25
7	2.017	2.089	-0.072	26
10	1.924	2.004	-0.080	27
36	2.009	2.103	-0.094	28
21	2.009	2.110	-0.101	29
28	2.428	2.531	-0.103	30
20	1.398	1.508	-0.110	31
12	1.851	1.965	-0.114	32
6	2.025	2.149	-0.124	33
27	2.111	2.264	-0.154	34
34	2.140	2.305	-0.165	35

Figure 7A.11

LINEAR LEAST-SQUARES CURVE FITTING PROGRAM

IOVARIABLES PASS 7   DEP VAR 1: LOG Y   MIN Y = 1.398D 00   MAX Y = 2.661D 00   RANGE Y = 1.263D 00

ALL 10 VARIABLES PLUS (X3 - .59)SQRD ARE PRESENT.
NO OBSERVATIONS ARE OMITTED.

IND.VAR(I)	NAME	COEF.B(I)	S.E. COEF.	T-VALUE	R(I)SQRD	MIN X(I)	MAX X(I)	RANGE X(I)	REL.INF.X(I)
0		-2.20939D 00							
1	X1	-2.00109D-02	1.620-02	1.8	0.7977	5.61D 00	1.475D 01	9.140D 00	0.22
2	X2	-1.57728D-02	4.700-03	3.4	0.5121	6.900D 01	8.800D 01	1.900D 01	0.24
3	X3--59	-8.56123D-01	2.660-01	3.2	0.5654	-1.700D-01	2.500D-01	4.200D-01	0.28
4	X4	-7.86364D-02	4.490-02	1.8	0.1503	-1.040D 00	2.910D 00	1.870D 00	0.12
5	X5	-4.24795D 00	1.260 00	3.4	0.8363	9.000D-02	2.600D 00	1.700D 00	0.57
6	X6	5.74715D-01	1.250-01	4.6	0.8875	6.180D-01	1.692D 00	1.700D-01	0.77
7	X7	1.26650D-02	1.210-02	1.0	0.6932	8.320D 01	9.550D 01	1.230D 01	0.12
8	X8	4.82144D-02	1.160-02	4.2	0.5006	6.500D-01	9.310D 00	8.660D 00	0.33
9	X9	1.14390D-03	6.470-04	1.8	0.2723	3.000D 00	1.850D 02	1.820D 02	0.16
10	X10	1.04327D-02	6.220-03	1.7	0.5920	1.000D-01	1.790D 01	1.780D 01	0.15
11	X3SQRD	-4.93395D 00	2.210 00	2.2	0.6553	0.0	6.250D-02	6.250D-02	0.24

NO. OF OBSERVATIONS   36
NO. OF IND. VARIABLES   11
RESIDUAL DEGREES OF FREEDOM   24
F-VALUE   23.3
RESIDUAL ROOT MEAN SQUARE   0.10239978
RESIDUAL MEAN SQUARE   0.01048572
RESIDUAL SUM OF SQUARES   0.25165717
TOTAL SUM OF SQUARES   2.94156556
MULT. CORREL. COEF. SQUARED   .9144

IDENT.	OBSV.	ORDERED BY COMPUTER INPUT WS DISTANCE	FITTED Y	RESIDUAL
7 14	1	2.	1.820	2.030	-0.211
7 21	2	2.	2.079	2.059	0.021
7 28	3	23.	2.047	2.067	-0.020
8 4	4	13.	1.544	1.613	-0.069
8 11	5	3.	2.204	2.091	0.113
8 18	6	3.	2.025	2.130	-0.105
8 25	7	3.	2.017	2.059	-0.042
9 8	8	17.	1.568	1.546	0.022
9 15	9	3.	1.924	2.015	-0.091
9 22	10	3.	2.121	1.999	0.121
9 29	11	4.	1.851	1.954	-0.103
10 6	12	4.	2.090	1.973	0.117
10 13	13	4.	1.826	1.854	-0.028
10 20	14	5.	2.149	2.110	0.039
11 3	15	13.	1.886	1.740	0.146
11 10	16	4.	2.097	2.132	-0.035
12 1	17	64.	1.716	1.711	0.005
12 8	18	59.	1.398	1.424	-0.026
12 15	19	21.	2.009	2.044	-0.036
12 22	20	3.	2.314	2.380	-0.067
12 29	21	3.	2.279	2.155	0.123
1 5	22	24.	2.431	2.441	-0.010
1 12	23	13.	2.591	2.481	0.111
1 19	24	18.	2.111	2.245	-0.135
1 26	25	15.	2.428	2.495	-0.067
2 2	26	15.	2.274	2.300	-0.025
2 9	27	9.	2.491	2.488	0.003
2 16	28	12.	2.415	2.331	0.084
2 23	29	12.	2.279	2.295	-0.016
3 1	30	5.	2.215	2.191	0.024
3 8	31	8.	2.140	2.218	-0.078
3 15	32	34.	1.924	2.015	-0.091
3 22	33	1.	2.305	2.197	0.108
4 12	34	11.	2.009	2.075	-0.067
4 21	35	11.	2.204	2.213	-0.008
4 28	36	5.	2.004	1.885	0.119

OBSV.	ORDERED BY RESIDUALS OBS. Y	FITTED Y	ORDERED RESID.	SEQ
16	1.886	1.740	0.146	1
23	2.279	2.155	0.123	2
11	2.271	2.191	0.121	3
38	2.000	1.885	0.117	4
13	2.090	1.973	0.113	5
5	2.204	2.091	0.113	6
25	2.591	2.481	0.111	7
35	2.305	2.197	0.108	8
31	2.415	2.331	0.084	9
26	2.661	2.619	0.042	10
15	2.149	2.110	0.039	11
33	2.215	2.191	0.024	12
9	1.568	1.546	0.022	13
2	2.079	2.059	0.021	14
3	2.467	2.447	0.020	15
19	1.716	1.711	0.005	16
30	2.491	2.488	0.003	17
37	2.204	2.213	-0.008	18
24	2.431	2.441	-0.010	19
32	2.279	2.295	-0.016	20
29	2.274	2.300	-0.025	21
20	1.398	1.424	-0.026	22
14	2.097	2.132	-0.028	23
17	2.009	2.044	-0.036	24
21	2.009	2.059	-0.042	25
7	2.017	2.059	-0.042	26
22	2.314	2.380	-0.067	27
36	2.009	2.075	-0.067	28
28	2.428	2.495	-0.067	29
4	1.544	1.613	-0.069	30
34	2.140	2.218	-0.078	31
10	1.924	2.015	-0.091	32
12	1.851	1.954	-0.103	33
6	2.025	2.130	-0.105	34
27	2.111	2.245	-0.135	35
1	1.820	2.030	-0.211	36

Figure 7A.12

141

CP VALUES FOR THE SELECTION OF VARIABLES

10VARIABLES PASS 7

NUMBER OF OBSERVATIONS   36
NUMBER OF VARIABLES IN FULL EQUATION   11
NUMBER OF VARIABLES IN BASIC EQUATION   0
REMAINDER OF VARIABLES TO BE CONSIDERED   11

P CP   VARIABLES IN EQUATION

11 11.1   X1 X2 X3 X4 X5 X6   X8 X9 X10 X3SQRD

12 12.0   X1 X2 X3 X4 X5 X6 X7 X8 X9 X10 X3SQRD

Figure 7A.13

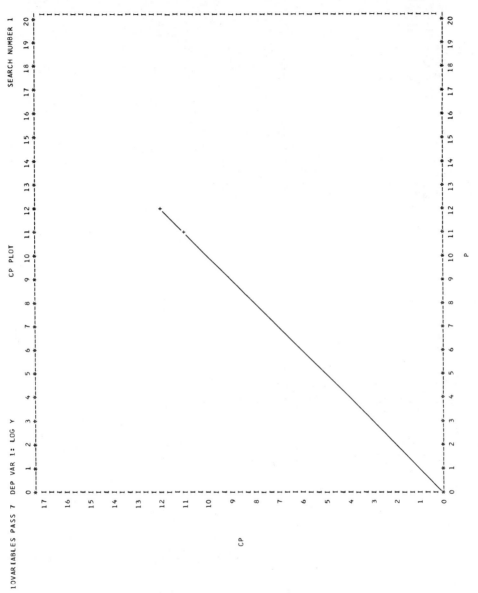

Figure 7A.14

143

IOVARIABLES PASS 8   DEP VAR 1: LOG Y   MIN Y = 1.398D 00   MAX Y = 2.661D 00   RANGE Y = 1.263D 00

9 VARIABLES PLUS (X3 - .59)SQRD ARE PRESENT, X7 IS OMITTED.
NO OBSERVATIONS ARE OMITTED.

IND.VAR(II)	NAME	COEF.B(II)	S.E. COEF.	T-VALUE	R(II)SQRD	MIN X(II)	MAX X(II)	RANGE X(II)	REL.INF.X(II)
0		3.31168D 00							
1	X1	-3.601D-02	1.620-02	2.0	0.7959	5.61CD 00	1.475D 01	9.140D 01	0.23
2	X2	-7.54401D-02	4.700-03	3.3	0.5099	6.900D 01	8.800D 01	1.900D 01	0.23
3	X3	-7.42934D-01	2.430-01	3.1	0.4796	-1.700D-01	2.500D-01	4.200D-01	0.25
4	X4	-7.98990D-02	4.500-02	1.8	0.1496	1.040D 00	2.910D 00	1.870D 00	0.12
5	X5	-3.90922D-02	1.220 00	3.2	0.8249	9.000D-02	2.600D-01	1.700D-01	0.53
6	X6	5.53110D-01	1.230-01	4.5	0.8844	6.180D-01	2.310D 00	1.692D 00	0.74
7	X8	4.52866D-01	1.130-02	4.0	0.4700	6.500D-01	9.310D 00	8.660D 00	0.31
8	X9	1.28348D-03	6.350-04	2.0	0.2402	3.000D 00	1.850D 02	8.200D 02	0.18
9	X10	1.32446D 00	5.630-03	2.4	0.4991	1.000D-01	1.790D 01	1.780D 01	0.19
10	X3SQRD	-6.23194D 00	1.830 00	3.4	0.4970	0.0	6.250D-02	6.250D-02	0.31

NO. OF OBSERVATIONS   36
NO. OF IND. VARIABLES   10
RESIDUAL DEGREES OF FREEDOM   25.4
F-VALUE
RESIDUAL ROOT MEAN SQUARE   0.10260554
RESIDUAL MEAN SQUARE   0.01052790
RESIDUAL SUM OF SQUARES   0.26319743
TOTAL SUM OF SQUARES   2.94156556
MULT. CORREL. COEF. SQUARED   .9105

-----ORDERED BY COMPUTER INPUT-----

IDENT.	OBSV.	WS DISTANCE	OBS.Y	FITTED Y	RESIDUAL
7 14	1	2.	1.820	2.013	0.194
7 21	2	2.	2.079	2.065	-0.015
7 28	3	2.	2.467	2.472	-0.005
8 11	4	12.	1.544	1.642	-0.097
8 18	5	3.	2.204	2.097	0.107
8 25	6	7.	2.025	2.133	-0.108
9 8	7	17.	1.568	1.536	-0.032
9 15	8	3.	1.924	2.069	-0.095
9 22	9	3.	2.121	1.985	-0.136
9 29	10	4.	1.851	2.020	-0.090
10 6	11	7.	2.090	1.979	0.111
10 13	12	4.	1.826	1.941	-0.039
10 20	13	5.	2.149	1.865	0.035
11 3	14	12.	1.886	2.115	0.147
11 10	15	4.	2.097	1.740	-0.048
12 1	16	58.	1.716	2.145	-0.012
12 8	17	57.	1.398	1.728	-0.011
12 15	18	3.	2.009	1.409	-0.065
12 22	19	3.	2.314	2.074	0.120
12 29	20	8.	2.279	2.370	0.000
1 5	21	2.	2.431	2.159	0.142
1 12	22	7.	2.591	2.431	-0.056
1 19	23	12.	2.661	2.449	0.053
1 26	24	14.	2.491	2.606	-0.001
2 2	25	16.	2.415	2.441	-0.058
2 9	26	13.	2.111	2.486	0.055
2 16	28	9.	2.274	2.332	-0.058
2 23	29	11.	2.492	2.362	0.053
3 1	30	8.	2.362	2.297	-0.001
3 8	31	2.	2.297	2.202	-0.018
3 15	32	8.	2.202	2.165	-0.025
3 22	34	2.	2.140	2.077	-0.068
4 12	36	7.	2.305	2.211	-0.007
4 21	37	10.	2.009	2.204	0.147
4 28	38	4.	2.004	1.857	

-----ORDERED BY RESIDUALS-----

OBSV.	OBS.Y	FITTED Y	ORDERED RESID.	SEQ
36	1.886	1.740	-0.147	1
38	2.004	1.857	0.147	2
25	2.591	2.449	0.142	3
11	2.121	1.985	0.136	4
35	2.305	2.180	0.125	5
23	2.279	2.159	0.120	6
13	2.204	2.097	0.111	7
5	2.661	2.606	0.107	8
26	2.415	2.362	0.055	9
31	2.149	2.115	0.053	10
15	1.568	1.536	0.035	11
9	2.079	2.065	0.032	12
2	2.215	2.079	0.015	13
33	2.431	2.202	0.013	14
24	2.491	2.492	0.000	15
30	2.204	2.472	-0.001	16
3	2.467	2.211	-0.005	17
37	1.398	1.409	-0.007	18
20	1.716	1.728	-0.011	19
19	1.728	1.716	-0.012	20
32	2.297	2.297	-0.018	21
34	1.826	2.165	-0.025	22
14	2.017	2.069	-0.038	23
17	2.314	2.370	-0.052	24
22	2.428	2.486	-0.056	25
28	2.274	2.332	-0.058	26
29	2.009	2.074	-0.058	27
21	2.074	2.077	-0.065	28
36	2.009	1.941	-0.068	29
12	1.851	2.020	-0.090	30
10	1.924	1.924	-0.095	31
12	1.544	1.642	-0.097	32
10	1.642	2.025	-0.108	33
6	2.025	2.133	-0.108	34
27	2.111	2.241	-0.130	35
1	1.820	2.013	-0.194	36

Figure 7A.15

10 VARIABLES PASS 8 DEP VAR 1: LOG Y

COMPONENT EFFECT OF EACH VARIABLE ON EACH OBSERVATION (IN UNITS OF Y)
(VARIABLES ORDERED BY THEIR RELATIVE INFLUENCE --- OBSERVATIONS ORDERED BY INFLUENCE OF MOST INFLUENTIAL VARIABLE)

SEQ.	OBSV.	6 X6	5 X5	7 X8	10 X3SQRD	3 X3	2 X2	1 X1	9 X10	8 X9	4 X4
1	3	0.39	-0.07	0.19	0.02	-0.10	-0.08	0.08	-0.07	0.00	0.00
2	26	0.33	-0.07	-0.00	-0.08	0.08	0.09	0.08	0.05	0.01	0.02
3	25	0.29	-0.11	0.05	-0.08	0.08	0.09	0.08	-0.01	-0.02	-0.02
4	27	0.27	-0.23	-0.13	-0.06	0.07	0.08	0.07	0.05	-0.01	0.01
5	28	0.25	-0.15	0.05	-0.10	0.09	0.06	0.07	0.03	-0.01	-0.03
6	30	0.23	-0.07	-0.10	-0.10	0.09	0.12	0.08	0.05	-0.01	-0.01
7	31	0.22	-0.11	-0.01	-0.00	0.04	0.11	0.08	0.08	-0.00	-0.02
8	24	0.21	-0.11	-0.01	-0.00	0.04	-0.12	0.07	0.01	-0.01	0.00
9	36	0.20	-0.07	-0.05	0.07	-0.07	0.0	0.01	0.02	0.00	-0.01
10	37	0.19	-0.15	-0.14	0.07	0.07	0.05	0.05	0.13	-0.01	0.02
11	29	0.17	-0.15	-0.08	-0.06	0.07	0.05	0.05	0.15	-0.02	0.05
12	7	0.16	-0.11	-0.01	0.02	0.03	-0.17	0.06	-0.04	-0.02	0.04
13	6	0.11	0.00	-0.03	0.01	0.04	-0.08	0.05	-0.09	-0.03	-0.04
14	21	0.07	-0.07	0.04	0.04	0.02	-0.05	0.05	-0.01	-0.02	-0.09
15	33	0.05	0.04	-0.17	-0.01	0.05	-0.08	0.00	-0.00	0.21	-0.00
16	34	0.04	0.00	-0.06	-0.02	0.03	-0.09	-0.03	0.06	0.01	0.03
17	35	0.03	0.00	0.04	-0.01	0.05	-0.08	-0.00	0.04	-0.02	0.03
18	5	0.03	0.00	0.05	0.04	0.02	-0.14	0.03	-0.06	-0.01	0.02
19	2	0.02	0.04	0.01	0.02	0.03	-0.09	-0.01	-0.01	-0.02	0.01
20	11	0.02	-0.07	-0.04	0.05	0.01	-0.11	0.05	-0.04	-0.00	-0.02
21	38	-0.01	-0.15	-0.10	0.07	-0.06	0.02	-0.02	0.03	-0.02	-0.02
22	32	-0.08	0.12	-0.11	-0.03	-0.06	0.08	0.00	0.07	-0.03	-0.03
23	23	-0.08	0.12	-0.03	0.04	0.02	0.02	0.01	-0.02	-0.00	-0.02
24	1	-0.10	0.04	0.03	0.07	-0.02	-0.03	0.01	-0.04	0.07	-0.06
25	22	-0.13	0.16	0.16	0.06	-0.00	-0.03	-0.01	-0.01	0.00	0.02
26	15	-0.13	0.00	0.16	0.06	0.03	-0.08	-0.01	-0.00	-0.01	0.01
27	10	-0.14	0.00	0.04	0.02	-0.04	0.05	-0.01	-0.07	-0.01	-0.01
28	17	-0.14	0.04	0.03	0.08	-0.05	0.06	-0.04	-0.04	-0.01	0.01
29	13	-0.15	0.04	-0.04	0.07	-0.07	0.08	-0.02	-0.03	-0.02	0.01
30	12	-0.17	0.04	-0.04	0.07	-0.06	0.03	-0.04	-0.01	-0.02	0.02
31	14	-0.24	0.04	-0.01	0.07	0.02	0.02	-0.06	-0.03	-0.03	0.02
32	4	-0.25	0.08	-0.12	0.04	-0.01	-0.02	-0.20	-0.04	0.01	0.04
33	9	-0.32	0.12	-0.07	0.07	-0.06	0.0	-0.10	-0.00	-0.02	-0.10
34	16	-0.32	0.08	0.01	0.05	-0.08	0.0	-0.12	-0.03	0.01	-0.02
35	19	-0.53	0.43	0.22	-0.15	-0.18	0.08	-0.12	-0.09	0.01	-0.02
36	20	-0.55	0.32	0.11	-0.31	-0.22	0.12	-0.12	-0.08	0.02	0.03

Figure 7A.16

145

LINEAR LEAST-SQUARES CURVE FITTING PROGRAM

IOVARTABLES PASS 8   DEP VAR 1: LOG Y

STANDARD DEVIATION ESTIMATED FROM RESIDUALS OF NEIGHBORING OBSERVATIONS (OBSERVATIONS 1 TO 4 APART IN FITTED Y ORDER).

NO.	CUMULATIVE STD DEV	ORDERED BY WSSD				ORDERED BY FITTED Y				
		WSSD	OBSV.	OBSV.	DEL RESIDUALS	WSSD	DEL RESIDUALS	FITTED Y	OBSV.	SEQ.
1	0.18	0.46	12	13	0.09	33.36	0.04	1.41	20	1
2	0.13	0.79	2	5	0.09	2.75	0.13	1.54	9	2
3	0.10	0.91	14	12	0.05	39.58	0.09	1.64	4	3
4	0.10	1.12	25	26	0.09	26.57	0.16	1.73	19	4
5	0.10	1.16	25	30	0.14	16.40	0.00	1.74	16	5
6	0.11	1.30	14	13	0.15	9.73	0.19	1.86	38	6
7	0.12	1.31	16	14	0.19	0.91	0.05	1.87	14	7
8	0.12	1.40	34	35	0.15	0.46	0.20	1.94	12	8
9	0.12	1.47	25	28	0.20	8.49	0.02	1.98	13	9
10	0.12	1.48	30	26	0.06	4.70	0.33	1.98	11	10
11	0.12	1.65	13	10	0.21	1.91	0.10	2.01	1	11
12	0.12	1.73	28	30	0.06	4.80	0.11	2.02	10	12
13	0.12	1.83	24	25	0.14	5.37	0.07	2.06	2	13
14	0.12	1.89	5	6	0.22	4.33	0.01	2.07	7	14
15	0.12	1.91	1	10	0.18	4.59	0.00	2.07	21	15
16	0.12	2.00	27	31	0.18	5.90	0.18	2.08	36	16
17	0.12	2.02	12	10	0.01	4.42	0.07	2.10	5	17
18	0.11	2.11	11	2	0.12	10.38	0.14	2.11	15	18
19	0.11	2.15	24	30	0.00	11.94	0.06	2.13	6	19
20	0.11	2.15	12	1	0.10	3.62	0.17	2.14	17	20
21	0.11	2.18	29	31	0.11	4.06	0.15	2.16	23	21
22	0.11	2.64	11	7	0.19	1.40	0.15	2.17	34	22
23	0.11	2.64	31	24	0.05	9.50	0.11	2.18	35	23
24	0.10	2.65	7	36	0.02	14.48	0.02	2.20	33	24
25	0.11	2.71	13	1	0.30	6.01	0.12	2.21	37	25
26	0.11	2.75	9	4	0.13	20.27	0.11	2.24	27	26
27	0.11	2.83	6	35	0.23	12.01	0.04	2.30	32	27
28	0.12	2.99	21	5	0.17	2.18	0.11	2.33	29	28
29	0.12	3.03	1	2	0.21	33.73	0.11	2.36	31	29
30	0.12	3.04	9	16	0.11	24.71	0.06	2.37	22	30
31	0.12	3.15	27	29	0.07	1.83	0.14	2.43	24	31
32	0.12	3.42	14	1	0.15	10.38	0.15	2.45	25	32
33	0.12	3.46	2	21	0.08	11.01	0.05	2.47	3	33
34	0.12	3.49	7	5	0.16	1.73	0.06	2.49	28	34
35	0.11	3.52	21	6	0.04	1.48	0.06	2.49	30	35
36	0.11	3.62	17	23	0.17	38.30	0.09	2.61	26	36

Figure 7A.17

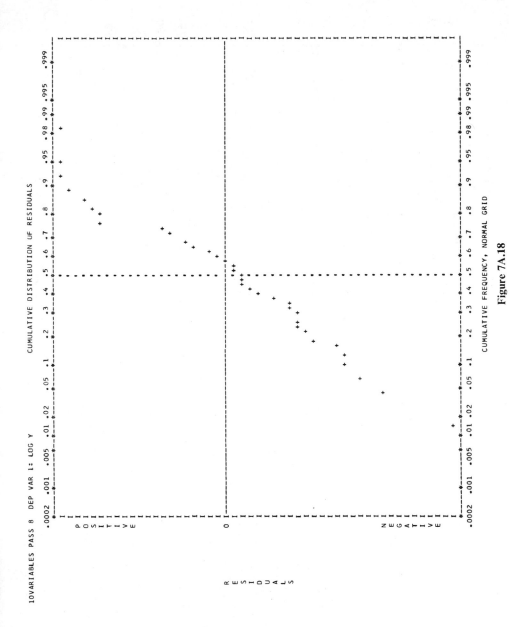

Figure 7A.18

147

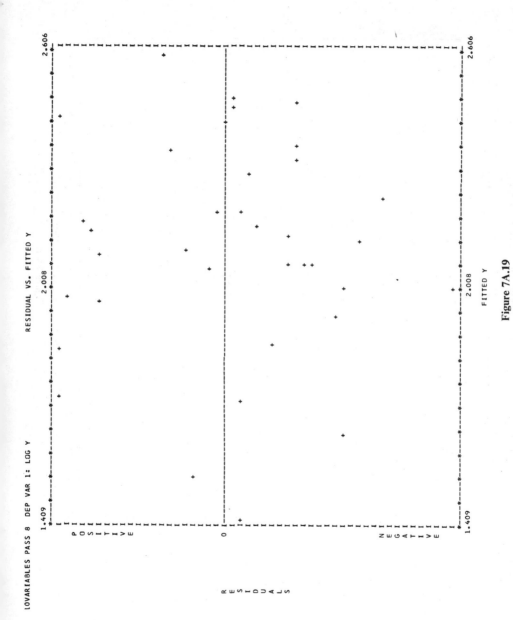

Figure 7A.19

LINEAR LEAST-SQUARES CURVE FITTING PROGRAM

1OVARIABLES PASS 9   DEP VAR 1: LOG Y   MIN Y = 1.398D 00   MAX Y = 2.661D 00   MAX Y = 2.661D 00   RANGE Y = 1.263D 00

ALL 10 VARIABLES PLUS (X3 - .59)SQRD ARE PRESENT.
OBSERVATION 1 IS OMITTED.

IND.VAR(I)	NAME	COEF.B(I)	S.E. COEF.	T-VALUE	R(I)SQRD	MIN X(I)	MAX X(I)	RANGE X(I)	REL.INF.X(I)
0		1.84594D 00							
1	X1	-3.95509D-02	1.500-02	2.6	0.8095	5.61CD 00	1.475D 01	9.140D 00	0.29
2	X2	-1.51374D-02	4.220-03	3.6	0.5115	6.9000-01	8.800D-01	1.900D-01	0.23
3	X3-.59	-8.9227D-02	2.390-01	3.8	0.1691	-1.7000-01	2.5000-01	4.2000-01	0.30
4	X4	-9.98450-02	4.290-02	0.9	0.0606	1.0000-02	1.7000-01	1.7000-01	0.06
5	X5	-4.43920D 00	1.130-02	3.9	0.8365	9.0000-02	2.6000-01	1.6920-01	0.60
6	X6	5.30659D-01	1.130-01	4.7	0.8893	6.1800-01	2.3100 01	1.2300 01	0.71
7	X7	1.77499D-02	1.110-02	1.6	0.6968	8.3200-01	9.5500 01	1.2300 01	0.17
8	X8	4.76476D-02	1.040-02	4.6	0.4993	6.5000-01	9.3100 01	8.6600 00	0.33
9	X9	1.06571D-03	5.810-04	1.8	0.2740	3.0000 00	1.8500 02	1.8200 02	0.15
10	X10	8.95563D-03	5.610-03	1.6	0.5913	1.0000-01	1.7900 01	1.7800 01	0.13
11	X3SQRD	-4.74122D 00	1.980 00	2.4	0.6479	0.0	6.2500-02	6.2500-02	0.23

NO. OF OBSERVATIONS		35
NO. OF IND. VARIABLES		11
RESIDUAL DEGREES OF FREEDOM		23
F-VALUE		28.7
RESIDUAL ROOT MEAN SQUARE		0.09178356
RESIDUAL MEAN SQUARE		0.00842422
RESIDUAL SUM OF SQUARES		0.19375711
TOTAL SUM OF SQUARES		2.85507933
MULT. CORREL. COEF. SQUARED		.9321

-------ORDERED BY COMPUTER INPUT-------

IDENT.	OBSV.	WS DISTANCE	OBS. Y	FITTED Y	RESIDUAL
7 21	2	2.	2.079	2.058	0.021
7 28	3	26.	2.467	2.433	0.034
8 4	4	18.	1.544	1.558	-0.014
8 11	5	3.	2.204	2.102	0.102
8 18	6	3.	2.025	2.133	-0.108
8 25	7	9.	2.017	2.058	-0.041
9 8	9	22.	1.568	1.580	-0.012
9 15	10	3.	1.924	2.026	-0.102
9 22	11	5.	2.121	2.028	0.093
9 29	12	5.	1.851	1.977	-0.126
10 6	13	8.	1.920	1.856	-0.013
10 13	14	6.	1.977	1.824	-0.030
10 20	15	6.	2.149	2.124	0.025
11 3	16	16.	1.886	1.758	0.129
11 10	17	4.	2.097	2.136	-0.039
12 1	19	80.	1.716	1.724	-0.008
12 8	20	71.	1.398	1.415	-0.017
12 15	21	3.	2.009	2.096	-0.088
12 22	22	10.	2.314	2.391	-0.078
12 29	23	3.	2.279	2.184	0.095
1 12	25	15.	2.591	2.495	0.096
1 19	26	16.	2.661	2.608	0.053
1 26	27	21.	2.111	2.239	-0.128
2 2	28	17.	2.428	2.501	-0.073
2 16	30	11.	2.491	2.487	0.005
2 23	31	11.	2.274	2.261	0.013
3 1	32	6.	2.415	2.326	0.089
3 8	33	9.	2.279	2.313	-0.034
3 15	34	3.	2.215	2.199	0.016
3 22	35	2.	2.140	2.225	-0.085
4 12	36	9.	2.305	2.195	0.111
4 21	37	14.	2.009	2.085	-0.076
4 28	38	7.	2.204	2.216	-0.011
			2.004	1.910	0.095

-------ORDERED BY RESIDUALS-------

OBSV.	OBS. Y	FITTED Y	ORDERED RESID.	SEQ
16	1.886	1.758	0.129	1
13	2.090	1.977	0.113	2
35	2.305	2.195	0.111	3
25	2.204	2.102	0.102	4
25	2.591	2.495	0.096	5
23	2.279	2.184	0.095	6
38	2.004	1.910	0.095	7
11	2.121	2.028	0.093	8
26	2.415	2.326	0.089	9
3	2.661	2.608	0.053	10
15	2.467	2.433	0.034	11
2	2.149	2.124	0.025	12
33	2.199	2.199	0.021	13
29	2.274	2.261	0.016	14
30	2.491	2.487	0.013	15
19	1.716	1.724	0.005	16
37	2.204	2.216	-0.008	17
9	1.568	1.580	-0.011	18
4	1.544	1.558	-0.012	19
24	2.431	2.447	-0.014	20
20	1.398	1.415	-0.015	21
14	1.826	1.856	-0.017	22
32	2.279	2.313	-0.030	23
7	2.017	2.058	-0.034	24
28	2.428	2.501	-0.039	25
36	2.009	2.085	-0.041	26
22	2.314	2.391	-0.073	27
34	2.140	2.225	-0.076	28
21	2.009	2.096	-0.078	29
10	1.924	2.026	-0.085	30
6	2.025	2.133	-0.088	31
12	1.851	1.924	-0.102	32
27	2.111	1.977	-0.108	33
			-0.126	34
			-0.128	35

Figure 7A.20

149

CP VALUES FOR THE SELECTION OF VARIABLES

10VARIABLES PASS 9

NUMBER OF OBSERVATIONS   35
NUMBER OF VARIABLES IN FULL EQUATION 11
NUMBER OF VARIABLES IN BASIC EQUATION 0
REMAINDER OF VARIABLES TO BE CONSIDERED 11

P	CP	VARIABLES IN EQUATION										
11	10.9	X1	X2	X3		X5	X6	X7	X8	X9	X10	X3SQRD
12	12.0	X1	X2	X3	X4	X5	X6	X7	X8	X9	X10	X3SQRD

Figure 7A.21

150

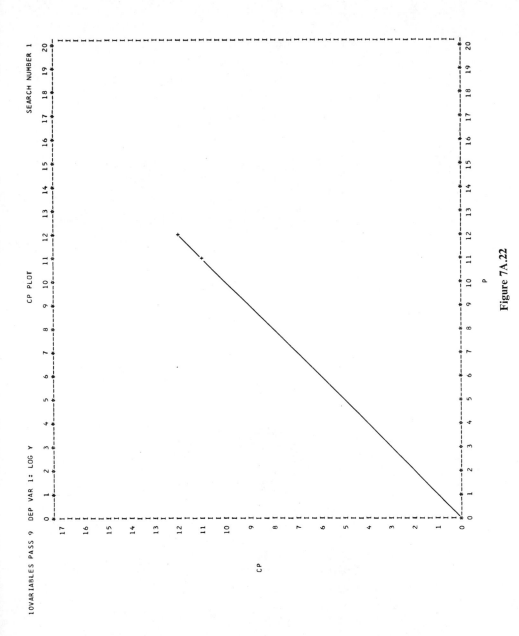

Figure 7A.22

151

LINEAR LEAST-SQUARES CURVE FITTING PROGRAM

10VARIABLES PASSIO   DEP VAR 1: LOG Y   MIN Y = 1.3980 00   MAX Y = 2.6610 00   RANGE Y = 1.2630 00

9 VARIABLES PLUS (X3 - .591SQRD ARE PRESENT, X4 IS OMITTED.
OBSERVATION 1 IS OMITTED.

IND.VAR(I)	NAME	COEF.B(I)	S.E. COEF.	T-VALUE	R(I)SQRD	MIN X(I)	MAX X(I)	RANGE X(I)	REL.INF.X(I)
0		1.655890 00							
1	X1	-4.234280-02	1.460-02	2.9	0.8015	5.610D 00	1.475D 01	9.140D 00	0.31
2	X2	-1.390970-02	3.990-03	3.5	0.4566	6.900D 00	8.800D 01	1.900D 01	0.21
3	X3	-8.901180-01	2.380-01	3.7	0.5655	-1.700D-01	2.500D-01	4.200D-01	0.30
5	X5	-4.189820 00	1.130-01	4.0	0.8362	-9.000D-02	2.600D-01	1.700D-01	0.60
6	X6	5.862530-02	1.120-01	4.6	0.8879	6.180D-01	2.310D 00	1.692D 00	0.70
7	X7	4.626990-02	1.030-02	4.5	0.4946	8.320D 01	1.230D 00	1.230D 00	0.18
8	X8	9.929490-04	5.740-04	1.7	0.2606	6.500D-01	9.310D 00	8.660D 00	0.32
9	X9	9.338910-03	5.580-03	1.7	0.5890	3.000D 00	1.850D 02	1.820D 02	0.14
10	X10	-4.305570 00	1.920 00	2.2	0.6269	1.000D-01	1.790D 01	1.780D 01	0.13
	X3SQRD					0.0	6.250D-02	6.250D-02	0.21

NO. OF OBSERVATIONS 35
NO. OF IND. VARIABLES 10
RESIDUAL DEGREES OF FREEDOM 24
F-VALUE 31.7
RESIDUAL ROOT MEAN SQUARE 0.09152321
RESIDUAL MEAN SQUARE 0.00837650
RESIDUAL SUM OF SQUARES 0.20103595
TOTAL SUM OF SQUARES 2.85507933
MULT. CORREL. COEF. SQUARED ...... .9296

------ORDERED BY COMPUTER INPUT------

IDENT.	OBSV.	WS DISTANCE	OBS. Y	FITTED Y	RESIDUAL
7 21	2	2.	2.079	2.054	0.025
7 28	3	25.	2.467	2.427	0.040
8 4	4	19.	1.544	1.528	0.016
8 11	5	3.	2.204	2.100	0.104
8 18	6	7.	2.025	2.120	-0.095
8 25	7	9.	2.017	2.051	-0.034
9 8	8	23.	1.568	1.618	-0.050
9 15	9	10.	1.924	2.025	-0.101
9 22	10	5.	2.121	2.029	0.091
9 29	11	12.	1.851	1.969	-0.117
10 6	12	5.	2.090	1.964	0.126
10 13	13	8.	1.826	1.845	-0.019
10 20	14	6.	2.149	2.121	0.028
11 3	15	16.	1.886	1.764	0.122
11 10	16	4.	2.097	2.127	-0.030
12 1	17	19.	1.716	1.734	-0.018
12 8	18	79.	1.398	1.409	-0.011
12 15	19	69.	2.009	2.144	-0.136
12 22	20	3.	2.314	2.375	-0.061
12 29	21	10.	2.431	2.438	-0.007
1 5	22	3.	2.279	2.193	0.085
1 12	23	15.	2.591	2.502	0.089
1 19	24	21.	2.661	2.598	0.063
1 26	25	11.	2.111	2.236	-0.126
2 2	26	17.	2.428	2.511	-0.083
2 9	27	5.	2.242	2.242	0.032
2 16	28	14.	2.491	2.490	0.001
2 23	29	10.	2.274	2.334	0.081
3 1	30	14.	2.415	2.327	-0.048
3 8	31	1.	2.279	2.215	0.001
3 15	32	8.	2.140	2.220	0.013
3 22	33	8.	2.009	2.120	0.116
3 29	34	1.	2.305	2.025	-0.080
4 12	35	14.	1.851	1.969	-0.090
4 21	37	7.	2.004	2.111	-0.012
4 28	38		2.004	1.919	0.085

------ORDERED BY RESIDUALS------

OBSV.	OBS. Y	FITTED Y	ORDERED RESID.	SEQ
13	2.090	1.964	0.126	1
16	1.886	1.764	0.122	2
35	2.305	2.189	0.116	3
5	2.204	2.100	0.104	4
11	2.121	2.029	0.091	5
25	2.591	2.502	0.089	6
23	2.279	2.193	0.085	7
38	2.004	1.919	0.085	8
31	2.415	2.334	0.081	9
26	2.661	2.598	0.063	10
29	2.467	2.427	0.040	11
15	2.274	2.242	0.032	12
2	2.121	2.121	0.028	13
4	1.544	1.528	0.025	14
33	2.215	2.202	0.016	15
30	2.491	2.490	0.013	16
24	2.431	2.438	-0.007	17
20	1.409	1.409	-0.011	18
37	2.204	2.216	-0.012	19
19	1.716	1.734	-0.018	20
14	1.826	1.845	-0.019	21
17	2.097	2.127	-0.030	22
7	2.017	2.051	-0.034	23
32	2.279	2.327	-0.048	24
9	1.568	1.618	-0.050	25
22	2.314	2.375	-0.061	26
34	2.140	2.220	-0.080	27
28	2.428	2.511	-0.083	28
36	2.009	2.099	-0.090	29
6	2.025	2.120	-0.095	30
10	1.924	2.025	-0.101	31
12	1.851	1.969	-0.117	32
27	2.111	2.236	-0.126	33
21	2.009	2.144	-0.136	34

Figure 7A.23

10VARIABLES PASS10   DEP VAR 1: LOG Y

COMPONENT EFFECT OF EACH VARIABLE ON EACH OBSERVATION (IN UNITS OF Y)
(VARIABLES ORDERED BY THEIR RELATIVE INFLUENCE --- OBSERVATIONS ORDERED BY INFLUENCE OF MOST INFLUENTIAL VARIABLE)

SEQ.	OBSV.	VARIABLES 5 X6	4 X5	7 X8	1 X1	3 X3	10 X3SQRD	2 X2	6 X7	8 X9	9 X10
1	3	0.36	-0.08	0.19	0.11	-0.13	0.02	-0.07	-0.05	0.00	-0.05
2	26	0.31	-0.08	-0.00	0.11	0.10	-0.06	0.08	-0.01	0.01	-0.03
3	25	0.27	-0.13	-0.05	0.10	0.10	-0.06	0.08	-0.01	-0.02	-0.01
4	27	0.25	-0.26	-0.13	0.09	0.09	-0.04	0.08	0.01	-0.01	-0.04
5	28	0.23	-0.17	0.15	0.10	0.11	-0.07	0.07	-0.04	-0.01	0.02
6	30	0.21	-0.08	0.05	0.10	0.11	-0.07	0.05	-0.04	-0.01	0.04
7	31	0.21	-0.13	-0.10	0.10	0.11	-0.07	0.11	-0.06	-0.00	0.06
8	24	0.20	-0.13	-0.01	0.10	0.05	0.00	0.10	0.01	-0.01	0.01
9	36	0.18	-0.08	-0.05	0.02	-0.08	0.05	-0.11	0.05	0.00	0.01
10	37	0.18	-0.17	-0.14	0.07	-0.07	0.05	-0.00	0.10	-0.01	0.09
11	29	0.15	-0.17	-0.08	0.07	0.09	-0.04	-0.04	-0.02	-0.01	0.10
12	7	0.14	-0.13	-0.01	0.09	0.03	0.02	-0.15	-0.01	-0.02	-0.03
13	6	0.10	0.01	-0.03	0.07	0.04	0.01	-0.07	-0.04	-0.02	-0.06
14	21	0.06	-0.08	0.04	0.07	0.02	0.03	-0.04	-0.04	-0.01	-0.01
15	33	0.05	0.05	-0.17	0.01	0.06	-0.01	-0.07	0.01	0.16	-0.00
16	34	0.03	-0.05	-0.06	-0.04	0.03	-0.02	-0.08	0.10	0.01	0.04
17	35	0.03	0.01	0.04	-0.00	0.06	-0.01	-0.07	0.01	-0.02	0.02
18	5	0.02	0.01	0.06	-0.04	0.02	0.03	-0.13	-0.02	-0.01	-0.04
19	2	0.02	0.05	0.01	-0.02	0.03	0.02	-0.08	0.04	-0.02	-0.04
20	11	-0.01	-0.08	-0.04	-0.07	-0.07	0.04	-0.10	0.04	-0.00	-0.03
21	38	-0.01	-0.17	-0.10	-0.03	0.07	-0.05	0.01	0.11	0.03	0.02
22	32	-0.01	0.14	-0.12	0.00	0.07	-0.02	0.07	-0.00	0.05	0.05
23	23	-0.08	-0.03	-0.03	-0.02	0.02	0.03	0.01	-0.01	-0.00	-0.01
24	22	-0.12	0.18	0.17	-0.02	-0.00	0.04	-0.03	0.01	0.05	-0.03
25	15	-0.13	0.01	0.16	-0.01	-0.00	0.04	-0.07	-0.01	-0.01	-0.02
26	10	-0.14	0.01	0.04	-0.05	-0.05	0.02	0.04	-0.01	-0.01	-0.02
27	17	-0.14	0.01	0.03	-0.02	-0.05	0.05	0.05	0.01	-0.01	-0.05
28	13	-0.14	0.05	-0.04	-0.06	-0.08	0.05	0.07	0.01	-0.01	-0.03
29	12	-0.16	0.05	-0.04	-0.01	-0.08	0.05	0.03	0.05	-0.02	-0.02
30	14	-0.22	-0.12	-0.01	-0.08	-0.07	0.05	0.01	-0.05	-0.02	-0.01
31	9	-0.24	0.14	-0.12	-0.27	-0.07	0.03	-0.01	-0.05	-0.02	-0.02
32	16	-0.30	0.09	0.02	-0.14	-0.10	0.04	-0.00	0.04	0.01	-0.03
33	19	-0.50	0.50	0.23	-0.16	-0.22	-0.10	0.07	-0.12	-0.02	-0.06
34	20	-0.52	0.36	0.11	-0.17	-0.27	-0.22	0.11	-0.08	0.01	-0.06

Figure 7A.24

153

IOVARIABLES PASSIO   DEP VAR 1: LOG Y

STANDARD DEVIATION ESTIMATED FROM RESIDUALS OF NEIGHBORING OBSERVATIONS (OBSERVATIONS 1 TO 4 APART IN FITTED Y ORDER).

NO.	CUMULATIVE STD DEV	ORDERED BY WSSD WSSD	OBSV.	OBSV.	DEL RESIDUALS	ORDERED BY FITTED Y WSSD	DEL RESIDUALS	FITTED Y	OBSV.	SEQ.
1	0.22	0.85	13	12	0.24	44.81	0.03	1.41	20	1
2	0.12	1.06	25	26	0.03	1.66	0.07	1.53	4	2
3	0.10	1.20	2	5	0.08	42.58	0.03	1.62	9	3
4	0.10	1.25	30	25	0.09	37.76	0.14	1.73	19	4
5	0.10	1.41	14	13	0.15	1.64	0.14	1.76	16	5
6	0.10	1.43	14	12	0.10	13.89	0.10	1.84	14	6
7	0.10	1.64	16	14	0.14	10.47	0.04	1.92	38	7
8	0.10	1.66	4	9	0.07	0.85	0.24	1.96	13	8
9	0.09	1.73	30	26	0.06	3.46	0.02	1.97	12	9
10	0.09	1.80	24	25	0.10	8.98	0.19	2.03	10	10
11	0.10	1.87	25	28	0.17	3.30	0.13	2.03	11	11
12	0.09	1.92	29	31	0.05	7.78	0.06	2.05	7	12
13	0.10	2.18	6	6	0.20	9.04	0.12	2.05	2	13
14	0.11	2.20	5	21	0.24	7.78	0.19	2.10	36	14
15	0.10	2.20	6	21	0.04	2.18	0.20	2.10	5	15
16	0.11	2.25	2	6	0.12	12.32	0.12	2.12	6	16
17	0.11	2.29	13	10	0.23	4.52	0.06	2.12	15	17
18	0.11	2.34	30	28	0.08	9.51	0.11	2.13	17	18
19	0.11	2.39	24	30	0.01	2.50	0.25	2.14	21	19
20	0.11	2.50	21	35	0.25	5.47	0.03	2.19	35	20
21	0.11	2.53	24	26	0.07	9.57	0.07	2.19	23	21
22	0.11	2.65	35	34	0.20	17.26	0.03	2.20	33	22
23	0.10	2.80	11	5	0.01	13.54	0.07	2.20	37	23
24	0.10	3.03	31	24	0.09	25.31	0.05	2.22	34	24
25	0.11	3.08	6	35	0.21	3.34	0.16	2.24	27	25
26	0.11	3.19	14	10	0.08	16.52	0.08	2.24	29	26
27	0.11	3.30	11	7	0.13	16.75	0.13	2.33	32	27
28	0.11	3.34	27	29	0.16	42.03	0.14	2.33	31	28
29	0.11	3.42	31	31	0.21	41.22	0.10	2.37	22	29
30	0.11	3.46	12	10	0.02	16.14	0.05	2.43	3	30
31	0.10	3.60	7	6	0.06	2.38	0.01	2.44	24	31
32	0.10	3.69	31	30	0.08	1.25	0.09	2.49	30	32
33	0.10	3.80	7	36	0.06	1.87	0.17	2.50	25	33
34	0.10	4.23	11	2	0.07	4.68	0.15	2.51	28	34
35	0.10	4.26	9	16	0.17	35.25	0.04	2.60	26	35

Figure 7A.25

154

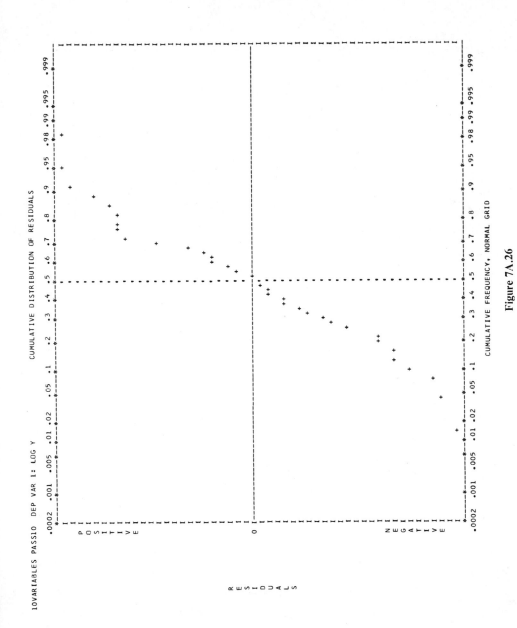

Figure 7A.26

155

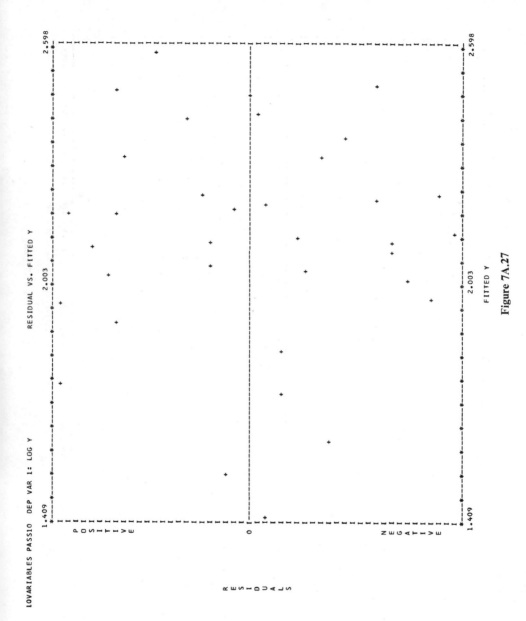

Figure 7A.27

LINEAR LEAST-SQUARES CURVE FITTING PROGRAM

6 VARIABLES PASS 2   DEP VAR 1: LOG Y

STANDARD DEVIATION ESTIMATED FROM RESIDUALS OF NEIGHBORING OBSERVATIONS (OBSERVATIONS 1 TO 4 APART IN FITTED Y ORDER).

NO.	CUMULATIVE STD DEV	------ORDERED BY WSSD------				------ORDERED BY FITTED Y------				SEQ.
		WSSD	OBSV.	OBSV.	DEL RESIDUALS	WSSD	DEL RESIDUALS	FITTED Y	OBSV.	
1	0.17	0.08	7	4	0.20	2.48	0.27	-0.41	26	1
2	0.12	0.21	30	17	0.08	1.07	0.08	-0.40	28	2
3	0.12	0.21	22	18	0.12	6.48	0.29	-0.37	27	3
4	0.11	0.29	10	13	0.08	5.01	0.03	-0.32	25	4
5	0.09	0.38	17	22	0.03	3.66	0.13	-0.26	29	5
6	0.11	0.45	27	29	0.25	7.96	0.19	-0.11	30	6
7	0.11	0.48	26	25	0.09	7.53	0.12	-0.07	24	7
8	0.12	0.53	22	19	0.22	0.38	0.03	-0.05	17	8
9	0.11	0.60	8	3	0.03	0.53	0.22	-0.02	22	9
10	0.12	0.63	1	4	0.24	6.53	0.14	0.03	19	10
11	0.11	0.67	18	20	0.05	7.36	0.01	0.06	21	11
12	0.11	0.70	7	1	0.04	5.06	0.04	0.08	23	12
13	0.10	0.78	14	3	0.06	4.04	0.07	0.10	18	13
14	0.11	0.80	7	16	0.20	6.14	0.12	0.20	31	14
15	0.10	0.86	16	4	0.00	34.51	0.17	0.22	20	15
16	0.10	0.87	12	7	0.12	1.13	0.28	0.74	11	16
17	0.10	0.88	14	8	0.03	14.05	0.09	0.75	12	17
18	0.10	0.88	30	22	0.10	9.51	0.02	0.89	15	18
19	0.10	0.93	12	1	0.16	0.70	0.04	0.89	7	19
20	0.10	0.97	19	18	0.09	2.89	0.24	0.90	1	20
21	0.10	0.98	1	14	0.11	0.86	0.00	0.91	16	21
22	0.11	1.02	28	29	0.33	3.98	0.19	0.93	4	22
23	0.10	1.07	28	27	0.08	3.82	0.06	0.93	5	23
24	0.11	1.13	11	12	0.28	6.31	0.05	1.02	14	24
25	0.11	1.17	3	9	0.08	3.17	0.03	1.04	6	25
26	0.11	1.35	16	6	0.18	0.60	0.03	1.09	8	26
27	0.11	1.61	17	19	0.19	1.17	0.08	1.18	3	27
28	0.11	1.83	8	9	0.12	3.88	0.03	1.20	9	28
29	0.11	1.84	5	8	0.04	2.85	0.12	1.25	2	29
30	0.11	2.01	4	8	0.15	0.29	0.08	1.33	10	30
31	0.11	2.09	30	19	0.12	6.50	0.19	1.35	13	31

Figure 7B.1

157

LINEAR LEAST-SQUARES CURVE FITTING PROGRAM

STACK LOSS PASS 13 DEP VAR 1: S.LOSS

COMPONENT EFFECT OF EACH VARIABLE ON EACH OBSERVATION (IN UNITS OF Y)
(VARIABLES ORDERED BY THEIR RELATIVE INFLUENCE --- OBSERVATIONS ORDERED BY INFLUENCE OF MOST INFLUENTIAL VARIABLE)

		VARIABLES		
		1	2	3
SEQ.	OBSV.	A.FLOW	W.TEMP	FLOWSQ
1	2	15.96	3.48	3.04
2	5	3.04	0.84	-0.22
3	6	3.04	1.37	-0.22
4	7	3.04	1.89	-0.22
5	8	3.04	1.89	-0.22
6	9	0.17	1.37	-0.35
7	10	0.17	-1.27	-0.35
8	11	0.17	-1.27	-0.35
9	12	0.17	-1.80	-0.35
10	13	0.17	-1.27	-0.35
11	14	0.17	-0.75	-0.35
12	20	-1.27	-0.22	-0.33
13	15	-5.57	-1.27	0.05
14	16	-5.57	-1.27	0.05
15	17	-5.57	-0.75	0.05
16	18	-5.57	-0.75	0.05
17	19	-5.57	-0.22	0.05

Figure 7C.1

LINEAR LEAST-SQUARES CURVE FITTING PROGRAM

STACK LOSS PASS 13   DEP VAR 1: S.LOSS

STANDARD DEVIATION ESTIMATED FROM RESIDUALS OF NEIGHBORING OBSERVATIONS (OBSERVATIONS 1 TO 4 APART IN FITTED Y ORDER).

NO.	CUMULATIVE STD DEV	ORDERED BY WSSD				ORDERED BY FITTED Y				
		WSSD	OBSV.	OBSV.	DEL RESIDUALS	WSSD	DEL RESIDUALS	FITTED Y	OBSV.	SEQ.
1	2.66	0.0	13	10	3.00	0.0	1.00	7.68	16	1
2	1.77	0.0	8	7	1.00	0.22	0.53	7.68	15	2
3	1.18	0.0	11	10	0.0	0.0	0.0	8.21	18	3
4	1.55	0.0	13	11	3.00	0.22	0.47	8.21	17	4
5	1.24	0.0	18	17	0.0	28.18	0.25	8.74	19	5
6	1.18	0.0	16	15	1.00	3.61	1.83	12.49	12	6
7	1.07	0.22	13	14	0.47	2.51	4.36	12.66	20	7
8	0.99	0.22	16	17	0.47	0.0	3.00	13.02	13	8
9	1.13	0.22	11	14	2.53	0.0	0.0	13.02	11	9
10	1.06	0.22	15	17	0.53	0.22	2.53	13.02	10	10
11	1.01	0.22	16	18	0.47	3.52	0.89	13.55	14	11
12	1.11	0.22	10	14	2.53	6.75	0.53	15.66	9	12
13	1.06	0.22	15	18	0.53	0.22	0.53	18.13	5	13
14	1.01	0.22	6	7	0.47	0.22	1.47	18.65	6	14
15	1.03	0.22	6	8	1.47	0.0	1.00	19.18	8	15
16	0.99	0.22	18	19	0.47	142.38	0.23	19.18	7	16
17	0.96	0.22	17	19	0.47	0.22	0.47	36.95	2	17

Figure 7C.2

159

LINEAR LEAST-SQUARES CURVE FITTING PROGRAM

STACK LOSS PASS 28

EQUATION 1 OF A MULTI-EQUATION PROBLEM

DATA READ WITH SPECIAL FORMAT

FORMAT CARD 1 (A6, 2I1, 4X, F3.0, 3X, F6.2, F4.0, F6.2)

BZERO = CALCULATED VALUE

DATA INPUT   3 INDEPENDENT VARIABLES   1 DEPENDENT VARIABLES

OBSV.	SEQ.	1-11-21	2-12-22	3-13-23	4-14-24	5-15-25	6-16-26	7-17-27	8-18-28	9-19-29	10-20-30
1	1	80.000	27.000	0.0	37.000						
2	1	62.000	23.250	0.0	18.750						
3	1	58.000	23.000	0.0	15.000						
4	1	58.000	18.000	0.0	12.800						
5	1	50.000	18.800	0.0	8.000						
6	1	56.000	20.000	0.0	15.000						

DATA TRANSFORMATIONS

POSITION	CODE	OPERATION	CONSTANT	LOCATION	OMIT	VARIABLE
1	8	ADD CONSTANT	-60.	1	0	1
2		NONE		2	0	2
3	9	MULTIPLY POSITIONS, 1X 1		3		3
4		NONE		4	0	4

DATA AFTER TRANSFORMATIONS   THE FITTED EQUATION HAS 3 INDEPENDENT VARIABLES, 1 DEPENDENT VARIABLES

| | A.FLOW | W.TEMP | FLOWSQ | S.LOSS | | | | | | |
OBSV.	1-11-21	2-12-22	3-13-23	4-14-24	5-15-25	6-16-26	7-17-27	8-18-28	9-19-29	10-20-30
1	2.00000D 01	2.70000D 01	4.00000D 02	3.70000D 01						
2	2.00000D 00	2.32500D 01	4.00000D 01	1.87500D 01						
3	-2.00000D 00	2.30000D 01	4.00000D 01	1.50000D 01						
4	-2.00000D 00	1.80000D 01	4.00000D 01	1.28000D 01						
5	-1.00000D 01	1.88000D 01	1.00000D 02	8.00000D 00						
6	-4.00000D 00	2.00000D 01	1.60000D 01	1.50000D 01						

SUMS OF VARIABLES
4.00000D 00 1.30050D 02 5.28000D 02 1.06550D 02

MEANS OF VARIABLES
6.66667D-01 2.16750D 01 8.80000D 01 1.77583D 01

ROOT MEAN SQUARES OF VARIABLES
1.02502D 01 3.38138D 00 1.57379D 00 1.00623D 01

SIMPLE CORRELATION COEFFICIENTS, R(I,I PRIME)

1	1.000	0.864	0.815	0.991
2	0.864	1.000	0.681	0.889
3	0.815	0.681	1.000	0.840
4	0.991	0.889	0.840	1.000

Figure 7C.3

160

LINEAR LEAST-SQUARES CURVE FITTING PROGRAM

STACK LOSS PASS 28   DEP VAR I: S.LOSS   MIN Y = 8.000D 00   MAX Y = 3.700D 01   RANGE Y = 2.900D 01

Y = B(0) + B(1)X1 + B(2)X2 + B(3)X1SQRD
Y = STACK LOSS
X1 = AIR FLOW — MEAN VALUE OF AIR FLOW(60.)
X2 = COOLING WATER TEMPERATURE
1ST DAY OBSERVATIONS WITH FLOW GREATER THAN 60 OMITTED(1,3,4,21)
CONTIGUOUS LINED OUT OBSERVATIONS AVERAGED(5-8, 10-14 AND 15-19).

IND.VAR(I)	NAME	COEF.B(I)	S.E. COEF.	T-VALUE	R(I)SQRD	MIN X(I)	MAX X(I)	RANGE X(I)	REL.INF.X(I)
0		7.62999D 00							
1	A.FLOW	7.69104D-01	1.81D-01	4.3	0.8429	-1.000D 01	2.000D 01	3.000D 01	0.80
2	W.TEMP	4.16159D-01	4.34D-01	1.0	0.7488	1.800D 01	2.700D 01	9.000D 00	0.13
3	FLOWSQ	6.76552D-03	8.09D-03	0.8	0.6669	4.000D 00	4.000D 02	3.960D 02	0.09

NO. OF OBSERVATIONS   6
NO. OF IND. VARIABLES   3
RESIDUAL DEGREES OF FREEDOM   2
F-VALUE   61.8
RESIDUAL ROOT MEAN SQUARE   1.64351794
RESIDUAL MEAN SQUARE   2.70115122
RESIDUAL SUM OF SQUARES   5.40230243
TOTAL SUM OF SQUARES   506.25208333
MULT. CORREL. COEF. SQUARED   .9893

------ORDERED BY COMPUTER INPUT------

IDENT.	OBSV.	WS DISTANCE	OBS. Y	FITTED Y	RESIDUAL
2	1	85.	37.000	36.955	0.045
5-8	2	1.	18.750	18.871	-0.121
9	3	2.	15.000	15.690	-0.690
10-14	4	3.	12.800	13.610	-0.810
15-19	5	25.	8.000	8.439	-0.439
20	6	5.	15.000	12.985	2.015

------ORDERED BY RESIDUALS------

OBSV.	OBS. Y	FITTED Y	ORDERED RESID.	SEQ
6	15.000	12.985	2.015	1
1	37.000	36.955	0.045	2
2	18.750	18.871	-0.121	3
5	8.000	8.439	-0.439	4
3	15.000	15.690	-0.690	5
4	12.800	13.610	-0.810	6

Figure 7C.4

161

STACK LOSS PASS 28   DEP VAR 1: S.LOSS   SEARCH NUMBER 1

SELECTION OF VARIABLES FOR BASIC EQUATION

CP = 1.5, VARIABLE ADDED, 1

CP = 2.7, VARIABLE ADDED 2

NUMBER OF OBSERVATIONS   6
NUMBER OF VARIABLES IN FULL EQUATION   3
NUMBER OF VARIABLES IN BASIC EQUATION   0
REMAINDER OF VARIABLES TO BE CONSIDERED   3

EQUATION   P   CP   VARIABLES IN EQUATION

           4   4.0   FULL EQUATION

                      NO BASIC SET OF VARIABLES

    1   2   1.5   BASIC SET PLUS   A.FLOW
                                         1

    2   3   2.7   BASIC SET PLUS   A.FLOW   W.TEMP
                                         1   2

    3   4   4.0   BASIC SET PLUS   A.FLOW   W.TEMP   FLOWSQ
                                         1   2   3

    4   3   2.9   BASIC SET PLUS   A.FLOW   FLOWSQ
                                         1   3

SWEEP NUMBER   8,   DELTA Z(K,K) = 0.

SEQ   EQUATION   P   ORDERED CP
 1   1   2   1.5
 2   2   3   2.7
 3   4   3   2.9
 4   3   4   4.0

Figure 7C.5

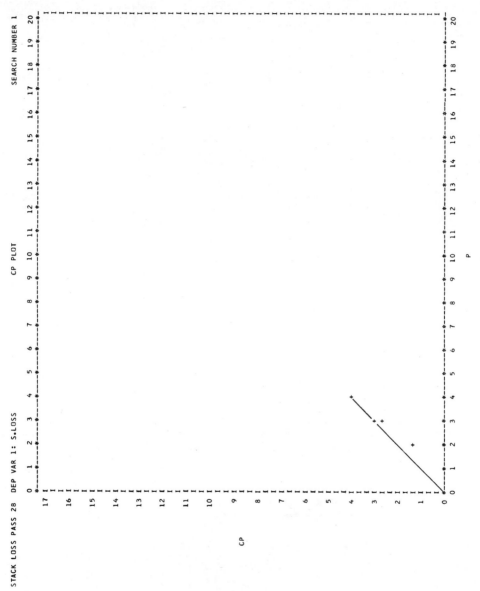

STACK LOSS PASS 28   DEP VAR 1: S.LOSS   CP PLOT   SEARCH NUMBER 1

Figure 7C.6

163

LINEAR LEAST-SQUARES CURVE FITTING PROGRAM

STACK LOSS PASS 30   DEP VAR 1: S.LOSS   MIN Y = 8.000D 00   MAX Y = 3.700D 01   RANGE Y = 2.900D 01

```
          Y = B(0) + B(I)X1
          Y = STACK LOSS
          X1 = AIR FLOW
```

1ST DAY OBSERVATIONS WITH FLOW GREATER THAN 60 OMITTED(1,3,4,21)
CONTIGUOUS LINED OUT OBSERVATIONS AVERAGED(5-8, 10-14 AND 15-19).

IND.VAR(I)	NAME	COEF.B(I)	S.E. COEF.	T-VALUE	R(I)SQRD	MIN X(I)	MAX X(I)	RANGE X(I)	REL.INF.X(I)
0		-4.1237D 01							
1	A.FLOW	9.72462D-01	6.71D-02	14.5	0.0	5.000D 01	8.000D 01	3.000D 01	1.01

```
NO. OF OBSERVATIONS            6
NO. OF IND. VARIABLES          1
RESIDUAL DEGREES OF FREEDOM    4
F-VALUE                      210.2
RESIDUAL ROOT MEAN SQUARE      1.53734343
RESIDUAL MEAN SQUARE           2.36342481
RESIDUAL SUM OF SQUARES        9.45369924
TOTAL SUM OF SQUARES         506.25208333
MULT. CORREL. COEF. SQUARED    .9813
```

-------ORDERED BY COMPUTER INPUT-------

IDENT.	OBSV.	WS DISTANCE	OBS. Y	FITTED Y	RESIDUAL
2	1	150.	37.000	36.559	0.441
5- 8	2	1.	18.750	19.055	-0.305
9	3	3.	15.000	15.165	-0.165
10-14	4	3.	12.800	15.165	-2.365
15-19	5	46.	8.000	7.385	0.615
20	6	9.	15.000	13.220	1.780

-------ORDERED BY RESIDUALS-------

OBSV.	OBS. Y	FITTED Y	ORDERED RESID.	SEQ
6	15.000	13.220	1.780	1
5	8.000	7.385	0.615	2
1	37.000	36.559	0.441	3
3	15.000	15.165	-0.165	4
2	18.750	19.055	-0.305	5
4	12.800	15.165	-2.365	6

Figure 7C.7

CHAPTER 8

Selection of Variables in Nested Data

8.1 Background of Example

As an example of cautious least-square fitting and selection of influential variables, we take the problem of estimating gasoline yields from various characteristics of the crude and a characteristic of the gasoline produced. The data in Table 8.1 appeared first in an article by N. H. Prater in the *Petroleum Refiner*, and again in a chapter on "Simple and Multiple Regression Analyses" by Hader and Grandage.

The data were obtained in a laboratory study of the distillation properties of various crude oils with respect to their yield of gasoline. The four independent variables measured were:

x_1, crude oil gravity, °API x_3, crude oil ASTM 10% point, °F
x_2, crude oil vapor pressure, psi x_4, gasoline ASTM end point, °F

Thus x_1, x_2, and x_3 are properties of the crude, and x_4 is a more or less independent property of the gasoline produced. The response, y, is gasoline yield, as percentage of crude. The crude oil ASTM 10% point and the gasoline ASTM end point are measurements of crude oil and gasoline volatility. Both measure the temperature at which a given amount of liquid has been vaporized. The end point is the temperature at which *all* of the liquid has been vaporized. As gasoline is distilled from a given crude, the volume of gasoline increases, as does its end point.

The purpose of this work was to obtain an equation for estimating gasoline yields, given the end point of the gasoline desired and the properties of available crude.

TABLE 8.1
PRATER'S DATA ON CRUDE OIL PROPERTIES AND GASOLINE YIELDS

Crude Oil			Gasoline	
Gravity °API x_1	Vapor Pressure, psi x_2	ASTM 10% Point, °F x_3	ASTM End Point, °F x_4	Yield, % y
38.4	6.1	220	235	6.9
40.3	4.8	231	307	14.4
40.0	6.1	217	212	7.4
31.8	0.2	316	365	8.5
40.8	3.5	210	218	8.0
41.3	1.8	267	235	2.8
38.1	1.2	274	285	5.0
50.8	8.6	190	205	12.2
32.2	5.2	236	267	10.0
38.4	6.1	220	300	15.2
40.3	4.8	231	367	26.8
32.2	2.4	284	351	14.0
31.8	0.2	316	379	14.7
41.3	1.8	267	275	6.4
38.1	1.2	274	365	17.6
50.8	8.6	190	275	22.3
32.2	5.2	236	360	24.8
38.4	6.1	220	365	26.0
40.3	4.8	231	395	34.9
40.0	6.1	217	272	18.2
32.2	2.4	284	424	23.2
31.8	0.2	316	428	18.0
40.8	3.5	210	273	13.1
41.3	1.8	267	358	16.1
38.1	1.2	274	444	32.1
50.8	8.6	190	345	34.7
32.2	5.2	236	402	31.7
38.4	6.1	220	410	33.6
40.0	6.1	217	340	30.4
40.8	3.5	210	347	26.6
41.3	1.8	267	416	27.8
50.8	8.6	190	407	45.7

166

TABLE 8.2
PRATER'S DATA ARRANGED IN ORDER OF DECREASING X_1

| Identi-fication Number | Crude Oil | | | Gasoline | |
| | Gravity, °API | Vapor Pressure, psi | ASTM 10% Point, °F | ASTM End Point, °F | Yield, % |
	x_1	x_2	x_3	x_4	y
1-1	50.8	8.6	190	205	12.2
2				275	22.3
3				345	34.7
4				407	45.7
2-1	41.3	1.8	267	235	2.8
2				275	6.4
3				358	16.1
4				416	27.8
3-1	40.8	3.5	210	218	8.0
2				273	13.1
3				347	26.6
4-1	40.3	4.8	231	307	14.4
2				367	26.8
3				395	34.9
5-1	40.0	6.1	217	212	7.4
2				272	18.2
3				340	30.4
6-1	38.4	6.1	220	235	6.9
2				300	15.2
3				365	26.0
4				410	33.6
7-1	38.1	1.2	274	285	5.0
2				365	17.6
3				444	32.1
8-1	32.2	5.2	236	267	10.0
2				360	24.8
3				402	31.7
9-1	32.2	2.4	284	351	14.0
2				424	23.2
10-1	31.8	0.2	316	365	8.5
2				379	14.7
3				428	18.0

8.2   Recognition of Nested Data

The first step in recognizing nested data is to sort the data on each variable. With larger sets of data, rearranging can be done easily on a card sorter, and the resultant groupings observed from a printout of the sorted cards. In this case, rearranging the data by hand in order of decreasing x_1, as in Table 8.2, reveals that we are dealing with *ten* crudes rather than thirty-two, and that gasoline ASTM end point, x_4, is indeed "nested" within crudes. Since the previous authors analyzed the data as though there were thirty-two independent observations with constant variance, we will have an opportunity to see how their conclusions differ from ours.

8.3   Data Identification and Entry

The data were identified for convenience by crude number and by observation number within crudes, and then entered on special forms for linear least-squares fitting by the computer. Indicator variables were included in order to separate and quantify the effects of the various crudes. Details concerning data entry and the various computer passes discussed in this chapter are given in Appendix 8A.

8.4   Fitting Equation, Ignoring Nesting

Pass 1 is the least-squares fit of the gasoline yields as a linear function of the four independent variables:

$$Y = b_0 + b_1 x_1 + b_2 x_2 + b_3 x_3 + b_4 x_4.$$

This corresponds to the treatment of the data by Prater* and by Hader and Grandage. Computer printouts of the results are shown in Figures 8A.3–8A.6. We will see the importance of recognizing nested data when these results are compared with subsequent fitted equations.

8.5   Fitting Equation to Nested Data

A number of computer passes were made to determine whether some other equations would fit the nested data better. The effect of gasoline end point on gasoline yield can be seen in Figure 8.1. A least-squares line has been drawn for each crude.

* There are a number of numerical errors in the Prater paper.

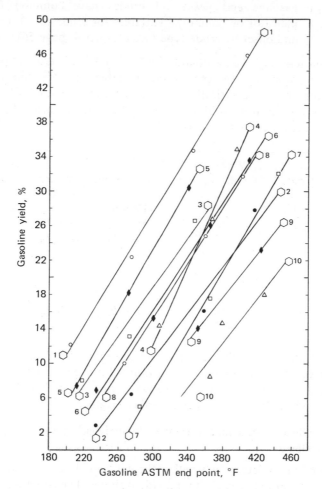

Figure 8.1 Gasoline yields from ten crudes.

Use of indicator (dummy) variables

Pass 2 was made under the assumption that a correlation between gasoline end point and yield could be represented by a set of ten straight lines with a common slope (each line representing a different crude). The equation used was:

$$Y = b_0 + b_1 x_1 + b_2 x_2 + \cdots + b_{10} x_{10},$$

where y = gasoline yield,

        x_1 = gasoline end point—all crudes have common b_1 slope,

$x_2, \ldots, x_{10}$ = indicator or "dummy" variables for crudes 2–10, to separate and offset by crude type (see Chapter 4, page 56).

The constant term, b_0, is an estimate of y for Crude 1 at zero end point, and the coefficient of each of the other crude variables indicates the magnitude by which that crude systematically differs from Crude 1. Thus, with $i = 2\text{–}10$, when a given x_i (indicator variable) is 1, the value of the associated b_i becomes the offset of the ith crude from Crude 1.

Crude	i, Where $x_i = 1$, All Other $x_i = 0$	Slope Intercept
1		$Y_1 = b_1 x_1 + b_0$
2	2	$Y_2 = b_1 x_1 + b_0 + b_2 x_2$
3	3	$Y_3 = b_1 x_1 + b_0 + b_3 x_3$
4	4	$Y_4 = b_1 x_1 + b_0 + b_4 x_4$
.	.	.
.	.	.
.	.	.
10	10	$Y_{10} = b_1 x_1 + b_0 + b_{10} x_{10}$

Computer printouts of Pass 2 are shown in Figures 8A.7–8A.9. As can be seen in Figure 8A.7, the fitted values for points 1 and 3 of Crude 4—observations *41* and *43*—have the largest residuals, suggesting that the slope for Crude 4 may differ from the slopes for other crudes.

Test for significance of added variables, "individual" versus "common" slopes

Pass 3 was run to determine whether the data could be represented significantly better by straight lines with individual slopes. The equation used was:

$$Y = b_0 + b_1 x_1 + b_2 x_2 + \cdots + b_{10} x_{10} + b_{11} x_{11} + \cdots + b_{19} x_{19}.$$

Variables $x_2, \ldots, x_{10}$ are indicator variables as before. Variables $x_{11}, \ldots, x_{19}$ are cross-product terms to allow each crude to have its own slope, for example, $x_{11} = x_1 x_2$, $x_{12} = x_1 x_3$. Transformation card entries were made as shown in Figure 8A.1 to include variables representing the cross products. The mean end point of all the data (332°F, obtained from Pass 1) was subtracted from each observation in order to obtain more meaningful t-values by reducing the correlations between independent variables.

The constant term, b_0, is then the value of Y for Crude 1 at 332°F, and Y is calculated for the various crudes as shown in the table. The results of the fit of the full equation are shown in Figure 8A.10.

Crude	i, Where $x_i = 1$, All Other $x_i = 0$	Slope	Intercept
1		$Y_1 = b_1 x_1 \qquad\qquad + b_0$	
2	2	$Y_2 = b_1 x_1 + b_{11} x_1 x_2 + b_0 + b_2 x_2$	
3	3	$Y_3 = b_1 x_1 + b_{12} x_1 x_3 + b_0 + b_3 x_3$	
4	4	$Y_4 = b_1 x_1 + b_{13} x_1 x_4 + b_0 + b_4 x_4$	
.	.	.	
.	.	.	
.	.	.	
10	10	$Y_{10} = b_1 x_1 + b_{19} x_1 x_{10} + b_0 + b_{10} x_{10}$	

The hypothesis that the additional individual slope coefficients do not improve the fit was tested by the F ratio of the mean square due to the extra variables divided by the residual mean square (Brownlee, Section 13.8, page 443, and Davies, 1960, Section 11B.32).

Source of Variation	Number of Coefficients	Degrees of Freedom	Residual Sum of Squares	Residual Mean Square
Common slope	$p \quad = 10$	21	74	
Individual slope	$p + q = 19$	12	30	2.5
Extra variables	$q = 9$	9	44	4.9

$$F(9, 12) = \frac{\text{RMS}_q}{\text{RMS}_{p+q}} = \frac{4.9}{2.5} = 1.9.$$

From F-tables, $F(0.95, 9, 12)$ is 2.8. Thus, if we use the 95% level as our criterion of significance, the calculated F-ratio value is nonsignificant. The data are therefore considered to be compatible with the hypothesis that the additional coefficients do not improve the fit.

Test for outliers

The question was then asked: If (as in Pass 2) all the lines are parallel, is point *4-3* far enough away from the Crude 4 line to be considered an "outlier"? Its measured yield from Figure 8A.7 is 3.4 units away from its calculated yield, contributing $(3.4)^2$ or 11.5 units to the sum of squares due to error. From Table XX (Fisher and Yates, page 86) we find that for a sample size of 32 the mean deviation of the largest member of a ranked sample is 2.07. The expected contribution to the sum of squares due to error would be $(2.07)^2$

times the mean square from Pass 2 (3.5) or 16.6 units. Since the contribution of point *4-3* to the sum of squares due to error is not abnormally large (11.5 versus 16.6), this point cannot be considered an outlier.

8.6 Fitting Equation Among Sets of Nested Data

From the computer printout of Pass 2, Figure 8A.7, we see that b_1 (the rate of change of gasoline yield per degree change in gasoline end point) was 0.1587. The standard error of this coefficient is 0.0057. Hence the slope might well be reported as 0.159 ± 0.006.

This average within-crude slope was used to adjust the ten crudes to the same end-point value, 332°F. This is done graphically in Figure 8.2, and the

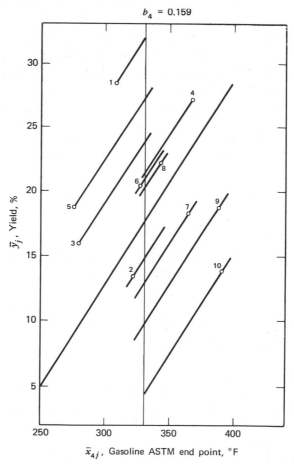

Figure 8.2 Graphical adjustment of yields to constant end point.

resulting gasoline yields are given in Figure 8A.11 as variable 4. This figure presents the data used to correlate the yield of gasoline at this constant end point with gravity, vapor pressure, and 10% point of the crudes, that is, with x_1, x_2, and x_3.

With three independent variables there are 2^3 equations to be considered. All seven combinations of variables were used, as shown in the table.

Pass	Crude Properties		
	Gravity	Vapor Pressure	10% Point
4	x_1	x_2	x_3
5	x_1	x_2	—
6	x_1	—	x_3
7	—	x_2	x_3
8	x_1	—	—
9	—	x_2	—
10	—	—	x_3

Results are summarized in Table 8.3.

8.7 Selecting Variables Based on Total Error, Recognizing Bias and Random Error

Since there are a number of combinations of variables from which to choose, Mallows' graphical method of plotting p versus C_p is used to review all the alternatives. The statistic C_p was defined in Chapter 6 as follows:

$$C_p = \frac{\text{RSS}}{s^2} - (N - 2p),$$

where RSS = residual sum of squares,

s^2 = best estimate of variance (normally the residual mean square of the equation with all variables present),

N = number of observations,

p = number of constants to be estimated; $p_{\max} = K + 1$ with b_0 present, $p_{\max} = K$ with b_0 absent.

For illustration, values for p and C_p have been calculated from computer Passes 4–10 and are given in Table 8.4. In practice, only Pass 4, the full equation, need be run in the curve fitting program, with the option set to search for the candidate equations.

TABLE 8.3
SUMMARY OF PASSES 4-10:
YIELD OF 332°F END-POINT GASOLINE VERSUS GRAVITY, VAPOR PRESSURE, AND 10% POINT OF CRUDE

	Pass 4 (x_1, x_2, x_3)				Pass 5 (x_1, x_2)				Pass 6 (x_1, x_3)				Pass 7 (x_2, x_3)			
	Value	b_i	$s(b_i)$	t_i	Value	b_i	$s(b_i)$	t_i	Value	b_i	$s(b_i)$	t_i	Value	b_i	$s(b_i)$	t_i
Degrees of freedom	6				7				7				7			
F-value	70				30				100				85			
R^2	0.971				0.89				0.966				0.961			
RMS	3.10				10.10				3.15				3.63			
RSS	18.6				70.7				22.1				25.4			
SSFE	630.6				578.5				627.1				623.8			
C_p	4.0				19.0				3.1				4.2			
Variable																
x_1, gravity		0.22	0.16	1.5		0.50	0.23	2.1		0.20	0.15	1.4				
x_2, vapor pressure		0.54	0.51	1.0		2.27	0.51	4.4*						0.47	0.55	0.9
x_3, 10% point		-0.16	0.04	4.1*						-0.19	0.02	8.9*		-0.18	0.04	5.0*

	Pass 8 (x_1)				Pass 9 (x_2)				Pass 10 (x_3)			
	Value	b_i	$s(b_i)$	t_i	Value	b_i	$s(b_i)$	t_i	Value	b_i	$s(b_i)$	t_i
Degrees of freedom	8				8				8			
F-value	10				35				175			
R^2	0.58				0.82				0.957			
RMS	33.76				14.5				3.52			
RSS	270.1				116.0				28.1			
SSFE	379.1				533.2				621.1			
C_p	81.0				31.0				3.0			
Variable												
x_1, gravity		1.13	0.34	3.3*								
x_2, vapor pressure						2.93	0.48	6.1*				
x_3, 10% point										-0.21	0.02	13.3*

* Significant at the 99% level.

174

TABLE 8.4
C_p OF EQUATIONS USED IN PASSES 4–10

$$C_p = \frac{RSS}{s^2} - (N - 2p), \quad N = 10, s^2 = 3.10^*$$

Pass	Variables	p	SSE	$\dfrac{SSE}{s^2}$	$N - 2p$	C_p
4	x_1, x_2, x_3	4	18.6	6.0	2	4.0
5	x_1, x_2	3	70.7	23.0	4	19.0
6	x_1, x_3	3	22.1	7.1	4	3.1
7	x_2, x_3	3	25.4	8.2	4	4.2
8	x_1	2	270.1	87.0	6	81.0
9	x_2	2	116.0	37.0	6	31.0
10	x_3	2	28.1	9.0	6	3.0
	None	1	649.2	209.0	8	201.0

* s^2 from Pass 4.

The C_p versus p plot of all of the equations is shown in Figure 8.3.

As all equations close to the zero-bias line, $C_p = p$, contain x_3, this is clearly the most important variable. The equation with x_1 and x_3 is on the line and so is judged to have no bias, or at least no more than the full equation in x_1, x_2, and x_3. The equation with x_3 alone has more bias (A to B) and less random error (B to C) than the equation with x_1 and x_3. Both have sensibly the same total squared error (bias plus random).

The value of each coefficient, as affected by the inclusion of other variables, is as follows:

Pass	b_1	b_2	b_3
4	0.22	0.54	−0.16
5	0.50	2.27	—
6	0.20	—	−0.19
7	—	0.47	−0.18
8	1.13	—	—
9	—	2.93	—
10	—	—	−0.21

Coefficient b_3 is relatively stable, regardless of the presence of other variables. With b_3 present, b_1 is relatively stable, b_2 more erratic. We do not have a good estimate of the influence of x_2, but it appears that we have determined clearly the influences of x_1 and x_3. Combining this information with that from the C_p plots, we conclude that the x_3 equation is the most economical one for

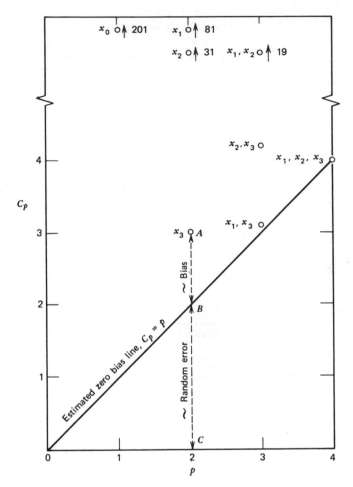

Figure 8.3 C_p versus p plot.

minimizing total error (bias plus random), and the x_1-x_3 equation the most economical for minimizing bias. We have chosen the x_3 equation from Pass 10 for our prediction equation.

8.8 Properties of Final Equation

From Pass 10, b_3, the rate of change of gasoline yield per degree of the 10% point of the crude, was found to be -0.212 ± 0.016. It is rather important to remember that the standard error of the coefficient is based on only 8 residual degrees of freedom and that b_3 is based on only ten crudes. The equation can be expected to hold only for crudes like these, and *not* for dissimilar ones.

The residual mean square of Pass 10 (3.52) is really composed of two parts. One is due to the fact that each adjusted y is the average of several within-crude values and hence suffers somewhat from the within-crude variation, and the other is due to whatever random factors vary from crude to crude. We have an estimate of the "within-crude" component from Pass 2; it is 3.53 with 21 degrees of freedom. We can get a rough estimate of the "among-crude" component as follows:

$$\text{RMS (Pass 14)} = s^2_{\text{among}} + \frac{s^2_{\text{within}}}{n},$$

where n is the average number of runs per crude, or 32/10 or 3.2. Thus:

$$3.52 = s^2_{\text{among}} + \frac{3.53}{3.2},$$

whence

$$s^2_{\text{among}} = 3.52 - 1.10 = 2.42.$$

All this work can be summarized in one equation *plus* the two separate variance-component estimates just given. The "nested" equation is

$$Y = 70.84 - 0.212x_3 + 0.159(x_4 - 332).$$

A detailed view of how this equation represents the thirty-two data points from which it was derived can be observed in the last two columns of Table 8.5, headed d_w and d_a. The d_w-values represent *within*-crude variations, and the d_a *among*-crude variations. The total deviation between an observation and its value predicted by the equation is the sum of a d_w and a d_a.

The estimated variance of Y for the equation given above, at $x_4 = 332°F$ end point, is:

$$\text{Est. Var. } (Y) = s^2 \left[\frac{1}{N} + \frac{(x_3 - \bar{x}_3)^2}{\Sigma(x_{3i} - \bar{x}_3)^2} \right]$$

(see Chapter 2, page 12). For $x_3 = \bar{x}_3$, Est. Var. $(\bar{Y})$ is s^2/N.

For this final equation, the effective variance is 3.52. Since there are ten crudes,

$$\text{Est. Var. } (\bar{Y}) = \frac{3.52}{10} = 0.352.$$

The resultant 95% confidence bounds at the center of the data are then

$$\bar{Y} \pm \{[2F(0.95, 2, N - 2)][\text{Est. Var. } (\bar{Y})]\}^{\frac{1}{2}} = \bar{y} \pm [2(4.46)(0.35)]^{\frac{1}{2}}$$
$$= \bar{y} \pm 1.77.$$

TABLE 8.5
COMPARISON OF OBSERVED AND PREDICTED YIELDS
(Prater Data)

Identification Number	Gasoline Yields, % on Crude		
	Observed	d_w (Pass 2)	d_a (Pass 14)
1-1	12.2	−0.2	2.0
2	22.3	−1.2	
3	34.7	0.1	
4	45.7	1.3	
2-1	2.8	3.2	0.8
2	6.4	0.4	
3	16.1	−3.0	
4	27.8	−0.6	
3-1	8.0	1.8	−2.0
2	13.1	−1.8	
3	26.6	0.0	
4-1	14.4	−3.1	−0.3
2	26.8	−0.3	
3	34.9	3.4	
5-1	7.4	−1.3	2.9
2	18.2	0.0	
3	30.4	1.3	
6-1	6.9	1.2	−3.0
2	15.2	−0.9	
3	26.0	−0.4	
4	33.6	0.1	
7-1	5.0	−0.6	0.3
2	17.6	−0.7	
3	32.1	1.3	
8-1	10.0	−0.1	−0.4
2	24.8	−0.1	
3	31.7	0.2	
9-1	14.0	1.2	−0.8
2	23.2	−1.2	
10-1	8.5	−1.2	0.5
2	14.7	2.8	
3	18.0	−1.6	

178

8.9   Comparison of Equations

It is interesting to compare both the fit and the usefulness of the equation just given with those of the former equation, which ignored nesting. As seen in Table 8.6, the new equation is simpler (only half the number of variables are required) and the fit is better (the residual mean square is one-third smaller). In addition, we now have estimates of two components of variance (among and within crudes). The earlier equation, ignoring these components, produced a confidence interval for Y that was 40% too small.

TABLE 8.6
COMPARISON OF EQUATIONS, FORMER AND PRESENT
(Prater Data)

	Equation	
Criterion	Former	Present
Number of independent variables	4	2
Overall fit		
C_p	4.0	3.0
Residual mean square	5.0	3.5
Components of variance		
Among crudes (9 df)	—	2.42
Within crudes (19 df)	—	3.53
95% confidence interval for Y at		
center of data	1.02	1.77

APPENDIX 8A

DATA PREPARATION AND COMPUTER PRINTOUTS OF EXAMPLE

Figure 8A.1 shows (1) the data entry form used for card keypunching; (2) the information given on the format card to tell the computer the locations and dimensions of the fields used for identification, observation numbers, sequence numbers, and variables on the data cards; and (3) a summary of the transformation card entries for the first six computer passes. Details of the control card and transformation entry form used for the first pass are shown in Figure 8A.2.

The order of the cards for the first two passes run in sequence was as follows:

> Pass 1   Control card (1 in columns 42 and 48)
>   Format card
>   Transformation cards (5)
>   Information cards (4)
>   Variable name cards (5)
>   Data cards
>   END card (the word "END" in columns 1–3)
>
> Pass 2   Control card (2 in columns 42 and 48)
>   Transformation card (5)
>   Information cards (6)

Both passes used the same format; otherwise, new format, data, and END cards would have been required for each pass.

Printouts of the various computer passes are given in the following figures:

Pass	Figures
1	8A.3–8A.6
2	8A.7–8A.9
3	8A.10
4	8A.11

ENTER WEIGHTING FACTOR, IF ANY, AS LAST ENTRY.
"END" CARD MUST BE LAST CARD.

IDENT.	OBSV. NO.	SEQ NO.	$X_{1-11-21}$	$X_{2-12-22}$	$X_{3-13-23}$	$X_{4-14-24}$	$X_{5-15-25}$	$X_{6-16-26}$	$X_{7-17-27}$	$X_{8-18-28}$	$X_{9-19-29}$	$X_{10-20-30}$	USER'S OPTION
CRUDE			GRAVITY	VAPOR PRS	10% PT	END PT		CRUDE 2,3,4,5,6,7,8,9,10				YIELD	
1	11		508	86	190	205						122	
	12					275						223	
	13					345						347	
	14					407						457	
2	21		413	118	267	235		1				28	
	22					275						64	
	23					358						161	
	24					416						278	
3	31		408	135	210	218		1				80	
	32					273						131	
	33					347						266	
4	41		403	48	231	307		1				144	
	42					367						268	
	43					395						349	
5	51		400	61	217	212		1				74	
	52					272						182	
	53					340						304	
6	61		384		220	235		1				69	
	62					300						152	

FORMAT CARD INFORMATION FOR PASSES 1 TO 6

```
          *a  *b  *c       *d    *e
Card:    (A6, I4, I2, 4F6.2, 26F1.0, 4X, F6.1)
Columns: 12345678901234567890123456789012345678 90---
```

* Explanation of Symbols Used

	Number with this type of Spacing	Type of Field	Number of Columns Each	Number of Columns to Right of Decimal Point
a	1	A = letter or number	6	0
b	1	I = integer	4	0
c	4	F = field	6	2
d	4	X = exclude	1	0
e	Statement must start and end with parenthesis.			

Figure 8A.1

181

Passes 1, 2 and 3

Position	Name	C	ARG	LOC	Variable Pass 1	Pass 2	Pass 3
1	Gravity	00		1	1		
2	Vapor Pressure	00		2	2		
3	10% Point	00		3	3		
4	End Point	00		4	4	1	
4	End Point-Mean E.P.	08*	-332.	4			1
5	(End Point-Mean E.P.)^2	09**	0404	5			
6							
7							
8							
9							
10							
11							

Crude

Position	Name	C	ARG	LOC	Variable Pass 1	Pass 2	Pass 3
12	2	00		12		2	2
13	3	00		13		3	3
14	4	00		14		4	4
15	5	00		15		5	5
16	6	00		16		6	6
17	7	00		17		7	7
18	8	00		18		8	8
19	9	00		19		9	9
20	10	00		20		10	10
21							

End Point x Crude

Position	Name	C	ARG	LOC	Variable Pass 1	Pass 2	Pass 3
22	End Point x 2	09**	0412	22			11
23	End Point x 3	09	0413	23			12
24	End Point x 4	09	0414	24			13
25	End Point x 5	09	0415	25			14
26	End Point x 6	09	0416	26			15
27	End Point x 7	09	0417	27			16
28	End Point x 8	09	0418	28			17
29	End Point x 9	09	0419	29			18
30	End Point x 10	09	0420	30			19
31	Gasoline Yield, %	00		40	5	11	20

* 08 denotes operation: $X \pm K$
** 09 denotes operation: $X(N) \cdot X(M)$

Figure 8A.1 (*continued*).

182

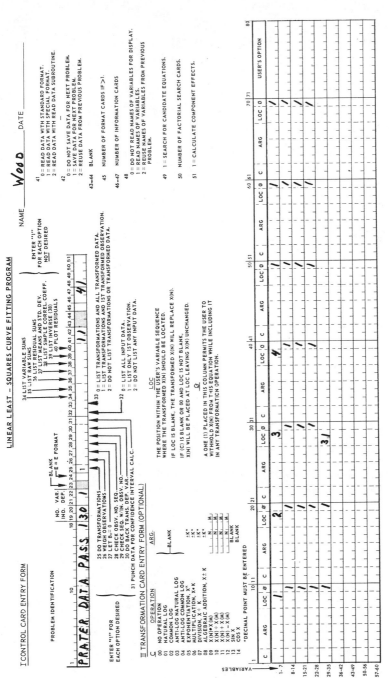

Figure 8A.2

183

PRATER DATA PASS 1

PROBLEM HAS ONE EQUATION

DATA READ WITH SPECIAL FORMAT

FORMAT CARD 1 (A6, I4, I2, 4F6.2, 30X, F6.1)

BZERO = CALCULATED VALUE

DATA INPUT   4 INDEPENDENT VARIABLES   1 DEPENDENT VARIABLES

OBSV.	SEQ.	1-11-21	2-12-22	3-13-23	4-14-24	5-15-25	6-16-26	7-17-27	8-18-28	9-19-29	10-20-30
11	1	50.800	8.600	190.000	205.000	12.200					
12	1	50.800	8.600	190.000	275.000	22.300					
13	1	50.800	8.600	190.000	345.000	34.700					
14	1	50.800	8.600	190.000	407.000	45.700					
21	1	41.300	1.800	267.000	235.000	2.800					
22	1	41.300	1.800	267.000	275.000	6.400					
23	1	41.300	1.800	267.000	358.000	16.100					
24	1	41.300	1.800	267.000	416.000	27.800					
31	1	40.800	3.500	210.000	218.000	8.000					
32	1	40.800	3.500	210.000	273.000	13.100					
33	1	40.800	3.500	210.000	347.000	26.600					
41	1	40.300	4.900	231.000	307.000	14.400					
42	1	40.300	4.800	231.000	367.000	26.800					
43	1	40.300	4.800	231.000	395.000	34.900					
51	1	40.000	6.100	217.000	212.000	7.400					
52	1	40.000	6.100	217.000	272.000	18.200					
53	1	40.000	6.100	217.000	340.000	30.400					
61	1	38.400	6.100	220.000	235.000	6.900					
62	1	38.400	6.100	220.000	300.000	15.200					
63	1	38.400	6.100	220.000	365.000	26.000					
64	1	38.400	6.100	220.000	410.000	33.600					
71	1	38.100	1.200	274.000	285.000	5.000					
72	1	38.100	1.200	274.000	365.000	17.600					
73	1	38.100	1.200	274.000	444.000	32.100					
81	1	32.200	5.200	236.000	360.000	24.800					
82	1	32.200	5.200	236.000	267.000	10.000					
83	1	32.200	5.200	236.000	402.000	31.700					
91	1	32.200	2.400	284.000	351.000	14.000					
92	1	32.200	2.400	284.000	424.000	23.200					
101	1	31.800	0.200	316.000	365.000	8.500					
102	1	31.800	0.200	316.000	379.000	14.700					
103	1	31.800	0.200	316.000	428.000	18.000					

SUMS OF VARIABLES
1.25600D 03 1.33800D 02 7.72800D 03 1.06270D 04 6.29100D 02

MEANS OF VARIABLES
3.92500D 01 4.18125D 00 2.41500D 02 3.32094D 02 1.96594D 01

ROOT MEAN SQUARES OF VARIABLES
5.63543D 00 2.61983D 00 3.75414D 01 6.97560D 01 1.07224D 01

SIMPLE CORRELATION COEFFICIENTS, R(I,I PRIME)

1	1.000	0.621	-0.700	-0.322	0.246
2	0.621	1.000	-0.906	-0.298	0.384
3	-0.700	-0.906	1.000	0.412	-0.315
4	-0.322	-0.298	0.412	1.000	0.712
5	0.246	0.384	-0.315	0.712	1.000

Figure 8A.3

184

LINEAR LEAST-SQUARES CURVE FITTING PROGRAM

PRATER DATA PASS 1   DEP VAR 1: GAS YD   MIN Y = 2.800D 00   MAX Y = 4.570D 01   RANGE Y = 4.290D 01

RESULTANT EQUATION:
Y = -6.8 + 0.227X1 + 0.554X2 - 0.149X3 + 0.155X4
Y = GASOLINE YIELD; X1 = GRAVITY OF CRUDE; X2 = VAPOR PRESSURE OF CRUDE;
X3 = 10 PERCENT POINT OF CRUDE; X4 = GASOLINE END POINT.

IND.VAR(I)	NAME	COEF.B(I)	S.E. COEF.	T-VALUE	R(I)SQRD	MAX X(I)	MIN X(I)	RANGE X(I)	REL.INF.X(I)
0		-6.8207D 00							
1	C GRAV	2.27246D-01	9.99D-02	2.3	0.4922	5.080D 01	3.180D 01	1.900D 01	0.10
2	C VP	5.53260D-01	3.70D-01	1.5	0.8284	8.400D 00	2.000D 00	8.400D 00	0.11
3	C 10PT	-1.49536D-01	2.92D-02	5.1	0.8662	3.160D 02	1.900D 02	1.260D 02	0.44
4	GAS EP	1.54650D-01	6.45D-03	24.0	0.2034	4.440D 02	2.050D 02	2.390D 02	0.86

NO. OF OBSERVATIONS   32
NO. OF IND. VARIABLES   4
RESIDUAL DEGREES OF FREEDOM   27
F-VALUE   171.7
RESIDUAL ROOT MEAN SQUARE   2.23444386
RESIDUAL MEAN SQUARE   4.99273937
RESIDUAL SUM OF SQUARES   134.80396290
TOTAL SUM OF SQUARES   3564.07718750
MULT. CORREL. COEF. SQUARED   .9622

ORDERED BY WS DISTANCE / COMPUTER INPUT

IDENT.	OBSV.	WS DISTANCE	OBS. Y	FITTED Y	RESIDUAL
1	11	92.	12.200	12.777	-0.577
1	12	30.	22.300	23.602	-1.302
1	13	15.	34.700	34.428	0.272
1	14	41.	45.700	44.016	1.684
2	21	48.	2.800	-0.022	2.822
2	22	19.	6.400	6.164	0.236
2	23	7.	16.100	19.000	-2.900
2	24	37.	27.800	27.970	-0.170
3	31	67.	8.000	6.700	1.300
3	32	21.	13.100	15.206	-2.106
3	33	6.	26.600	26.650	-0.050
4	41	4.	14.400	17.930	-3.530
4	42	6.	26.800	27.209	-0.409
4	43	19.	34.900	31.539	3.361
5	51	72.	7.400	5.983	1.417
5	52	20.	18.200	15.262	2.938
5	53	3.	30.400	25.779	4.621
6	61	47.	6.900	8.728	-1.828
6	62	7.	15.200	18.780	-3.580
6	63	7.	26.000	28.833	-2.833
6	64	31.	33.600	35.792	-2.192
7	71	16.	5.000	5.604	-0.604
7	72	10.	17.600	17.976	-0.376
7	73	65.	32.100	30.194	1.906
8	81	4.	24.800	23.760	1.040
8	82	21.	10.000	9.377	0.623
8	83	24.	31.700	30.255	1.445
9	91	11.	14.000	13.640	0.360
9	92	49.	23.200	24.929	-1.729
10	101	32.	8.500	9.710	-1.210
10	102	37.	14.700	11.876	2.824
10	103	70.	18.000	19.453	-1.453

ORDERED BY RESIDUALS

OBSV.	OBS. Y	FITTED Y	ORDERED RESID.	SEQ
53	30.400	25.779	4.621	1
43	34.900	31.539	3.361	2
52	18.200	15.262	2.938	3
102	14.700	11.876	2.824	4
21	2.800	-0.022	2.822	5
73	32.100	30.194	1.906	6
14	45.700	44.016	1.684	7
83	31.700	30.255	1.445	8
51	7.400	5.983	1.417	9
31	8.000	6.700	1.300	10
82	24.800	23.760	1.040	11
81	10.000	9.377	0.623	12
91	14.000	13.640	0.360	13
13	34.700	34.428	0.272	14
22	6.400	6.164	0.236	15
33	26.600	26.650	-0.050	16
24	27.800	27.970	-0.170	17
72	17.600	17.976	-0.376	18
42	26.800	27.209	-0.409	19
11	12.200	12.777	-0.577	20
71	5.000	5.604	-0.604	21
101	8.500	9.710	-1.210	22
12	22.300	23.602	-1.302	23
103	18.000	19.453	-1.453	24
92	23.200	24.929	-1.729	25
61	6.900	8.728	-1.828	26
32	13.100	15.206	-2.106	27
64	33.600	35.792	-2.192	28
63	26.000	28.833	-2.833	29
23	16.100	19.000	-2.900	30
41	14.400	17.930	-3.530	31
62	15.200	18.780	-3.580	32

Figure 8A.4

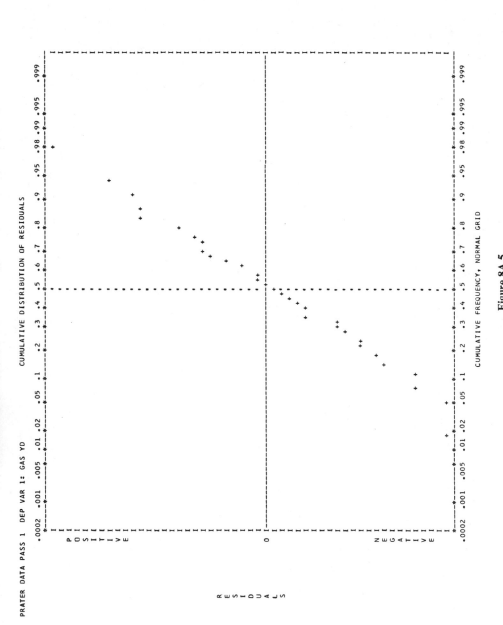

Figure 8A.5

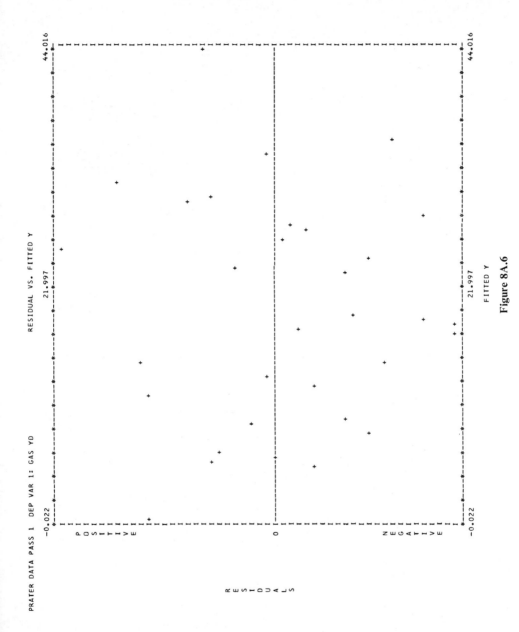

Figure 8A.6

LINEAR LEAST-SQUARES CURVE FITTING PROGRAM

PRATER DATA PASS 2   DEP VAR 1: GAS YD   MIN Y = 2.800D 00 MAX Y = 4.570D 01 RANGE Y = 4.290D 01

RESULTANT EQUATION:
Y = -20.2 + 0.159X1 - 17.5X2 - 8.3X3 - 11.0X4 - 4.8X5 - 11.4X6
Y = -20.2 + 0.159X7 - 19.5X8 - 22.7X9 - 28.1X10

Y = GASOLINE YIELD; X1 = GASOLINE END POINT; X2 ---- X10 = CRUDES 2 TO 10
Y FOR CRUDE 1 = B(0) + B(1)X1; Y FOR CRUDE 2 = (B(0) + B(2)X2) + B(1)X1;
Y FOR CRUDE 3 = (B(0) + B(3)X3) + B(1)X1; ETC.

IND-VAR(I)	NAME	COEF+B(I)	S.E. COEF.	T-VALUE	R(I)SQRD	MIN X(I)	MAX X(I)	RANGE X(I)	REL.INF.X(I)
0	GAS EP	-2.01637D 01	5.720-03	27.8	0.2843	2.050D 02	4.440D 02	2.390D 02	0.88
1	CRUDE2	1.58730D-01	1.330 00	13.2	0.4304	0.0	1.000D 00	1.000D 00	0.41
2	CRUDE3	-8.27475D 00	1.440 00	5.7	0.3376	0.0	1.000D 00	1.000D 00	0.19
3	CRUDE4	-1.10303D 01	1.460 00	7.5	0.2920	0.0	1.000D 00	1.000D 00	0.26
4	CRUDE5	-4.76735D 00	1.450 00	3.3	0.3804	0.0	1.000D 00	1.000D 00	0.11
5	CRUDE6	-1.13952D 01	1.330 00	8.5	0.4326	0.0	1.000D 00	1.000D 00	0.27
6	CRUDE7	-1.94863D 01	1.470 00	13.2	0.4000	0.0	1.000D 00	1.000D 00	0.45
7	CRUDE8	-1.21139D 01	1.450 00	8.4	0.3815	0.0	1.000D 00	1.000D 00	0.28
8	CRUDE9	-2.27440D 01	1.690 00	13.5	0.3404	0.0	1.000D 00	1.000D 00	0.53
9	CRDE10	-2.81133D 01	1.510 00	18.6	0.4312	0.0	1.000D 00	1.000D 00	0.66

NO. OF OBSERVATIONS   32
NO. OF IND. VARIABLES   10
RESIDUAL DEGREES OF FREEDOM   21
F-VALUE   98.9
RESIDUAL ROOT MEAN SQUARE   1.87885466
RESIDUAL MEAN SQUARE   3.53009485
RESIDUAL SUM OF SQUARES   74.13199186
TOTAL SUM OF SQUARES   3564.07718750
MULT. CORREL. COEF. SQUARED   .9792

IDENT.	OBSV.	----ORDERED BY COMPUTER INPUT---- WS DISTANCE	OBS. Y	FITTED Y	RESIDUAL
1	11	122.	12.200	12.376	-0.176
1	12	30.	22.300	23.487	-1.187
1	13	8.	34.700	34.598	0.102
1	14	46.	45.700	44.439	1.261
2	21	139.	2.800	-0.376	3.176
2	22	95.	6.400	6.977	-0.427
2	23	122.	18.100	19.148	-1.048
2	24	115.	27.800	28.354	-0.554
3	31	77.	8.000	6.165	1.835
3	32	47.	13.100	14.895	-1.795
3	33	24.	26.600	26.641	-0.041
4	41	39.	14.400	17.536	-3.136
4	42	43.	26.800	27.060	-0.260
4	43	63.	34.900	31.504	3.396
5	51	114.	7.400	8.720	-1.320
5	52	37.	18.200	18.243	-0.043
5	53	12.	30.400	30.037	0.363
6	61	101.	10.900	5.743	5.157
6	62	41.	15.200	16.060	-0.860
6	63	42.	26.000	26.377	-0.377
6	64	77.	33.600	33.520	0.080
7	71	110.	5.000	5.588	-0.588
7	72	101.	17.600	18.286	-0.686
7	73	183.	32.100	30.826	1.274
8	81	70.	24.800	24.865	-0.065
8	82	75.	10.000	10.103	-0.103
8	83	137.	31.700	31.532	0.168
9	91	101.	14.000	12.806	1.194
9	92	196.	23.200	24.394	-1.194
10	101	196.	8.500	9.659	-1.159
10	102	204.	14.700	11.881	2.819
10	103	254.	18.000	19.659	-1.659

OBSV.	----ORDERED BY RESIDUALS---- OBS. Y FITTED Y	ORDERED RESID.	SEQ
43	34.900 31.504	3.396	1
21	2.800 -0.376	3.176	2
102	14.700 11.881	2.819	3
31	8.000 6.165	1.835	4
53	30.400 29.037	1.363	5
73	32.100 30.826	1.274	6
14	45.700 44.439	1.261	7
91	14.000 12.806	1.194	8
61	10.900 5.743	1.157	9
6	6.400 5.973	0.427	10
22	6.400 31.532	0.168	11
83	31.700 34.598	0.102	12
13	34.700 33.520	0.080	13
64	33.600 26.641	-0.041	14
33	26.600 18.243	-0.043	15
52	18.200 24.865	-0.065	16
82	24.800 10.103	-0.103	17
81	10.000 12.376	-0.176	18
11	12.200 31.532	-0.260	19
42	26.800 27.060	-0.377	20
63	26.000 26.377	-0.377	21
24	27.800 28.354	-0.554	22
71	5.000 5.588	-0.588	23
72	17.600 18.286	-0.686	24
62	15.200 16.060	-0.860	25
101	8.500 9.659	-1.159	26
12	22.300 23.487	-1.187	27
92	23.200 24.394	-1.194	28
51	7.400 8.720	-1.320	29
103	18.000 19.659	-1.659	30
32	13.100 14.895	-1.795	31
23	18.100 19.148	-3.048	32
41	14.400 17.536	-3.136	

Figure 8A.7

188

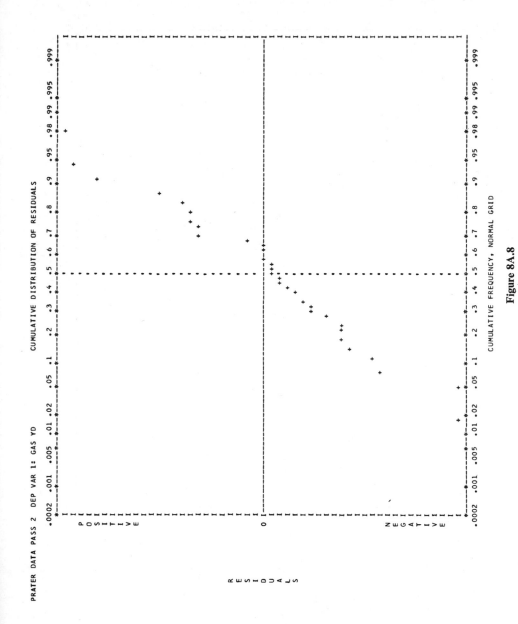

Figure 8A.8

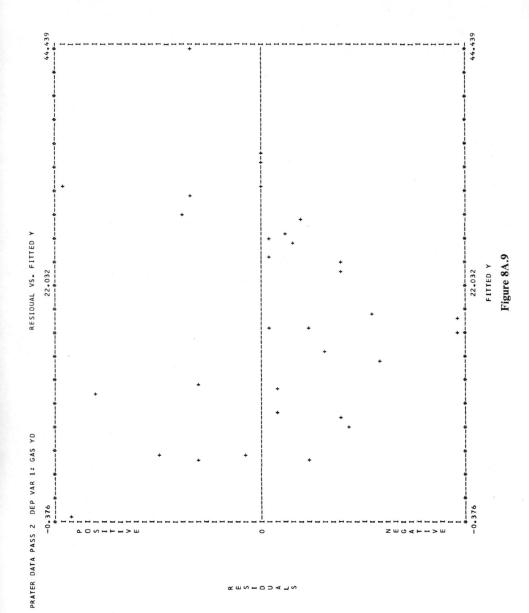

Figure 8A.9

190

LINEAR LEAST-SQUARES CURVE FITTING PROGRAM

PRATER DATA PASS 3   DEP VAR 1: GAS YD   MIN Y = 2.800D 00   MAX Y = 4.570D 01   RANGE Y = 4.290D 01

Y = GASOLINE YIELD;
X1 = GASOLINE END POINT − 332 ; X2 --- X10 = CRUDE 2 THROUGH 10;
X11 = (END POINT − 332)(CRUDE 2);   X19 = (END POINT − 332)(CRUDE 10)

IND.VAR(I)	NAME	COEF.B(I)	S.E. COEF.	T-VALUE	R(I)SQRD	MIN X(I)	MAX X(I)	RANGE X(I)	REL.INF.X(I)
0	GAS EP	3.27296D 01	1.05D-02	15.9	0.8484	-1.27D 02	-1.12D 02	2.39D 02	0.93
1	CRUDE2	1.66858D-01	1.16D 00	15.5	0.4622	0.0	1.000D 00	1.000D 00	0.42
2	CRUDE3	-9.12314D 00	1.54D 00	5.9	0.6085	0.0	1.000D 00	1.000D 00	0.21
3	CRUDE4	-1.29302D 00	1.38D 00	9.4	0.5128	0.0	1.000D 00	1.000D 00	0.30
4	CRUDE5	-3.76118D 00	1.60D 00	2.4	0.6356	0.0	1.000D 00	1.000D 00	0.09
5	CRUDE6	-1.16137D 00	1.15D 00	10.1	0.4572	0.0	1.000D 00	1.000D 00	0.27
6	CRUDE7	-2.00631D 00	1.32D 00	15.2	0.4692	0.0	1.000D 00	1.000D 00	0.47
7	CRUDE8	-1.23281D 01	1.25D 00	9.8	0.4079	0.0	1.000D 00	1.000D 00	0.49
8	CRUDE9	-2.11241D 01	2.21D 00	9.6	0.2738	0.0	1.000D 00	1.000D 00	0.62
9	CRDE10	-2.65446D 00	1.54D 00	11.3	0.6314	0.0	1.000D 00	1.000D 00	0.62
10	XICD 2	-2.11448D 01	1.50D 00	2.0	0.4765	-9.700D 01	8.400D 01	8.400D 01	0.13
11	XICD 3	-2.05327D-02	2.03D-02	1.0	0.6148	-1.1400 02	1.5000 01	1.2900 02	0.06
12	XICD 4	6.19353D-02	2.71D-02	2.3	0.3922	-2.5000 01	6.3000 01	8.8000 00	0.13
13	XICD 5	-1.28238D-02	2.05D-02	0.6	0.6678	-1.2000 02	8.0000 00	1.2800 02	0.04
14	XICD 6	-1.33198D-02	1.60D-02	0.8	0.4361	-9.700 01	7.8000 00	1.7500 02	0.05
15	XICD 7	3.55534D-03	1.76D-02	0.2	0.4762	-9.700D 01	1.1200 02	1.5900 02	0.01
16	XICD 8	-6.38203D-03	1.94D-02	0.3	0.3180	-4.700 01	7.0000 00	1.3500 02	0.01
17	XICD 9	-4.08302D-02	3.25D-02	1.3	0.7172	-6.5000 01	9.2000 01	9.200D 01	0.02
18	XICD10	-3.78597D-02	3.56D-02	1.1	0.8270	0.0	9.6000 01	9.600D 01	0.09

NO. OF OBSERVATIONS   32
NO. OF IND. VARIABLES   19
RESIDUAL DEGREES OF FREEDOM   12
F-VALUE   73.6
RESIDUAL ROOT MEAN SQUARE   1.58978901
RESIDUAL MEAN SQUARE   2.52742910
RESIDUAL SUM OF SQUARES   30.32914923
TOTAL SUM OF SQUARES   3564.07718750
MULT. CORREL. COEF. SQUARED   .9915

--------ORDERED BY COMPUTER INPUT--------

IDENT.	OBSV.	WS DISTANCE	OBS. Y	FITTED Y	RESIDUAL
1	11	187.	12.200	11.539	0.661
1	12	45.	23.020	23.219	-0.199
1	13	11.	34.700	34.899	-0.199
1	14	71.	45.700	45.244	0.456
2	21	212.	2.800	1.604	1.196
2	22	142.	6.400	7.032	-0.632
2	23	112.	16.100	18.296	-2.196
2	24	185.	27.800	26.168	1.632
3	31	181.	8.000	6.925	1.075
3	32	75.	13.100	14.973	-1.873
3	33	38.	26.600	25.801	0.799
4	41	71.	14.400	14.080	0.320
4	42	78.	26.800	27.807	-1.007
4	43	112.	34.900	34.213	0.687
5	51	53.	7.400	7.407	-0.007
5	52	14.	18.200	18.188	0.012
5	53	153.	14.000	14.000	-0.006
6	61	60.	6.900	6.223	0.677
6	62	61.	15.200	16.203	-1.003
6	63	116.	26.000	26.183	-0.183
6	64	163.	33.600	33.092	0.508
7	71	150.	5.000	4.657	0.343
7	72	276.	17.600	18.290	-0.690
7	73	66.	32.100	31.753	0.347
8	81	104.	24.800	24.895	-0.095
8	82	83.	18.000	17.971	0.029
8	83	167.	10.000	9.971	0.065
9	91	262.	14.000	14.000	0.000
9	92	248.	23.200	23.200	-0.000
10	101	261.	8.500	10.422	-1.922
10	102	342.	14.700	12.228	2.472
10	103		18.000	18.549	-0.549

--------ORDERED BY RESIDUALS--------

OBSV.	OBS. Y	FITTED Y	ORDERED RESID.	SEQ
102	14.700	12.228	2.472	1
24	27.800	26.168	1.632	2
21	8.000	6.604	1.196	3
31	8.000	6.925	1.075	4
33	26.600	25.801	0.799	5
43	34.900	34.213	0.687	6
61	6.900	6.223	0.677	7
11	12.200	11.539	0.661	8
64	33.600	33.092	0.508	9
14	45.700	45.244	0.456	10
73	32.100	31.753	0.347	11
71	5.000	4.657	0.343	12
41	14.400	14.080	0.320	13
83	10.000	9.971	0.065	14
82	18.000	18.290	0.029	15
52	18.200	18.188	0.012	16
91	14.000	14.000	0.000	17
92	23.200	23.219	-0.000	18
53	14.000	14.000	-0.006	19
51	7.400	7.407	-0.007	20
81	24.800	24.895	-0.095	21
63	26.000	26.183	-0.183	22
13	34.700	34.899	-0.199	23
123	23.219	34.899	-0.199	24
72	17.600	18.290	-0.549	25
72	6.400	7.032	-0.632	26
12	22.300	23.219	-0.687	27
62	16.203	18.290	-0.690	28
42	15.200	16.203	-1.003	29
32	26.800	27.807	-1.007	30
101	13.100	14.973	-1.873	31
23	16.100	18.296	-2.196	32

Figure 8A.10

LINEAR LEAST-SQUARES CURVE FITTING PROGRAM

PRATER DATA PASS 4

EQUATION 1 OF A MULTI-EQUATION PROBLEM

DATA READ WITH SPECIAL FORMAT

FORMAT CARD 1 (A6,I4,I2,3F6.2,6X,F6.3)

BZERO = CALCULATED VALUE

DATA INPUT   3 INDEPENDENT VARIABLES   1 DEPENDENT VARIABLES

OBSV.	SEQ.	1-11-21	2-12-22	3-13-23	4-14-24	5-15-25	6-16-26	7-17-27	8-18-28	9-19-29	10-20-30
100	1	50.800	8.600	190.000	32.550						
200	1	41.300	1.800	267.000	15.036						
300	1	40.800	3.500	210.000	24.275						
400	1	40.300	4.800	231.000	21.519						
500	1	40.000	6.100	217.000	27.782						
600	1	38.400	6.100	220.000	21.154						
700	1	38.100	1.200	274.000	13.063						
800	1	32.200	5.200	236.000	20.436						
900	1	32.200	2.400	284.000	9.806						
1000	1	31.800	0.200	316.000	4.436						

DATA TRANSFORMATIONS

POSITION	CODE	OPERATION
1		NONE
2		NONE
3		NONE
4		NONE

	CONSTANT LOCATION	OMIT VARIABLE
1	0	1
2	0	2
3	0	3
4	0	4

DATA AFTER TRANSFORMATIONS   THE FITTED EQUATION HAS 3 INDEPENDENT VARIABLES, 1 DEPENDENT VARIABLES

OBSV.	GRAV. 1-11-21	VAP PR 2-12-22	10 PCT 3-13-23	YIELD 4-14-24	5-15-25	6-16-26	7-17-27	8-18-28	9-19-29	10-20-30
100	5.08000D 01	8.60000D 00	1.90000D 02	3.25500D 01						
200	4.13000D 01	1.80000D 00	2.67000D 02	1.50360D 01						
300	4.08000D 01	3.50000D 00	2.10000D 02	2.42750D 01						
400	4.03000D 01	4.80000D 00	2.31000D 02	2.15190D 01						
500	4.00000D 01	6.10000D 00	2.17000D 02	2.77820D 01						
600	3.84000D 01	6.10000D 00	2.20000D 02	2.11540D 01						
700	3.81000D 01	1.20000D 00	2.74000D 02	1.30630D 01						
800	3.22000D 01	5.20000D 00	2.36000D 02	2.04360D 01						
900	3.22000D 01	2.40000D 00	2.84000D 02	9.80600D 00						
1000	3.18000D 01	2.00000D-01	3.16000D 02	4.43600D 00						

SUMS OF VARIABLES
3.85900D 02 3.99000D 01 2.44500D 03 1.90057D 02

MEANS OF VARIABLES
3.85900D 01 3.99000D 00 2.44500D 02 1.90057D 01

ROOT MEAN SQUARES OF VARIABLES
5.71809D 00 2.62444D 00 3.91869D 01 8.49313D 00

SIMPLE CORRELATION COEFFICIENTS, R(I,I PRIME)

1	1.000	-0.613	-0.712	0.764
2	0.613	1.000	-0.897	0.906
3	-0.712	-0.897	1.000	-0.978
4	0.764	0.906	-0.978	1.000

192

Figure 8A.11

CHAPTER 9

Nonlinear Least Squares, a Complex Example

9.1 Introduction

In this chapter we use both linear and nonlinear least-squares programs to arrive at our final equation. The data give the amounts of heat released over six different time periods by fourteen samples of cement of differing chemical compositions. The equation is *nonlinear* in time, while two of its coefficients are *linear* and *quadratic* functions of composition.

Indicator variables are used first in a nonlinear equation to allow for systematic differences between individual cements while ascertaining the relationship between observations of those cements over a period of time. The dependence of the coefficients of the indicator variables on composition is then investigated by the C_p-search technique in the linear least-squares program. The results are returned to the nonlinear iterative program to fit the final equation involving both time and composition.

In the end, statistics of the fit and interpretation of the results are compared with those obtained by others.

Statistical problems discussed include:
1. Using an equation which is nonlinear in time.
2. Handling biased data caused by using differences instead of raw observations.
3. Developing potential equation forms.
4. Finding outliers.
5. Using indicator variables to separate the effects of different samples.
6. Choosing between common and individual rate equations.
7. Finding high correlation between a variable that can be measured immediately and one that requires a long time before it can be measured.

8. Handling composition variables when the components must add to unity.

9. Selecting influential variables.

10. Studying the component effect of each sample on each variable.

11. Determining whether "far out" observations control the form of the equation or only extend the range of applicability.

12. Checking on the possibility of nested data.

13. Specifying components of error to measure lack-of-fit.

14. Judging the reasonableness of the final equation.

9.2   Background of Example

In the construction of massive structures such as dams, the amount and the rate of heat evolved during the hardening of cement become extremely important. A large rise in temperature during hardening may result in excessive initial expansion and later, when the cement is eventually cooled to the surrounding temperatures, in contraction and cracking. Because of structure thickness, it is often necessary to build cooling ducts or refrigeration coils within the structure to carry away excess heat. However, an inordinate number of cooling ducts weakens the structure, and refrigeration over a long period of time increases costs. Thus there is a considerable incentive to determine the effect of cement composition on the amount and the rate at which heat evolves during hardening.

In a study of this problem, Woods, Steinour, and Starke of the Riverside Cement Company made experimental cements covering a wide range of compositions. Table 9.1 shows both the analysis of the oxides and the calculated compounds of the kilned "clinkers" used to make each cement.

Samples prepared from each cement were aged for 3, 7, 28, 90, 180, and 365 days under uniform conditions. The cumulative heat given off by each sample was determined indirectly by measuring the amount of heat evolved when the samples were dissolved in acid. It was assumed that Hess's law prevailed, namely, that the change in the heat content of a system in passing from one state to another is independent of the path. Using this assumption, the experimenters determined (1) the acid heat of solution of each cement in the initial state (unreacted with water), (2) the acid heat of solution of each sample after reaction with water and aging for various lengths of time, and (3) the difference between the first and the second quantity to obtain the amount of heat evolved during hardening over each period of time. The calculated cumulative heats of hardening are given in Table 9.2.

The experimenters used *linear* least squares to fit a separate equation for each of the six time periods, with four composition proportions as independent variables. A single equation, including time as a variable, would have

TABLE 9.1
CLINKER ANALYSIS AND CALCULATED CLINKER COMPOUNDS

		Clinker Analysis, Weight Per Cent					
Analysis	Cement	Silica, SiO₂	Alumina, Al₂O₃	Ferric Oxide, Fe₂O₃	Lime, CaO	Magnesia, MgO	Total
1	22	27.68	3.76	1.98	64.97	2.48	100.87
2	23	25.96	3.48	5.06	63.15	2.32	99.97
3	92	21.86	5.75	2.77	65.02	5.04	100.44
4	88	24.60	5.85	2.80	64.18	2.40	99.83
5	96	25.04	3.86	2.11	66.57	2.36	99.94
6	85	22.32	6.17	2.85	66.47	2.43	100.24
7	94	20.93	4.64	5.74	66.26	2.08	99.65
8	24	23.54	4.83	7.21	62.03	2.24	99.85
9	89	21.96	4.65	6.06	64.07	2.32	99.06
10	90	21.44	8.81	1.19	66.64	2.48	100.64
11	25	22.48	5.00	7.46	62.72	2.24	99.90
12	95	21.34	6.07	2.93	67.03	2.56	99.93
13	91	21.94	5.57	2.68	67.71	2.44	100.34
14	70	25.72	4.12	6.06	61.05	2.08	99.03

		Published Calculated Clinker Compounds, Weight Per Cent					
Analysis	Cement	4CaO-Al₂O₃-Fe₂O₃	3CaO-Al₂O₃	3CaO-SiO₂	2CaO-SiO₂	MgO	Total
1	22	6	7	26	60	2.5	101.5
2	23	15	1	29	52	2.3	99.3
3	92	8	11	56	20	5.0	100.0
4	88	8	11	31	47	2.4	99.4
5	96	6	7	52	33	2.4	100.4
6	85	9	11	55	22	2.4	99.4
7	94	17	3	71	6	2.1	99.1
8	24	22	1	31	44	2.2	100.2
9	89	18	2	54	22	2.3	98.3
10	90	4	21	47	26	2.5	100.5
11	25	23	1	40	34	2.2	100.2
12	95	9	11	66	12	2.6	100.6
13	91	8	10	68	12	2.4	100.4
14	70	18	1	17	61	2.1	99.1

TABLE 9.2
CUMULATIVE HEAT FROM HARDENING CEMENT WITH 40% WATER, CALORIES PER GRAM

	Period of Aging					
Cement	3 Days	7 Days	28 Days	90 Days	180 Days	365 Days
22	51.2	53.1	72.5	81.2	78.5	85.5
23	40.6	45.7	61.0	71.9	74.3	76.0
92	74.8	84.7	92.2	100.6	104.3	110.4
88	53.2	63.1	73.7	77.7	87.6	90.6
96	71.9	78.8	88.8	92.6	95.9	103.5
85	80.6	90.3	101.0	103.7	109.2	109.8
	74.0	88.6	101.7	104.5	107.0	110.3
94	77.3	87.5	92.4	100.5	102.7	108.0
24	42.1	46.8	61.5	70.0	72.5	71.6
89	64.7	77.9	85.8	90.1	93.1	97.0
90	89.5	96.6	108.8	114.1	115.9	122.7
25	58.8	62.8	74.2	78.1	83.8	83.1
95	84.8	94.5	103.7	107.5	113.3	115.4
91	84.0	91.3	99.9	106.2	109.4	116.3
70	28.3	36.3	49.8	—	—	62.6

been more compact, more precise, more comprehensive, and hence more informative.

Hald (Section 20.3) and Draper and Smith (Chapter 6 and Appendix B) have used observations of thirteen of the fourteen cements at 180 days to illustrate the standard arithmetic of "multiple regression." No contribution to understanding the interrelationships of composition, time, and heat of hardening was claimed or expected.

Let us review in detail how the data were taken. Although ultimately we may wish to have our equations in terms of the cumulative heats of hardening, *measurements* were made in terms of the acid heat of solution of each aged sample. These values in turn were subtracted from the *initial* acid heat of solution of the dry cement from which the samples were made. This gave an estimate of the amount of heat given off by the exothermic reactions of aging and hardening. However, as this subtraction was repeated at each of the six time periods, any random fluctuation in a cement's *initial* measurement affected all six differences equally. Thus, the six differences for each cement are not statistically independent, although the seven measurements are. Least-squares fitting is mathematically simpler and easier to interpret if we fit an equation directly to the statistically independent observed heats of acid solution. This equation, in turn, can be modified to calculate the desired cumulative heats of hardening. We will pursue this course of action.

TABLE 9.3
Observed Acid Heat of Solution, Calories per Gram

NO.	CEMENT	0 DAYS	3 DAYS	7 DAYS	28 DAYS	90 DAYS	180 DAYS	365 DAYS
1	22	588.7	537.5	535.6	516.2	507.5	510.2	503.2
2	23	575.7	535.1	530.0	514.7	503.8	501.4	499.7
3	92	623.4	548.8	538.9	531.4	523.0	519.3	513.2
3	92	624.1						
4	88	595.5	542.3	532.4	521.8	517.8	507.9	504.9
5	96	603.2						
5	96	604.5	532.6	525.7	515.7	511.9	508.6	501.0
6	85	618.8	538.2	528.5	517.8	515.1	509.6	509.0
6	85		544.8	530.2	517.1	514.3	511.8	508.5
7	94	610.0	532.7	522.5	517.6	509.5	507.3	502.0
7	94	611.7						
8	24	573.0	530.9	526.2	511.5	503.0	500.5	501.4
9	89	599.0	534.3	521.1	513.2	508.9	505.9	502.0
10	90	633.4	543.9	536.8	524.6	519.3	517.5	510.7
11	25	585.9	527.1	523.1	511.7	507.8	502.1	502.8
12	95	625.6	540.8	531.1	521.9	518.1	512.3	510.2
13	91	624.0	540.0	532.7	524.1	517.8	514.6	507.7
14	70	564.7	536.4	528.4	514.9			502.1

Before we try to fit an equation to the observed values, let us see how much can be learned about their measurement. We realize from the start that this may be only an academic exercise because of the small number of replicates. Nonetheless, such a search is often worthwhile. The observed values are given in Table 9.3.

9.3 Replicates

Clinker mixtures were prepared from a finely ground commercial kiln feed by adding technical oxides and carbonates to bring the oxides to the desired values. Each mixture was fired in a furnace, cooled, and pulverized. Before storage, a portion of each was withdrawn for chemical analysis. There were no replicate clinker mixtures or replicate chemical analyses.

The fourteen cements were prepared by grinding the same amount of gypsum (3.2%) with each clinker mixture. Gypsum, $CaSO_4 \cdot 2H_2O$, is used commercially to delay setting, increase strength, and improve dimensional stability.

Duplicate initial heat of solution measurements were made on three cements to estimate the precision of the calorimeter. The results are shown in the table.

DUPLICATE INITIAL HEAT OF SOLUTION MEASUREMENTS

Observation	Cements		
	92	94	96
1	623.4	610.0	603.2
2	624.1	611.7	604.5
Difference	0.7	1.7	1.3
d^2	0.49	2.89	1.69

$\Sigma\, d^2/2 = 5.07/2 = 2.535$
Variance $= 2.535/3 = 0.845$
Standard deviation $= 0.92$ with 3 df

A portion of each cement was then mixed with water, and the resulting cement paste put into ten glass vials. All ten vials were then placed in a constant-temperature cabinet until required for the heat of solution measurements after aging 3, 7, 28, 90, 180, and 365 days.

To test the cumulative precision of the preparation and curing of the cement pastes, their preparation for admission to the calorimeter, and the heat of solution measurements, a second set of cement paste samples was prepared at a different time from Cement 85. The heat of solution measurements of these duplicate samples are given in the table. The 3-day difference

HEAT OF SOLUTION MEASUREMENTS ON DUPLICATE AGED SAMPLES

Batch	Days					
	3	7	28	90	180	365
1	538.2	528.5	517.8	515.1	509.6	509.0
2	544.8	530.2	517.1	514.3	511.8	508.5
Difference	6.6	1.7	−0.7	−0.8	2.2	−0.5
d^2	43.56	2.89	0.49	0.64	4.84	0.25
$\Sigma\, d^2 =$	52.67					

is so much larger than the differences for the other five periods that we use Cochran's g-test to judge whether all six differences could come from the same random normal source:

$$g(1, 6) = \frac{d^2 \max}{\Sigma d^2} = \frac{43.56}{52.67} = 0.827.$$

From Dixon and Massey (Table A-17) we find the value

$$g(0.95, 1, 6) = 0.781.$$

Thus at the 95% confidence level, the difference in the pair of duplicate observations at 3 days is *not* compatible with the others.

From the five acceptable pairs, we compute

$$\sum d^2/2 = 9.11/2 = 4.55,$$

$$\text{Variance} = 4.55/5 = 0.910, \text{ and the}$$

$$\text{Standard deviation} = 0.95 \text{ with 5 df.}$$

The standard deviation of 0.95 for the duplicate aged sample determinations and the standard deviation of 0.92 for the duplicate initial heat of solution measurements are remarkably close.

Although one of the 3-day observations of Cement 85 is an outlier, we will not know which it is until we determine the relationship between the various heat of solution measurements and time.

9.4 Potential Equations

The usual trial plots of the heat of solution, z, versus time, t, were made: z vs. t, z vs. $\log t$, $\log z$ vs. t, and $\log z$ vs. $\log t$. Only the last gave tolerably straight lines, and even these showed upward curvature at small t and large z. When a constant, E, was subtracted from z, we were able to obtain straight-line plots on log-log paper [equivalent to $\log (z - E)$ vs. $\log (t)$]. Two such plots are shown in Figure 9.1. A different value of E was required for each cement. The constant E appears to represent the acid heat of solution after hardening for an infinite length of time, that is, the ultimate or equilibrium heat of solution. (Cement used in aqueducts by the ancient Romans still contains unreacted cells or granules; water has not broken through all of the crystalline barriers.)

The resulting equation, $\log (z_t - E) = A - B \log (t)$, is nonlinear in that the value of E is unknown and there is no transformation which will linearize

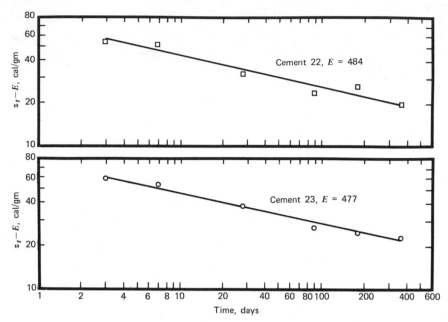

Figure 9.1 Acid heat of solution minus constant versus time.

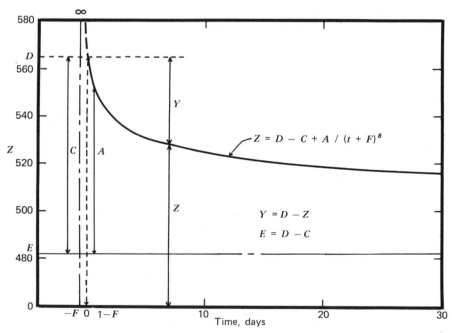

Figure 9.2 Heat of solution versus time.

it in all coefficients. Since we have observations at time zero, a time correction, F, may also be needed. Hence a convenient form of the equation is:

$$(9.1) \qquad Z - E = \frac{A}{(t + F)^B},$$

where t = time from moment of adding water, measured in days,

Z = estimated heat evolved when sample is dissolved in acid at time t (heat of acid solution measured in calories per gram), and $z =$ observed value of same quantity,

E = heat of acid solution after infinite time, that is, the ultimate heat of solution,

A = a constant characteristic of the sample, the remaining heat of hardening at time $(1 - F)$,

B = an exponent, and

F = time correction.

Since

$$(9.2) \qquad E = D - C,$$

where D = initial heat of solution at time zero, and C = cumulative heat of hardening after infinite time, that is, the ultimate heat of hardening, then

$$(9.3) \qquad Z = D - C + \frac{A}{(t + F)^B}.$$

At $t = 0$ by definition, $Z = D$, the initial heat of solution, and

$$D = D - C + \frac{A}{(0 + F)^B},$$

so that

$$(9.4) \qquad F = \left(\frac{A}{C}\right)^{1/B}.$$

Substituting for F in 9.3 gives the equation for the heat of solution in terms of four constants: A, B, C, and D,

$$(9.5) \qquad Z = D - C + \frac{A}{[t + (A/C)^{1/B}]^B}.$$

The equation can also be written in terms of B, C, D, and F:

$$(9.6) \qquad Z = D - C\left[1 - \left(\frac{F}{t + F}\right)^B\right].$$

Both forms of the equation will be helpful in interpreting the final fit of the data. The basic equation, 9.3, is illustrated in Figure 9.2.

The cumulative heat of hardening, Y, has been defined as

$$Y = D - Z.$$

Substituting $D - Y$ for Z in equations 9.3, 9.5, and 9.6, we have:

(9.7)
$$Y = C - \frac{A}{(t + F)^B},$$

(9.8)
$$Y = C - \frac{A}{[t + (A/C)^{1/B}]^B},$$

and

(9.9)
$$Y = C\left[1 - \left(\frac{F}{t + F}\right)^B\right].$$

We will return to these equations later.

9.5   Nonlinear Fit—Observations of Individual Cements versus Time

The data from each of the fourteen cements were used separately to fit an equation of the form 9.5. The results of each fit are given in Table 9.4.

TABLE 9.4
Acid Heat of Solution:
Nonlinear Estimation of Coefficients—Individual Equations

Cement Number	Con- stant, A	Ex- ponent, B	Ultimate Heat of Hardening, C	Initial Heat of Solution, D	Number of Obser- vations	Degrees of Freedom	Residual Sum of Squares
22	72.41	0.221	104.74	588.67	7	3	56.35
23	79.13	0.209	100.44	575.66	7	3	17.13
92	95.90	0.103	161.34	623.74	8	4	8.24
88	104.63	0.102	147.46	595.50	7	3	19.66
96	86.45	0.098	149.23	603.84	8	4	13.99
85	64.23	0.506	112.51	618.81	13	9	35.70
94	75.21	0.116	145.62	610.85	8	4	14.33
24	63.69	0.309	83.91	572.96	7	3	20.82
89	56.62	0.431	99.80	599.01	7	3	11.98
90	69.46	0.166	147.33	633.40	7	3	8.83
25	54.31	0.192	101.80	585.90	7	3	10.84
95	56.12	0.245	128.18	625.60	7	3	6.83
91	90.19	0.097	165.21	623.98	7	3	6.71
70	69.89	0.315	73.58	564.68	5	1	0.98
Totals					105	49	232.39

The Computer Nonlinear Least-Squares Curve Fitting Program (SHARE Library Number 360D-13.6.007 and VIM Library Number G2 CAL NLWOOD) was used to estimate the coefficients of the equation. This program, as mentioned previously, is a modification of the University of Wisconsin's GAUSHAUS program, which utilizes the Marquardt "maximum neighborhood" nonlinear estimation technique mentioned in Chapter 2. Details are given in the User's Manual.

9.6 Fit with Indicator Variables

In order to determine whether all of the cements can be represented by an equation with a single value of B, we write:

$$(9.10) \qquad Z = D' - C' + \frac{A'}{(t + F')^B},$$

where Z = estimated acid heat of solution at time t (in days),

$D' = D_1 v_1 + D_2 v_2 + \cdots + D_k v_k$ ($v_1, \ldots, v_k$ are indicator variables to separate the initial heats of solution of the k cements, $D_1, \ldots, D_k$),

$C' = C_1 v_1 + C_2 v_2 + \cdots + C_k v_k$,

$A' = A_1 v_1 + A_2 v_2 + \cdots + A_k v_k$, and

$F' = (A'/C')^{1/B}$.

Thus, there is a D, C, and A coefficient for each of the fourteen cements and one B coefficient—43 coefficients in all.

The starting values of the coefficients were guessed from plots such as Figure 9.1, but taken purposely wide of their marks to test the ability of the computer program to converge: 0.8 for B, 100 for each C, 80 for each A, and the average initial heat of solution for each cement's D coefficient. Computer printouts of the result of fitting this equation are shown in Figures 9A.1–9A.4.

Only nine iterations were required to obtain the final fit. The criterion used for ending the search for the "best" values of the coefficients was that the relative change in each coefficient from the previous iteration be less than 0.000001. The resulting change in the sum of squares of the 105 residuals from the previous iteration was less than 0.01.

The method appears to be rather robust in regard to the approximation of the starting values. The starting value of 0.8 for the exponent B shifted to 0.22; the starting value of 100 for C (the ultimate heat of hardening) shifted to a range of values from 83 to 137; and the starting value of 80 for A shifted to a range of 52–78. The values chosen for D (the initial acid heat of solution) remained essentially unchanged.

The multiple correlation coefficient squared, a measure of how well the fitting equation accounts for the total variation of the observed values of the dependent variable, is 0.9976:

$$R_z^2 = 1 - \frac{\text{Residual sum of squares}}{\text{Total sum of squares}} = 1 - \frac{299}{122,985}$$
$$= 1 - 0.00243 = 0.9976.$$

The residual versus cumulative frequency plot indicates that the distribution of the residuals is roughly normal. (The center portion is close to a straight line.) Individual hand plots of the residuals of each cement versus time show that the fit is not biased for individual cements and that, at time periods above zero, the residuals are apparently randomly distributed.

The residual versus fitted Z plot looks peculiar at first sight, in that there is a larger spread of residuals at the lower fitted Z-values than at the higher values—the initial heats of solutions. In these equations the observations on aged samples have little influence on the fit of the observations at time zero. We rationalized this by supposing that the steepness of the curve was making back-extrapolation uninformative. But as we show later, when the data on composition are used, first weight percent and then mol percent, the residuals become more evenly dispersed over all values of fitted Y.

The statistics of the fit of equation 9.5 with an individual exponent for each cement, and the fit of equation 9.10 with a common exponent, are shown in the table.

EFFECT OF COMMON VERSUS INDIVIDUAL EXPONENTS

B	Number of Coefficients	Degrees of Freedom	Residual Sum of Squares	Residual Mean Square
Common	$p = 43$	62	299.3	4.8
Individual	$p + q = 56$	49	232.4	4.7
	$q = 13$	13	66.9	5.2

$$F(13, 49) = \frac{\text{RMS}_q}{\text{RMS}_{p+q}} = \frac{5.2}{4.7} = 1.1.$$

From F-tables, $F(0.95, 12, 40)$ is 2.0 and $F(0.95, 15, 60)$ is 1.8; so the individual exponents did not significantly improve the fit. Therefore, we conclude that all the cements can be represented by an equation with a single exponent.

Similar tests were made on coefficients D, C, and A. In each case, the use of individual coefficients for each cement significantly improved the fit over that

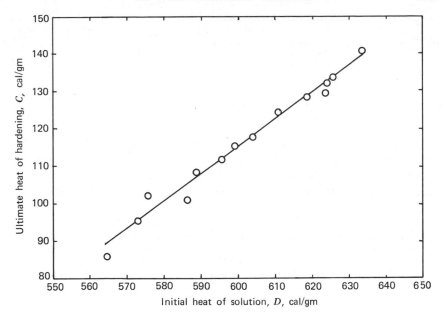

Figure 9.3   Ultimate heat of hardening versus initial heat of solution.

obtained with common coefficients. Hence, individual coefficients are needed for D, C, and A.

Earlier we had observed that one of the duplicate observations of sample 85 at 3 days was undoubtedly an outlier, but were unable to identify it. From the list of ordered residuals, Figure 9A.2, it is now apparent that the second observation is the culprit. It has the largest residual of all of the observations.

In Pass 2, Figures 9A.5–9A.7, the same equation is used as in Pass 1, but observation *85-3-2* is omitted. The residual sum of squares is reduced considerably, from 299.35 to 253.10, a decrease of 46.25 units. From the squared difference of the duplicate observations of *85-3*, we find that 43.56/2 or only 21.78 units can be directly attributed to removing the second observation. The remaining 24.47 units (46.25–21.78) is the indirect effect of the deletion. R_z^2 is increased from 0.9976 to 0.9979.

Observation *85-3-2* will be omitted in the remaining analysis.

A plot of coefficients C_i versus D_i ($i = 1, \dots, 14$), Figure 9.3, reveals an unexpected *linear* relationship between the initial heat of solution of each cement and its ultimate heat of hardening. As seen in Figure 9A.8, the equation $C = GD - H$ represents the data with an R_c^2 of 0.98. Using

(9.11) $C' = GD' - H,$

we can rewrite the nonlinear equation as

(9.12)
$$Z = H + (1 - G)D' + \frac{A'}{(t + F')^B},$$

where now

$$F' = \left(\frac{A'}{GD' - H}\right)^{1/B}$$

and B, G, and H are *common* to all cements.

The nonlinear fit of equation 9.12 is shown in Figure 9A.9. The residual root mean square is 1.99 with 73 degrees of freedom, and R_z^2 is 0.9976. The residual versus cumulative frequency plot and the residual versus fitted Y plot (not shown) are similar to the plots of the nonlinear fit with individual C coefficients.

EFFECT OF COMMON G AND H VERSUS INDIVIDUAL C COEFFICIENTS

Equation with	Number of Coefficients	Degrees of Freedom	Residual Sum of Squares	Residual Mean Square
Common G and H	$p = 31$	73	290.7	4.0
Individual C	$p + q = 43$	61	253.1	4.1
	$q = 12$	12	37.6	3.1

$$\frac{\text{RMS}_q}{\text{RMS}_{p+q}} = \frac{3.1}{4.1} = 0.8; \quad F(0.95, 12, 60) = 1.9$$

The statistics of the fit of both equations are shown in the table. The new equation with *twelve* fewer coefficients fits as well as the old one. Thus, we conclude that equation 9.12 can be used to calculate the ultimate heat of hardening from the fitted value of the initial heat of solution.

9.7 Fit with Composition

Before composition variables can replace the indicator variables representing the different cements in equation 9.12, we will need to determine the *forms* of the equations and to estimate the starting values of their coefficients.

In Figure 9A.9 we saw how the indicator variables were used in the nonlinear equation to estimate coefficients D and A for each cement. We now try to find the dependence of these coefficients on the composition of the clinker from which each cement was made.

The proportions (or fractions) found by chemical analysis of the oxides of each clinker should total 100%. Since we do not know which of the oxides of each clinker is in error, we correct each proportionally by dividing by the clinker's total weight per cent and multiplying by 100.

We have chosen equations of the type first given by Scheffé (1958; also see Gorman and Hinman, 1962) to represent the dependence of the heat of solution on the five composition variables. These equations are symmetrical in all components. They do not arbitrarily omit one component and hence do not conceal real effects. The equations are polynomials of the second order whose terms have understandable meanings.

In order to reduce the covariances between the various coefficients, we subtract from each oxide the minimum value observed for it in all the samples. The resulting composition is reproportioned so that the sum of all five oxide proportions is again 100%. This is done by dividing each adjusted value by 100 minus the sum of the minimum observations. The minimum values and the resulting divisor are as follows:

Oxide	Minimum Concentration, %
Silica, SiO_2	21.00
Alumina, Al_2O_3	3.48
Ferric oxide, Fe_2O_3	1.18
Lime, CaO	61.60
Magnesia, MgO	2.08
Sum	= 89.34
Divisor = (100 − Sum)	= 10.66

The transformed composition variables then are:

$$x_1 = (\% \text{ silica} \quad - 21.00)/10.66$$
$$x_2 = (\% \text{ alumina} \quad - \ 3.48)/10.66$$
$$x_3 = (\% \text{ ferric oxide} - \ 1.18)/10.66$$
$$x_4 = (\% \text{ lime} \quad - 61.60)/10.66$$
$$x_5 = (\% \text{ magnesia} \quad - \ 2.08)/10.66$$

Reproportioning in composition space is somewhat analogous to the usual technique of subtracting the mean from variables before taking their cross products and squared terms.

Because of the limited number of cements (fourteen), the number of cross products we can fit at one time is limited. Magnesia (x_5) theoretically remains unreacted, so we thought it safe to study later the potential influence of its cross products. The following set of equations can be fitted using *linear* least squares:

(9.13)
$$D, A = b_1 x_1 + b_2 x_2 + b_3 x_3 + b_4 x_4 + b_5 x_5$$
$$+ b_{12} x_1 x_2 + b_{13} x_1 x_3 + b_{14} x_1 x_4$$
$$+ b_{23} x_2 x_3 + b_{24} x_2 x_4 + b_{34} x_3 x_4.$$

Computer printouts of the fit of these equations are given in Figures 9A.10–9A.16.

Since there is no b_0 term, the basic equations must contain the five corrected oxides. Thus, after fitting the above equations, C_p searches are made to determine which cross-product terms are influential. The following statistics are obtained:

Dependent Variable	Full Equations		Result of C_p Search		
	R_y^2	RRMS	Most Influential Cross Products	p	C_p
D	0.998	2.2	x_1x_3, x_2x_4	7	3.4
A	0.990	1.6	x_1x_2, x_1x_4	7	6.4

The fit of each equation is relatively good. With D, the search indicates that the value of C_p can be reduced from 11 with the full equation to 3.4 by using only the cross products x_1x_3 and x_2x_4. With A, C_p can be reduced from 11 to 6.4 by using cross products x_1x_2 and x_1x_4.

Plots of p (the number of coefficients in each equation) versus C_p are shown in Figures 9A.13 and 9A.16. Each plot represents the results of calculating the fit of 2^6 or 64 combinations of cross products added to the basic equation. Only equations with C_p's less than the full equation ($C_p = 11 = p$) are represented. Although we are not compelled to select the variable combinations with the lowest C_p-values, in this case (as seen in Figures 9A.17 and 9A.18) they appear reasonable. We have for the selected equations:

Dependent Variable	R_y^2	RRMS
D	0.997	1.55
A	0.978	1.57

The fit of the two dependent variables and the subsequent C_p searches took 6 seconds of computer time.

The cross products involving x_5—x_1x_5, x_2x_5, x_3x_5, x_4x_5—are now added to the reduced linear equations to determine whether any will provide a fit with a lower C_p-value. None does.

We are now ready to replace the indicator variable estimates of D' and A' in equation 9.12 with those of composition. The fourteen coefficients in the composition equations:

$$(9.14) \quad D = d_1x_1 + d_2x_2 + d_3x_3 + d_4x_4 + d_5x_5 + d_{13}x_1x_3 + d_{24}x_2x_4$$

and

$$(9.15) \quad A = a_1x_1 + a_2x_2 + a_3x_3 + a_4x_4 + a_5x_5 + a_{12}x_1x_2 + a_{14}x_1x_4$$

and the earlier estimates of B, G, and H are used to start the nonlinear iterations.

Results of this fit (Pass 9) are shown in Figure 9A.19. After eight iterations the relative change in each coefficient was less than 0.000001. The residual root mean square is 1.98, measured with 87 degrees of freedom. This matches closely the value of 1.99 found with the previous nonlinear fit using indicator variables. Replacing the *specific* indicator variables with the more *general* composition variables only reduced R_z^2 from 0.9976 to 0.9972. The cumulative distribution plot (Figure 9A.20) is similar to that obtained with the indicator variables (Figure 9A.6). The residual versus fitted Z plot (Figure 9A.21), however, shows a more nearly symmetrical dispersion of residuals at the larger Z-values (time $= 0$) than that obtained with the indicator variables (Figure 9A.7).

In order to determine with security just how much lack of fit there is in this equation, we would need some measure of the cumulative error caused by variations in kiln feed preparation, furnace firing, chemical analysis, gypsum addition, water addition, storage conditions, storage time, and calorimeter measurements. Unfortunately there were no replicate clinker samples. We have one replicate cement sample (85) from which to estimate the cumulative error of water addition, storage conditions, and storage time, and three replicate calorimeter measurements, but clearly these are not sufficient to provide the needed error estimates.

We have come to the end of the path on which we started. We do not think that further improvement is possible by empirical fitting. But now that a good fit has been obtained, we may hope for further theoretical developments by those able to make them.

In order to simplify the use of the equation given by Pass 9, Figure 9A.19, let us convert the coded composition to the composition actually observed by adding back the minimum observed value of each variable and multiplying by 10.66. The components of the nonlinear equation now contain only these variables:

$$t = \text{time measured in days,}$$
$$x_1 = \% \text{ silica in the clinker,}$$
$$x_2 = \% \text{ alumina,}$$
$$x_3 = \% \text{ ferric oxide,}$$
$$x_4 = \% \text{ lime, and}$$
$$x_5 = \% \text{ magnesia.}$$

In the *heat of solution* equation:

(9.16)
$$Z = H + (1 - G)D + \frac{A}{(t + F)^B},$$

and in the corresponding *cumulative heat of hardening* equation:

(9.17)
$$Y = -H + GD - \frac{A}{(t + F)^B},$$

where

$$F = \left(\frac{A}{GD - H}\right)^{1/B},$$

the coefficients B, G, and H, common to all cements, remain unchanged:

$$B = 0.20, \quad G = 0.726, \quad \text{and} \quad H = 320.$$

The decoded equations for D and A are:

(9.18)   Initial heat of solution, $D =$

$$
\begin{aligned}
& 4.2 \ \times \ (\% \text{ silica}) \\
& -56.4 \ \times \ (\% \text{ alumina}) \\
& +14.4 \ \times \ (\% \text{ ferric oxide}) \\
& + \ 6.4 \ \times \ (\% \text{ lime}) \\
& +12.1 \ \times \ (\% \text{ magnesia}) \\
& - \ 0.46 \times \ (\% \text{ silica} \times \% \text{ ferric oxide}) \\
& + \ 1.00 \times \ (\% \text{ alumina} \times \% \text{ lime})
\end{aligned}
$$

and

(9.19)   Sample constant, $A =$

$$
\begin{aligned}
& 50.0 \ \times \ (\% \text{ silica}) \\
& -28.7 \ \times \ (\% \text{ alumina}) \\
& -13.4 \ \times \ (\% \text{ ferric oxide}) \\
& + \ 8.0 \ \times \ (\% \text{ lime}) \\
& - \ 7.9 \ \times \ (\% \text{ magnesia}) \\
& + \ 0.77 \times \ (\% \text{ silica} \times \% \text{ alumina}) \\
& - \ 0.98 \times \ (\% \text{ silica} \times \% \text{ lime}).
\end{aligned}
$$

A useful relationship is obtained from $C = GD - H$, that is, Ultimate heat of hardening $= 0.726(\text{Initial heat of solution}) - 320$, or simply

(9.20) $C = 0.726(D - 441).$

Using this relationship, we find

(9.21)   Ultimate heat of hardening, $C = -320$

$$
\begin{aligned}
& + \ 3.0 \ \times \ (\% \text{ silica}) \\
& -40.9 \ \times \ (\% \text{ alumina}) \\
& +10.4 \ \times \ (\% \text{ ferric oxide}) \\
& + \ 4.6 \ \times \ (\% \text{ lime}) \\
& + \ 8.8 \ \times \ (\% \text{ magnesia}) \\
& - \ 0.33 \times \ (\% \text{ silica} \times \% \text{ ferric oxide}) \\
& + \ 0.73 \times \ (\% \text{ alumina} \times \% \text{ lime}).
\end{aligned}
$$

The component effects of each of the oxides on the initial heat of solution, D, and the sample constant, A, are given in Tables 9.5 and 9.6. In both tables we see that Cement 90 was highly influential in determining the effect of alumina, x_2, and alumina interactions, while Cement 92 was most

TABLE 9.5

COMPONENT EFFECT TABLE

INITIAL HEAT OF SOLUTION

IDENT.	SILICA PCT	B1X1	FERRIC OXIDE PCT	B3X3	CROSS X1X3	PRODUCT B6X1X3	U, TOTAL	ALUMINA PCT	B2X2	LIME PCT	B4X4	CROSS X2X4	PRODUCT B7X2X4	V, TOTAL	MAGNESIA PCT	B5X5	INITIAL HEAT OF SOLUTION
14 70	26.0	108.7	6.1	88.4	159.	-72.9	124.2	4.2	-234.7	61.6	392.8	256.	256.3	414.5	2.1	25.4	564.2
8 24	23.6	98.7	7.2	104.3	170.	-78.1	124.9	4.8	-272.8	62.1	395.9	301.	300.3	423.4	2.2	27.1	575.4
2 23	26.0	108.7	5.1	73.1	131.	-60.3	121.5	3.5	-196.3	63.2	402.5	220.	219.8	426.0	2.3	28.1	575.6
11 25	22.5	94.2	7.5	107.9	168.	-77.1	125.0	5.0	-282.3	62.8	400.1	314.	314.0	431.8	2.2	27.1	584.0
1 22	27.4	114.9	2.0	28.4	54.	-24.7	118.5	3.7	-210.2	64.4	410.4	240.	239.9	440.1	2.5	29.7	588.4
4 88	24.6	103.2	2.8	40.5	69.	-31.7	112.0	5.9	-330.5	64.3	409.7	377.	376.5	455.7	2.4	29.1	596.7
9 89	22.2	92.8	6.1	88.4	136.	-62.2	119.0	4.7	-264.8	64.7	412.2	304.	303.4	450.8	2.3	28.3	598.1
5 96	25.1	104.9	2.1	30.5	53.	-24.3	111.1	3.9	-217.8	66.6	424.5	257.	257.1	463.7	2.4	28.6	603.4
7 94	21.0	87.9	5.8	83.2	121.	-55.5	115.7	4.7	-262.6	66.5	423.7	310.	309.4	470.5	2.1	25.3	611.4
6 85	22.3	93.2	2.8	41.1	63.	-29.0	105.3	6.2	-347.2	66.3	422.6	408.	407.9	483.3	2.4	29.3	617.9
13 91	21.9	91.5	2.7	38.6	58.	-26.8	103.3	5.6	-313.1	67.5	430.0	375.	374.4	491.3	2.4	29.4	624.0
3 92	21.8	91.1	2.8	39.8	60.	-27.5	103.4	5.7	-322.9	64.7	412.5	371.	370.4	460.0	5.0	60.7	624.1
12 95	21.4	89.4	2.9	42.4	63.	-28.7	103.0	6.1	-342.6	67.1	427.4	407.	407.2	492.0	2.6	31.0	626.1
10 90	21.3	89.2	1.2	17.1	25.	-11.6	94.7	8.8	-493.7	66.2	422.0	580.	579.3	507.5	2.5	30.8	633.0

EFFECT OF SILICA AND FERRIC OXIDE

EFFECT OF ALUMINA AND LIME

TABLE 9.6

COMPONENT EFFECT TABLE

SAMPLE-CONSTANT A

IDENT.	SILICA PCT	B1X1	ALUMINA PCT	B2X2	FERRIC OXIDE PCT	B3X3	LIME PCT	B4X4	MAGNESIA PCT	B5X5	CROSS PRODUCTS X1X2	B6X1X2	X1X4	B7X1X4	A
14 70	26.0	1296.0	4.2	-119.6	6.1	-82.1	61.6	493.8*	2.1	-16.7	108.	82.9	1601.	-1574.3	80.1**
3 92	21.8	1086.1	5.7	-164.5	2.8	-37.0	64.7	518.5	5.0	-39.9*	125.	95.6	1409.	-1385.3	73.5
1 22	27.4	1369.4**	3.7	-107.1	2.0	-26.3	64.4	515.9	2.5	-19.5	102.	78.5	1767.	-1737.8*	72.9
4 88	24.6	1229.7	5.9	-168.4	2.8	-37.6	64.3	514.9	2.4	-19.1	144.	110.8	1584.	-1557.6	72.6
2 23	26.0	1295.8	3.5	-100.0**	5.1	-67.9	63.2	505.9	2.3	-18.4	90.	69.3*	1640.	-1612.8	71.9
8 24	23.6	1176.5	4.8	-139.0	7.2	-96.9	62.1	497.6	2.2	-17.8	114.	87.5	1465.	-1440.0	67.8
10 90	21.3	1063.1	8.8	-251.6*	1.2	-15.9**	66.2	530.3	2.5	-20.2	186.	143.0**	1411.	-1387.0	61.8
6 85	22.3	1111.1	6.2	-176.9	2.8	-38.1	66.3	531.1	2.4	-19.3	137.	105.1	1477.	-1451.7	61.3
11 25	22.5	1122.9	5.0	-143.8	7.5	-100.2*	62.8	502.8	2.2	-17.8	113.	86.4	1413.	-1389.1	61.2
12 95	21.4	1065.7	6.1	-174.6	2.9	-39.3	67.1	537.2	2.6	-20.3	130.	99.5	1432.	-1408.4	59.7
13 91	21.9	1091.1	5.6	-159.5	2.7	-35.8	67.5	540.5**	2.4	-19.3	121.	93.1	1476.	-1450.8	59.3
5 96	25.1	1250.3	3.9	-111.0	2.1	-28.3	66.6	533.5	2.4	-18.8	97.	74.2	1669.	-1640.9	59.0
9 89	22.2	1106.2	4.7	-134.9	6.1	-82.1	64.7	518.0	2.3	-18.6	104.	79.8	1434.	-1409.8	58.7
7 94	21.0	1048.1*	4.7	-133.8	5.8	-77.3	66.5	532.6	2.1	-16.6**	98.	75.0	1397.	-1373.1**	54.9*
AVERAGE	23.3		5.2		4.1		64.9		2.5		121.		1514.		

* MINIMUM VALUE

** MAXIMUM VALUE

TABLE 9.7
FITTED VALUES OF THE HEAT OF SOLUTION

NO.	CEMENT	INITIAL	3 DAYS	7 DAYS	28 DAYS	90 DAYS	180 DAYS	365 DAYS	ULTIMATE
1	22	588.4	539.5	530.7	519.0	511.2	507.4	504.0	481.4
2	23	575.6	534.9	526.5	514.9	507.3	503.5	500.2	477.9
3	92	624.1	550.1	541.1	529.1	521.3	517.4	514.0	491.2
4	88	596.7	541.6	532.9	521.1	513.4	509.6	506.2	483.7
5	96	603.4	532.9	525.6	516.0	509.7	506.6	503.8	485.5
6	85	617.9	538.7	531.1	521.1	514.6	511.4	508.5	489.5
7	94	611.4	531.8	525.0	516.0	510.2	507.3	504.7	487.7
8	24	575.4	531.8	523.7	512.8	505.6	502.0	498.9	477.8
9	89	598.1	531.2	524.0	514.4	508.1	505.0	502.3	484.0
10	90	633.0	543.3	535.6	525.5	518.9	515.7	512.8	493.6
11	25	584.0	529.2	521.7	511.8	505.3	502.0	499.2	480.2
12	95	626.1	539.7	532.3	522.6	516.2	513.0	510.2	491.7
13	91	624.0	538.8	531.4	521.8	515.4	512.3	509.5	491.1
14	70	564.2	536.9	528.3	515.9	507.5	503.3	499.6	474.7

influential in determining the effect of magnesia, x_5. The remaining cements have relatively uniform distributions of the other components; silica, ferric oxide, and lime.

In order to determine whether Cements 90 and 92 merely extended the range of the effect of composition, or indeed controlled the form of the equation, two additional runs (not shown) were made, in each of which one of these cements was omitted. The results show that their influence is consistent with that of the remaining cements, and therefore their presence only extends the effective range of applicability.

Fitted values of the initial heat of solution, the heat of solution at specific time periods, and the ultimate heat of solution of each cement are given in Table 9.7.

Estimates of the cumulative heat of hardening are calculated by subtracting the fitted values of the heat of solution at specific time periods from the fitted value of the initial heat of solution. These values and the ultimate heats of hardening are given in Table 9.8.

Table 9.9 displays the residuals from the nonlinear fit of the heat of solution data. Cement 22 at 7 days has the largest residual, 4.9. The fit does not appear to be biased for any of the individual cements. In no case are there fewer than two of the seven residuals of a given cement with the same sign.

TABLE 9.8

FITTED VALUES OF THE CUMULATIVE HEAT OF HARDENING

NO.	CEMENT	3 DAYS	7 DAYS	28 DAYS	90 DAYS	180 DAYS	365 DAYS	ULTIMATE
1	22	49.0	57.7	69.4	77.2	81.0	84.4	107.0
2	23	40.7	49.1	60.6	68.3	72.0	75.4	97.7
3	92	74.1	83.1	95.0	102.9	106.7	110.2	133.0
4	88	55.1	63.9	75.6	83.3	87.1	90.5	113.1
5	96	70.6	77.8	87.5	93.7	96.8	99.6	117.9
6	85	79.2	86.8	96.8	103.3	106.5	109.4	128.4
7	94	79.7	86.5	95.4	101.3	104.1	106.7	123.7
8	24	43.7	51.7	62.6	69.8	73.4	76.6	97.6
9	89	66.9	74.2	83.7	90.0	93.1	95.8	114.1
10	90	89.7	97.4	107.5	114.1	117.3	120.2	139.4
11	25	54.8	62.2	72.2	78.7	81.9	84.8	103.8
12	95	86.4	93.8	103.5	109.9	113.0	115.8	134.4
13	91	85.3	92.6	102.3	108.6	111.7	114.5	132.9
14	70	27.2	35.8	48.2	56.6	60.8	64.6	89.4

Table 9.10 in turn displays the residuals obtained by subtracting the heats of hardening (calculated from the nonlinear equation) from those published by the experimenters. Errors in the published values have been confounded; the errors in the initial and the aged measurements in some cases cancel each other, in other cases are additive. Cement 88 at 90 days now has the largest residual, 5.6, while the residual of Cement 22 at 7 days is reduced to 4.6. Apparently there is more bias—Cement 85 has no negative residuals and Cement 70 only one, while Cements 24 and 91 have only one positive residual.

Comparing these tables of residuals reinforces the lesson that directly measured quantities should be used whenever possible, rather than differences that propagate errors through subsets of data.

The percentage of the ultimate heat of hardening evolved by each cement when observations were made is given in Table 9.11. As previously noted, the majority of the cements had already given off 50% of their potential heat of hardening by the first observation at 3 days. Thus the data provide little information on the early evolution of heat.

TABLE 9.9
RESIDUALS FROM HEAT OF SOLUTION DATA

NO.	CEMENT	0 DAYS	3 DAYS	7 DAYS	28 DAYS	90 DAYS	180 DAYS	365 DAYS
1	22	0.3	−2.0	4.9	−2.8	−3.7	2.8	−0.8
2	23	0.1	0.2	3.5	−0.2	−3.5	−2.1	−0.5
3	92−1	−0.7	−1.3	−2.2	2.3	1.7	1.9	−0.8
3	92−2	0.0						
4	88	−1.2	0.7	−0.5	0.7	4.4	−1.7	−1.3
5	96−1	−0.2						
5	96−2	1.1	−0.3	0.1	−0.3	2.2	2.0	−2.8
6	85−1	0.9	−0.5	−2.6	−3.3	0.5	−1.8	0.5
6	85−2			−0.9	−4.0	−0.3	0.4	0.0
7	94−1	−1.4	0.9	−2.5	1.6	−0.7	0.0	−2.7
7	94−2	0.3						
8	24	−2.4	−0.9	2.5	−1.3	−2.6	−1.5	2.5
9	89	0.9	3.1	−2.9	−1.2	0.8	0.9	−0.3
10	90	0.4	0.6	1.2	−0.9	0.4	1.8	−2.1
11	25	1.9	−2.1	1.4	−0.1	2.5	0.1	3.6
12	95	−0.5	1.1	−1.2	−0.7	1.9	−0.7	−0.0
13	91	−0.0	1.2	1.3	2.3	2.4	2.3	−1.8
14	70	0.5	−0.5	0.1	−1.0			2.5

In the equation for the cumulative heat of hardening,

$$(9.9) \qquad Y = C\left[1 - \left(\frac{F}{t + F}\right)^{B}\right],$$

the proportion of the heat of hardening given off at time t can be calculated from the term

$$1 - \left(\frac{F}{t + F}\right)^{B}.$$

Since B is constant, the *rate* at which heat is evolved is controlled by F, the *amount* by C.

TABLE 9.10
RESIDUALS FROM PUBLISHED HEAT OF HARDENING DATA

NO.	CEMENT	3 DAYS	7 DAYS	28 DAYS	90 DAYS	180 DAYS	365 DAYS
1	22	2.2	-4.6	3.1	4.0	-2.5	1.1
2	23	-0.1	-3.4	0.4	3.6	2.3	0.6
3	92	0.7	1.6	-2.8	-2.3	-2.4	0.2
4	88	-1.9	-0.8	-1.9	-5.6	0.5	0.1
5	96	1.3	1.0	1.3	-1.1	-0.9	3.9
6	85-1	1.4	3.5	4.2	0.4	2.7	0.4
6	85-2		1.8	4.9	1.2	0.5	0.9
7	94	-2.4	1.0	-3.0	-0.8	-1.4	1.3
8	24	-1.6	-4.9	-1.1	0.2	-0.9	-5.0
9	89	-2.2	3.7	2.1	0.1	-0.0	1.2
10	90	-0.2	-0.8	1.3	0.0	-1.4	2.5
11	25	4.0	0.6	2.0	-0.6	1.9	-1.7
12	95	-1.6	0.7	0.2	-2.4	0.3	-0.4
13	91	-1.3	-1.3	-2.4	-2.4	-2.3	1.8
14	70	1.1	0.5	1.6			-2.0

Since $F = (A/C)^{1/B}$, we deduce the following:

1. If two cements have equivalent initial heats of solution and hence equal ultimate heats of hardening, the one with the larger A- or F-value will evolve its heat more slowly.

2. If two cements have equivalent values of A, the one with the smaller ultimate heat of hardening not only will have less heat to evolve but also will do so at a slower rate.

The "half-life" of each cement's heat of hardening, $t_{(1/2)}$, can be calculated from

$$(9.22) \qquad \frac{Y_{0.5}}{C} = \frac{1}{2} = 1 - \left(\frac{F}{t_{(1/2)} + F}\right)^B,$$

or

$$(9.23) \qquad t_{(1/2)} = [(2)^{1/B} - 1]F = 32.0F.$$

TABLE 9.11
PERCENTAGE OF HEAT OF HARDENING GIVEN OFF

NO.	CEMENT	3 DAYS	7 DAYS	28 DAYS	90 DAYS	180 DAYS	365 DAYS
14	70	30.	40.	54.	63.	68.	72.
2	23	42.	50.	62.	70.	74.	77.
8	24	45.	53.	64.	72.	75.	78.
1	22	46.	54.	65.	72.	76.	79.
4	88	49.	56.	67.	74.	77.	80.
11	25	53.	60.	70.	76.	79.	82.
3	92	56.	62.	71.	77.	80.	83.
9	89	59.	65.	73.	79.	82.	84.
5	96	60.	66.	74.	79.	82.	84.
6	85	62.	68.	75.	80.	83.	85.
13	91	64.	70.	77.	82.	84.	86.
12	95	64.	70.	77.	82.	84.	86.
10	90	64.	70.	77.	82.	84.	86.
7	94	64.	70.	77.	82.	84.	86.

The fitted values of A, C, A/C, F, and the "half-life" of each cement are given in Table 9.12. Cement samples 91 and 92 have essentially the same ultimate heat of hardening, C, but with different values of A the half-life of one is about three times as long as the other. The reverse situation can be seen in the comparison of Cements 25 and 90. Although 90 has a slightly higher A-value, the C-value of 25 is sufficiently lower than that of 90 to provide a half-life more than four times as long. In this case a lower C-value is more effective in reducing the heat of hardening than a lower A-value. We see from Table 9.8 that at 3 days the difference in the cumulative heat of hardening between Cements 91 and 92 is only 11 calories per gram, or 15%, whereas the difference between Cements 25 and 90 is 35 calories per gram, or 63%. The values of C and A chosen for specifications will depend on whether a low-early or a low-later heat of hardening is desired. This may determine when refrigeration can be terminated. Cement 85, for example, has a lower heat of hardening value than Cement 94 at 3 days, but from then on 94 evolves less heat than 85.

Figure 9.4 shows, by a single point on an *A-C* grid, the two basic responses for each cement. From this plot we can see which cements give nearly the same total heats, which have almost the same *A*-values, and, by their closeness to the lines A/C = constant, which have nearly the same proportional rate constants, *F*, or half-lives.

From the preceding discussion we realize that not only is considerable

TABLE 9.12

A, *C*, *A/C*, *F*, AND HALF-LIFE VALUES

NO.	CEMENT	A	C	A/C	F, DAYS	HALF-LIFE, DAYS
14	70	80.07	89.4	0.90	0.573	18.34
2	23	71.89	97.7	0.74	0.213	6.81
8	24	67.76	97.6	0.69	0.159	5.08
1	22	72.88	107.0	0.68	0.144	4.61
4	88	72.57	113.1	0.64	0.107	3.42
11	25	61.24	103.8	0.59	0.070	2.23
3	92	73.46	133.0	0.55	0.050	1.60
9	89	58.74	114.1	0.51	0.035	1.12
5	96	59.02	117.9	0.50	0.030	0.97
6	85	61.32	128.4	0.48	0.024	0.77
13	91	59.26	132.9	0.45	0.017	0.54
12	95	59.73	134.4	0.44	0.017	0.54
10	90	61.83	139.4	0.44	0.017	0.53
7	94	54.85	123.7	0.44	0.017	0.53

flexibility available in the combination of *C* and *A* to arrive at a given heat of hardening at a specific time, but also there are numerous combinations of composition whereby the specified values of *C* and *A* can be attained.

We have analyzed only the heat of solution data from these experiments. The choice of composition for a particular application will depend, of course, on an aggregate of physical properties, as well as on availability and cost.

9.8   Possible Nesting within Cements

Since the data were taken on fourteen cements (with an average of 7.5 observations on each, spread over time), it is natural to inquire whether there may be nesting, as in the data used in Chapter 8. We can imagine that some random impulse affecting a cement might well make all its z-values high (or low) by about the same amount. Such a factor would produce an "among-cement component of variance": we call it σ_1^2. This component is at least conceptually distinct from the component we will call σ_0^2, produced by random

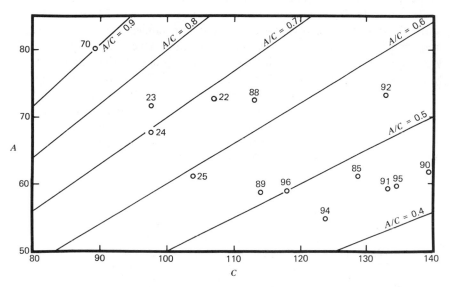

Figure 9.4   Distribution of cements on A versus C grid. Cement number is attached to each point.

variation of measurements taken at different times on the *same* cement.

Statistical habit would lead us to compare the fourteen curves at their *average* values to get an estimate of among-cement random variability with minimal contamination by σ_0^2, but this does not seem appropriate here. The fourteen curves are, more probably, randomly displaced in *two* directions. They may be displaced at their starting points in the z direction, that is, by their D_i, by any random factors that influence initial heats of solution. We are not thinking here of random fluctuations in composition; these will be taken care of (insofar as they appear as changes in oxide concentrations) by the seventeen-coefficient (nonlinear) fit. There may be variations in molecular chemical composition, which are not reflected in the oxide content but nevertheless influence the heat of solution and so move a whole curve up or down by a small amount.

Quite independent of these among-cement variations in heats of solution, there may be random variations in the recorded time of the reactions. There is ample evidence in the uniformity of these data that the experimenters were very careful workers indeed. But it was probably not realized that for five or six of these cements half the total heat of hardening evolves in a day or less and hence that heats of solution at early times should have been measured and time-recorded to the nearest hour. These time errors would produce a cement-to-cement component of variance in F (or equivalently in A).

To return to D, we have a good estimate of σ_0^2, "with 73 degrees of freedom" in our mean square of 4.0 from the 31-coefficient fit (Pass 4). But we require an estimate of Var. (D_1) based on this value. This appears in the nonlinear fit as a separate value for each D_{1i}. Since all the standard errors estimated for D_{1i} are close to 1.8 and average 1.84, we take the estimated variance of D_i based on within-cement random variation to be 3.4.

We now require estimates of all D_i which (contrary to those found in the 31-coefficient fit) have minimal disturbance from among-cement sources. The best we have are the values from the final 17-coefficient equation (Pass 9). We call these D_{2i} and show them in Table 9.13, beside the corresponding D_{1i} for each cement.

TABLE 9.13

COMPARISON OF FITTED VALUES OF D

Pass 4: Cements Separated and Fitted by Indicator Variables

Pass 9: Cements Fitted by Composition Variables

i	Cement	Pass 4 D_{1i}	Pass 9 D_{2i}	$D_{1i} - D_{2i}$
1	22	588.5	588.4	−0.1
2	23	575.0	575.6	−0.6
3	92	624.1	624.1	0.0
4	88	595.5	596.7	−1.2
5	96	603.9	603.4	0.5
6	85	618.8	617.9	0.9
7	94	610.7	611.4	−0.7
8	24	573.0	575.4	−2.4
9	89	598.9	598.1	0.8
10	90	633.1	633.0	0.1
11	25	586.6	584.0	2.6
12	95	625.6	626.1	−0.5
13	91	624.0	624.0	0.0
14	70	565.0	564.2	0.8

$\mathrm{SS}(D) = 17.42; \quad \mathrm{MS}(D) = 17.42/7 = 2.49$

We see from Table 9.13 that the residual sum of squares for $(D_{1i} - D_{2i})$ is 17.42. We feel that this should reflect about $(14 - 7)$ or 7 degrees of freedom, since we have used 7 composition coefficients in fitting the 14 D_i. The residual mean square is then $17.4/7$ or 2.5. Since this value is *smaller* than the 3.4 we expected if only within-cement random factors were operating, we have no evidence whatever for an among-cement random component. This test must be of low power, both because of the small number of degrees of freedom available, and because of the inexact assumption that mean squares from nonlinear least-squares fits provide the usual unbiased, chi-square distributed estimates of the corresponding variances. All we can say is that we see no sign whatever of any cement-to-cement random variation in the heat of solution, in excess of that expected from the observed within-cement variation.

9.9 Comparison with Previous Linear Least-Square Fits

As the reader will recall, the experimenters used the derived heats of hardening data at specific time periods to fit a series of equations using *linear* least squares. They considered that magnesia was unreactive and then chose the postulated clinker compounds given in Table 9.1 as the independent variables. They used an equation without a b_0 term, namely,

$$(9.24) \qquad Y = b_1x_1 + b_2x_2 + b_3x_3 + b_4x_4.$$

Finally, they chose the first observation of each sample (except for 85-3, where they used the average) as being "most probable."

The accompanying table shows the results of such fits at the specific time periods (for direct comparison, we have corrected the oxide contents to 100% before converting algebraically to the postulated compounds):

Time Periods, Days	Number of Observations	Number of Coefficients	Degrees of Freedom	RSS	RMS	RRMS
3	14	4	10	155.1	15.5	3.9
7	14	4	10	103.8	10.4	3.2
28	14	4	10	142.9	14.3	3.8
90	13	4	9	113.7	12.6	3.6
180	13	4	9	36.4	4.0	2.0
365	14	4	10	56.5	5.6	2.4
Totals	82	24	58	608.4		

$$\text{Pooled RMS} = \frac{\text{Total RSS}}{\text{Total df}} = \frac{608.4}{58} = 10.5$$

Pooled residual root mean square $= 3.2$

Not only do the residual mean square values vary from time period to time period, but also the pooled RMS of 10.5 is considerably larger than the 3.9 RMS of the nonlinear fit. A larger number of coefficients might be expected to fit the data better (24 vs. 17), but in this case the cross-product terms are needed and the additional coefficients do not include them. There is a further substantial drawback in the inability to interpolate or extrapolate easily to other time periods.

By contrast, a direct fit of all the data (except one observation that is demonstrably defective) has produced equations which provide some aid in understanding the chemical and physical relationships involved. Specifically, we have found that:

1. The ultimate heat of hardening, C, is a linear function of the initial heat of solution, D.
2. The initial heat of solution in turn is a quadratic function of composition (however, see Section 9.10).
3. Both the heat of hardening, Y, and the heat of solution, Z, are nonlinear functions of time.
4. The rate constant F, $(A/C)^{1/B}$, is a complex function of composition, since A and C are quadratic in composition.

A side point of interest in the original analysis was the selection of independent variables. The experimenters calculated the composition of the clinker compounds from the analysis of the oxides, using assumptions which were widely accepted at that time. Since one set of values is derived algebraically from the other, the fit of the equations would have been identical but the coefficients and their interpretation would have been different. No data to challenge these assumptions were taken in this series of experiments. Since then the development of more definitive analytical techniques (see Hansen) has altered these views, and they undoubtedly will continue to change as more is learned about the chemistry of cement. Thus the experimenters and others since may have been misled because the fundamental units of the variables were not used in the early data analysis. This is not a rare occurrence, but the lesson involved is a hard one to learn.

9.10   Fit Using Composition in Mol Percents

The previous fit was made using the weight percents of the clinker oxides. In chemical reactions, it is often more meaningful to chemists and physicists to use mol percents as units of measurement. In order to determine if any different conclusions might be drawn, we repeat the entire sequence of fitting using mol percents.

The number of mols of any component is the weight percent of that component divided by its molecular weight. Its mol fraction, in turn, is the number of mols of that component divided by the sum of the number of mols of all components. Mol percent is the mol fraction multiplied by 100. For example, the mol percent of silica in sample 22 is calculated as follows:

$$\text{Mol}\% \text{ silica} = \frac{\dfrac{W\% \text{ silica}}{MW \text{ silica}}}{\dfrac{W\% \text{ silica}}{MW \text{ silica}} + \dfrac{W\% \text{ alumina}}{MW \text{ alumina}} + \dfrac{W\% \text{ F.O.}}{MW \text{ F.O.}} + \dfrac{W\% \text{ lime}}{MW \text{ lime}} + \dfrac{W\% \text{ magnesia}}{MW \text{ magnesia}}} \times 100$$

$$= \frac{\dfrac{27.68}{60.06}}{\dfrac{27.68}{60.06} + \dfrac{3.76}{101.94} + \dfrac{1.98}{159.68} + \dfrac{64.97}{56.08} + \dfrac{2.48}{40.32}} \times 100$$

$$= 26.64.$$

Table 9.14 gives the analysis of the clinker oxides in mol percents.

TABLE 9.14

CLINKER ANALYSIS, MOL PERCENT

ANALYSIS	CEMENT	SILICA	ALUMINA	FERRIC OXIDE	LIME	MAGNESIA
1	22	26.64	2.13	0.72	66.96	3.55
2	23	25.70	2.03	1.88	66.96	3.42
3	92	21.13	3.28	1.01	67.32	7.26
4	88	24.26	3.40	1.04	67.78	3.53
5	96	24.33	2.21	0.77	69.27	3.42
6	85	21.92	3.57	1.05	69.91	3.55
7	94	20.95	2.74	2.16	71.05	3.10
8	24	23.81	2.88	2.74	67.19	3.37
9	89	22.17	2.77	2.30	69.27	3.49
10	90	20.97	5.08	0.44	69.79	3.73
11	25	22.77	2.98	2.84	68.03	3.38
12	95	21.00	3.52	1.08	70.64	3.75
13	91	21.43	3.21	0.98	70.83	3.55
14	70	26.00	2.45	2.30	66.10	3.13

When the minimum concentration of each oxide is used, the transformed composition variables of equation 9.13 are:

$$x_1 = (\text{mol \% silica} \qquad -20.90)/7.43$$
$$x_2 = (\text{mol \% alumina} \qquad -2.03)/7.43$$
$$x_3 = (\text{mol \% ferric oxide} \;\; -0.44)/7.43$$
$$x_4 = (\text{mol \% lime} \qquad -66.10)/7.43$$
$$x_5 = (\text{mol \% magnesia} \qquad -3.10)/7.43$$

The C_p searches of the full equations in D and A are made again using these transformed variables. The results (not shown) indicate that in this case D can be estimated quite well ($R_D^2 = 0.9938$, RRMS = 2.07) with only the five composition variables while the estimate of $A(R_A^2 = 0.9942$, RRMS = 1.24) requires the addition of all the cross product terms.

The resulting 19 variable-equation nonlinear fit, Figures 9.22–9.25, has a residual sum of squares of 339.9, a residual root mean square of 2.0, an R_z^2 of 0.9972, and a F-value of 1600 with 85 degrees of freedom.

The residuals of the observations at time 0 (the larger fitted Y-values) have a greater dispersion than before (compare Figure 9.25 with Figure 9.21). Hence, since the total residual sum of squares is smaller, the residuals of the observations at the other time periods are slightly smaller than in the fit using weight percents. None of the residuals changed sign.

The untransformed coefficients for equations 9.16 and 9.17 are as follows:

		S. E. Coef.
B = 0.195		0.03
G = 0.72		0.08
H = 315.0		51.0
(9.25)	Initial heat of solution, $D =$	
	$-0.4 \times$ (mol % silica)	0.3
	$+5.9 \times$ (mol % alumina)	0.9
	$-7.6 \times$ (mol % ferric oxide)	0.7
	$+8.4 \times$ (mol % lime)	0.1
	$+7.8 \times$ (mol % magnesia)	0.4
(9.26)	Sample constant, $A =$	
	$+79.0 \times$ (mol % silica)	13.0
	$-375.0 \times$ (mol % alumina)	107.0
	$+592.0 \times$ (mol % ferric oxide)	82.0
	$+11.0 \times$ (mol % lime)	2.0
	$-13.0 \times$ (mol % magnesia)	7.0
	$+6.0 \times$ (mol % silica × mol % alumina)	1.0
	$-6.2 \times$ (mol % silica × mol % ferric oxide)	0.9
	$-1.5 \times$ (mol % silica × mol % lime)	0.2
	$-14.0 \times$ (mol % alumina × mol % ferric oxide)	2.0
	$+3.0 \times$ (mol % alumina × mol % lime)	1.0
	$-6.0 \times$ (mol % ferric oxide × mol % lime)	1.0

Other than the new coefficients in terms of mol percents (which we will leave for cement chemists to interpret), no additional information was gained. The equation for D, the initial heat of solution, and its correlation with C, the ultimate heat of hardening, is fortunately simpler. The equation for A, however, is more complex. The overall fit using mol percents is as good as that using weight percents. Under these conditions we are more secure about estimating the initial heats of solution and the ultimate heats of hardening, less secure about estimating the time constants F.

9.11   Conclusions

The experimenters collected their data in order to learn more about the effects of chemical composition on the rate and the amount of heat evolved during the setting of Portland cement. *The aims of the data analyst should be the same.*

We have found an equation (9.16) that describes all the data (one point dropped) acceptably. We have shown:

1. How the amount of heat evolved varies with time (equation 9.1).

2. The prediction of the ultimate heat of hardening, C, from the initial acid heat of solution, D (equation 9.20).

3. The dependence of the initial heat on chemical composition (equations 9.18 and 9.25).

4. A table of constants, F, that characterize the rate of evolution of heat for each cement (Table 9.12). F is the time required for 0.128 of the total heat to be evolved; half of the ultimate heat is evolved in time 32.0F.

5. The dependence of F on A (a sample constant) and C, and, in turn, their dependence on chemical composition (equations 9.19, 9.21, and 9.26).

We have tried to meet the requirements given in Chapter 2 for a good method of fitting. All the data have been used (except one point shown to be defective); we have employed 17 coefficients to describe 104 observations with an R_z^2 of 0.9972 and an F-value of 1800; we have estimated and studied the residual error and then have used it to estimate the standard errors of the coefficients in our equation; finally, we have tested the data for nesting and found none.

The experimenters used *linear* least squares to fit a separate equation for each of the six time periods. The single *nonlinear* equation (9.16), with time as a variable, is more compact, more precise, more comprehensive, and hence more informative.

The data have served to introduce nonlinear least squares, to exemplify a massive use of indicator variables, and to permit discovery of an unexpected relation (that between the initial heat of solution and the ultimate heat of hardening). Finally, the example justifies our initial contention that much can be learned from sets of multifactor data taken by careful experimenters, even though the standard statistical requirements of balanced conditions were not met.

APPENDIX 9A

COMPUTER PRINTOUT OF NONLINEAR EXAMPLE

Pass	Figures	Program	Conditions
1	9A.1–9A.4	Nonlinear	Indicator variables, B common
2	9A.5–9A.7	Nonlinear	Indicator variables, 85-3-2 omitted
3	9A.8	Linear	C as a function of D
4	9A.9	Nonlinear	Indicator variables, B, G, and H common
5	9A.10–9A.13	Linear	D as a function of wt.% composition C_p search
6	9A.14–9A.16	Linear	A as a function of wt.% composition, C_p search
7	9A.17	Linear	Selected equation, D vs. wt.% composition
8	9A.18	Linear	Selected equation, A vs. wt.% composition
9	9A.19–9A.21	Nonlinear	Wt.% composition variables, B, G, and H common
10	9A.22–9A.25	Nonlinear	Mol% composition variables, B, G, and H common

NON-LINEAR ESTIMATION   CEMENT HEAT, PASS 1

MODEL NUMBER   2

$Z = D^* - C^* + (A^* / L T + F^*)EXP B)$
Z = ACID HEAT OF SOLUTION
B = EXPONENT COMMON TO ALL CEMENTS
D^* = INITIAL ACID HEAT OF SOLUTION = $D1V1 + D2V2 + --- + D14V14$
A^* = CONSTANT = $A1V1 + A2V2 + --- + A14V14$
C^* = ULTIMATE CUMULATIVE HEAT OF HARDENING = $C1V1 + C2V2 + --- + C14V14$
F^* = TIME CORRECTION TERM = $(A^*/C^*)EXP(1/B)$
T = TIME, MEASURED IN DAYS, AFTER WATER IS ADDED TO THE CEMENT
$V1$ --- $V14$ = INDICATOR VARIABLES FOR EACH SAMPLE OF CEMENT
NO OBSERVATIONS OMITTED.

NUMBER OF COEFFICIENTS   43

STARTING LAMBA   0.100

STARTING NU   10.000

MAX NO. OF ITERATIONS   20

DATA FORMAT (6X, 4X, 2X, 2F6.1, 6X, 14F1.0)

OBSV. NO.	IDENT.	SEQ.	1-11-21	2-12-22	3-13-23	4-14-24	5-15-25	6-16-26	7-17-27	8-18-28	9-19-29	10-20-30
1	22 0	1	588.700	0.0	1.000	0.0	0.0	0.0	0.0	0.0	0.0	0.0
		2	537.500	3.000	1.000	0.0	0.0	0.0	0.0	0.0	0.0	0.0
2	22 3	1	535.000	7.000	1.000	0.0	0.0	0.0	0.0	0.0	0.0	0.0
		2	0.0	0.0	0.0	0.0	0.0	0.0	0.0	0.0	0.0	0.0
3	22 7	1	516.200	28.000	1.000	0.0	0.0	0.0	0.0	0.0	0.0	0.0
		2	0.0	0.0	0.0	0.0	0.0	0.0	0.0	0.0	0.0	0.0
4	2228	1	507.500	90.000	1.000	0.0	0.0	0.0	0.0	0.0	0.0	0.0
		2	0.0	0.0	0.0	0.0	0.0	0.0	0.0	0.0	0.0	0.0
5	2290	1	510.200	180.000	1.000	0.0	0.0	0.0	0.0	0.0	0.0	0.0
		2	0.0	0.0	0.0	0.0	0.0	0.0	0.0	0.0	0.0	0.0
6	2218	1	503.200	365.000	1.000	0.0	0.0	0.0	0.0	0.0	0.0	0.0
		2	0.0	0.0	0.0	0.0	0.0	0.0	0.0	0.0	0.0	0.0
7	2236	1	575.700	0.0	0.0	1.000	0.0	0.0	0.0	0.0	0.0	0.0
		2	0.0	0.0	0.0	1.000	0.0	0.0	0.0	0.0	0.0	0.0
8	23 0	1	535.100	3.000	0.0	1.000	0.0	0.0	0.0	0.0	0.0	0.0
		2	0.0	0.0	0.0	1.000	0.0	0.0	0.0	0.0	0.0	0.0
9	23 3	1	530.000	7.000	0.0	1.000	0.0	0.0	0.0	0.0	0.0	0.0
		2	0.0	0.0	0.0	1.000	0.0	0.0	0.0	0.0	0.0	0.0
10	23 7	1	514.700	28.000	0.0	1.000	0.0	0.0	0.0	0.0	0.0	0.0
		2	0.0	0.0	0.0	1.000	0.0	0.0	0.0	0.0	0.0	0.0
11	2328	1	503.800	90.000	0.0	1.000	0.0	0.0	0.0	0.0	0.0	0.0
		2	0.0	0.0	0.0	1.000	0.0	0.0	0.0	0.0	0.0	0.0
12	2390	1	501.400	180.000	0.0	1.000	0.0	0.0	0.0	0.0	0.0	0.0
		2	0.0	0.0	0.0	1.000	0.0	0.0	0.0	0.0	0.0	0.0
13	2318	1	499.700	365.000	0.0	1.000	0.0	0.0	0.0	0.0	0.0	0.0
		2	0.0	0.0	0.0	1.000	0.0	0.0	0.0	0.0	0.0	0.0
14	2336	1	623.600	0.0	0.0	0.0	1.000	0.0	0.0	0.0	0.0	0.0
		2	0.0	0.0	0.0	0.0	1.000	0.0	0.0	0.0	0.0	0.0
15	92 0	1	624.100	0.0	0.0	0.0	1.000	0.0	0.0	0.0	0.0	0.0
		2	0.0	0.0	0.0	0.0	1.000	0.0	0.0	0.0	0.0	0.0
16	92 3	1	548.800	3.000	0.0	0.0	1.000	0.0	0.0	0.0	0.0	0.0
		2	0.0	0.0	0.0	0.0	1.000	0.0	0.0	0.0	0.0	0.0
17	92 7	1	538.900	7.000	0.0	0.0	1.000	0.0	0.0	0.0	0.0	0.0
		2	0.0	0.0	0.0	0.0	1.000	0.0	0.0	0.0	0.0	0.0
18	9228	1	531.400	28.000	0.0	0.0	1.000	0.0	0.0	0.0	0.0	0.0
		2	0.0	0.0	0.0	0.0	1.000	0.0	0.0	0.0	0.0	0.0
19	9228	1	0.0	0.0	0.0	0.0	1.000	0.0	0.0	0.0	0.0	0.0
		2	0.0	0.0	0.0	0.0	1.000	0.0	0.0	0.0	0.0	0.0

Figure 9A.1

#	ID	c	A	B										
20	9290	1	523.000	90.000	0.0	0.0	1.000	0.0	0.0	0.0	0.0	0.0	0.0	0.0
21	9218	2	519.300	180.000	0.0	0.0	1.000	0.0	0.0	0.0	0.0	0.0	0.0	0.0
22	9236	1	513.200	0.0	0.0	0.0	1.000	0.0	0.0	0.0	0.0	0.0	0.0	0.0
23	88 0	2	595.500	365.000	0.0	0.0	1.000	0.0	0.0	0.0	0.0	0.0	0.0	0.0
24	88 3	1	542.300	0.0	0.0	0.0	0.0	1.000	0.0	0.0	0.0	0.0	0.0	0.0
25	88 7	2	532.400	3.000	0.0	0.0	0.0	1.000	0.0	0.0	0.0	0.0	0.0	0.0
26	8828	1	521.800	7.000	0.0	0.0	0.0	1.000	0.0	0.0	0.0	0.0	0.0	0.0
27	8890	2	517.800	28.000	0.0	0.0	0.0	1.000	0.0	0.0	0.0	0.0	0.0	0.0
28	8818	1	507.900	90.000	0.0	0.0	0.0	1.000	0.0	0.0	0.0	0.0	0.0	0.0
29	8836	2	504.900	180.000	0.0	0.0	0.0	1.000	0.0	0.0	0.0	0.0	0.0	0.0
30	96 0	1	604.500	365.000	0.0	0.0	0.0	0.0	1.000	0.0	0.0	0.0	0.0	0.0
31	96 0	2	603.200	0.0	0.0	0.0	0.0	0.0	1.000	0.0	0.0	0.0	0.0	0.0
32	96 3	1	532.600	3.000	0.0	0.0	0.0	0.0	1.000	0.0	0.0	0.0	0.0	0.0
33	96 7	2	525.700	7.000	0.0	0.0	0.0	0.0	1.000	0.0	0.0	0.0	0.0	0.0
34	9628	1	515.700	28.000	0.0	0.0	0.0	0.0	1.000	0.0	0.0	0.0	0.0	0.0
35	9690	2	511.900	90.000	0.0	0.0	0.0	0.0	1.000	0.0	0.0	0.0	0.0	0.0
36	9618	1	508.600	180.000	0.0	0.0	0.0	0.0	1.000	0.0	0.0	0.0	0.0	0.0
37	9636	2	501.000	365.000	0.0	0.0	0.0	0.0	1.000	0.0	0.0	0.0	0.0	0.0
38	85 0	1	618.800	0.0	0.0	0.0	0.0	0.0	0.0	1.000	0.0	0.0	0.0	0.0
39	85 3	2	538.200	3.000	0.0	0.0	0.0	0.0	0.0	1.000	0.0	0.0	0.0	0.0
40	85 3	1	544.800	3.000	0.0	0.0	0.0	0.0	0.0	1.000	0.0	0.0	0.0	0.0
41	85 7	2	528.500	7.000	0.0	0.0	0.0	0.0	0.0	1.000	0.0	0.0	0.0	0.0
42	85 7	1	530.200	7.000	0.0	0.0	0.0	0.0	0.0	1.000	0.0	0.0	0.0	0.0
43	8528	2	517.800	28.000	0.0	0.0	0.0	0.0	0.0	1.000	0.0	0.0	0.0	0.0
44	8528	1	517.100	28.000	0.0	0.0	0.0	0.0	0.0	1.000	0.0	0.0	0.0	0.0
45	8590	2	515.100	90.000	0.0	0.0	0.0	0.0	0.0	0.0	1.000	0.0	0.0	0.0
46	8590	1	514.300	90.000	0.0	0.0	0.0	0.0	0.0	0.0	1.000	0.0	0.0	0.0
47	8518	2	509.600	180.000	0.0	0.0	0.0	0.0	0.0	0.0	1.000	0.0	0.0	0.0
48	8518	1	511.800	180.000	0.0	0.0	0.0	0.0	0.0	0.0	1.000	0.0	0.0	0.0
49	8536	2	509.000	365.000	0.0	0.0	0.0	0.0	0.0	0.0	1.000	0.0	0.0	0.0
50	8536	1	508.500	365.000	0.0	0.0	0.0	0.0	0.0	0.0	1.000	0.0	0.0	0.0
51	94 0	2	610.000	0.0	0.0	0.0	0.0	0.0	0.0	0.0	0.0	1.000	0.0	1.000
52	94 0	1	611.700	0.0	0.0	0.0	0.0	0.0	0.0	0.0	0.0	1.000	0.0	1.000

Figure 9A.1 (*continued*)

#	ID	idx											
53	94 3	1	532.700	3.000	0.0	0.0	0.0	0.0	0.0	0.0	0.0	1.000	0.0
		2	0.0	0.0	0.0	0.0	0.0	0.0	0.0	0.0	0.0	0.0	0.0
54	94 7	1	522.500	7.000	0.0	0.0	0.0	0.0	0.0	0.0	0.0	1.000	0.0
		2	0.0	0.0	0.0	0.0	0.0	0.0	0.0	0.0	0.0	0.0	0.0
55	9428	1	517.600	28.000	0.0	0.0	0.0	0.0	0.0	0.0	0.0	1.000	0.0
		2	0.0	0.0	0.0	0.0	0.0	0.0	0.0	0.0	0.0	0.0	0.0
56	9490	1	509.500	90.000	0.0	0.0	0.0	0.0	0.0	0.0	0.0	1.000	0.0
		2	0.0	0.0	0.0	0.0	0.0	0.0	0.0	0.0	0.0	0.0	0.0
57	9418	1	507.300	180.000	0.0	0.0	0.0	0.0	0.0	0.0	0.0	1.000	0.0
		2	0.0	0.0	0.0	0.0	0.0	0.0	0.0	0.0	0.0	0.0	0.0
58	9436	1	502.000	365.000	0.0	0.0	0.0	0.0	0.0	0.0	0.0	1.000	1.000
		2	0.0	0.0	0.0	0.0	0.0	0.0	0.0	0.0	0.0	0.0	0.0
59	24 0	1	573.000	0.0	0.0	0.0	0.0	0.0	0.0	0.0	0.0	0.0	1.000
		2	0.0	0.0	0.0	0.0	0.0	0.0	0.0	0.0	0.0	0.0	0.0
60	24 3	1	530.900	3.000	0.0	0.0	0.0	0.0	0.0	0.0	0.0	0.0	1.000
		2	0.0	0.0	0.0	0.0	0.0	0.0	0.0	0.0	0.0	0.0	0.0
61	24 7	1	526.200	7.000	0.3	0.0	0.0	0.0	0.0	0.0	0.0	0.0	1.000
		2	0.0	0.0	0.0	0.0	0.0	0.0	0.0	0.0	0.0	0.0	0.0
62	2428	1	511.500	28.000	0.0	0.0	0.0	0.0	0.0	0.0	0.0	0.0	1.000
		2	0.0	0.0	0.0	0.0	0.0	0.0	0.0	0.0	0.0	0.0	0.0
63	2490	1	503.000	90.000	0.0	0.0	0.0	0.0	0.0	0.0	0.0	0.0	1.000
		2	0.0	0.0	0.0	0.0	0.0	0.0	0.0	0.0	0.0	0.0	0.0
64	2418	1	500.500	180.000	0.0	0.0	0.0	0.0	0.0	0.0	0.0	0.0	0.0
		2	0.0	0.0	0.0	0.0	0.0	0.0	0.0	0.0	0.0	0.0	0.0
65	2436	1	501.400	365.000	0.0	0.0	0.0	0.0	0.0	0.0	0.0	0.0	0.0
		2	0.0	0.0	0.0	0.0	0.0	0.0	0.0	0.0	0.0	0.0	0.0
66	89 0	1	599.000	0.0	0.0	0.0	0.0	0.0	0.0	0.0	0.0	0.0	0.0
		2	0.0	0.0	0.0	0.0	0.0	0.0	0.0	0.0	0.0	0.0	0.0
67	89 3	1	534.300	3.000	0.0	0.0	0.0	0.0	0.0	0.0	0.0	0.0	0.0
		2	0.0	0.0	0.0	0.0	0.0	0.0	0.0	0.0	0.0	0.0	0.0
68	89 7	1	521.100	7.000	0.0	0.0	0.0	0.0	0.0	0.0	0.0	0.0	0.0
		2	0.0	0.0	0.0	0.0	0.0	0.0	0.0	0.0	0.0	0.0	0.0
69	8928	1	513.200	28.000	0.0	0.0	0.0	0.0	0.0	0.0	0.0	0.0	0.0
		2	0.0	0.0	0.0	0.0	0.0	0.0	0.0	0.0	0.0	0.0	0.0
70	8990	1	508.900	90.000	0.0	0.0	0.0	0.0	0.0	0.0	0.0	0.0	0.0
		2	0.0	0.0	0.0	0.0	0.0	0.0	0.0	0.0	0.0	0.0	0.0
71	8918	1	505.000	180.000	0.0	0.0	0.0	0.0	0.0	0.0	0.0	0.0	0.0
		2	0.0	0.0	0.0	0.0	0.0	0.0	0.0	0.0	0.0	0.0	0.0
72	8936	1	502.000	365.000	0.0	0.0	0.0	0.0	0.0	0.0	0.0	0.0	0.0
		2	0.0	0.0	0.0	0.0	0.0	0.0	0.0	0.0	0.0	0.0	0.0
73	90 0	1	633.400	0.0	0.0	0.0	0.0	0.0	0.0	0.0	0.0	0.0	0.0
		2	0.0	0.0	0.0	0.0	0.0	0.0	0.0	0.0	0.0	0.0	0.0
74	90 3	1	543.900	3.000	0.0	0.0	0.0	0.0	0.0	0.0	0.0	0.0	0.0
		2	0.0	0.0	0.0	0.0	0.0	0.0	0.0	0.0	0.0	0.0	0.0
75	90 7	1	536.800	7.000	0.0	0.0	0.0	0.0	0.0	0.0	0.0	0.0	0.0
		2	0.0	0.0	0.0	0.0	0.0	0.0	0.0	0.0	0.0	0.0	0.0
76	9028	1	524.600	28.000	0.0	0.0	0.0	0.0	0.0	0.0	0.0	0.0	0.0
		2	0.0	0.0	0.0	0.0	0.0	0.0	0.0	0.0	0.0	0.0	0.0
77	9090	1	519.300	90.000	0.0	0.0	0.0	0.0	0.0	0.0	0.0	0.0	0.0
		2	0.0	0.0	0.0	0.0	0.0	0.0	0.0	0.0	0.0	0.0	0.0
78	9018	1	517.500	180.000	0.0	0.0	0.0	0.0	0.0	0.0	0.0	0.0	0.0
		2	0.0	0.0	0.0	0.0	0.0	0.0	0.0	0.0	0.0	0.0	0.0
79	9036	1	510.700	365.000	0.0	0.0	0.0	0.0	0.0	0.0	0.0	0.0	0.0
		2	0.0	0.0	0.0	0.0	0.0	0.0	0.0	0.0	0.0	0.0	0.0
80	25 0	1	585.900	0.0	1.000	0.0	0.0	0.0	0.0	0.0	0.0	0.0	0.0
		2	0.0	0.0	0.0	0.0	0.0	0.0	0.0	0.0	0.0	0.0	0.0
81	25 3	1	527.100	3.000	1.000	0.0	0.0	0.0	0.0	0.0	0.0	0.0	0.0
		2	0.0	0.0	0.0	0.0	0.0	0.0	0.0	0.0	0.0	0.0	0.0
82	25 7	1	523.100	7.000	1.000	0.0	0.0	0.0	0.0	0.0	0.0	0.0	0.0
		2	0.0	0.0	0.0	0.0	0.0	0.0	0.0	0.0	0.0	0.0	0.0
83	2528	1	511.700	28.000	1.000	0.0	0.0	0.0	0.0	0.0	0.0	0.0	0.0
		2	0.0	0.0	0.0	0.0	0.0	0.0	0.0	0.0	0.0	0.0	0.0
84	2590	1	507.800	90.000	1.000	0.0	0.0	0.0	0.0	0.0	0.0	0.0	0.0
		2	0.0	0.0	0.0	0.0	0.0	0.0	0.0	0.0	0.0	0.0	0.0
85	2518	1	502.100	180.000	1.000	0.0	0.0	0.0	0.0	0.0	0.0	0.0	0.0
		2	0.0	0.0	0.0	0.0	0.0	0.0	0.0	0.0	0.0	0.0	0.0

Figure 9A.1 (*continued*)

229

No.	Code													
86	2536	1	502.800	365.000	0.0	0.0	0.0	0.0	0.0	0.0	0.0	0.0	0.0	0.0
87	95 0	2	625.600	0.0	1.000	0.0	0.0	0.0	0.0	0.0	0.0	0.0	0.0	0.0
88	95 3	1	540.800	0.0	0.0	1.000	0.0	0.0	0.0	0.0	0.0	0.0	0.0	0.0
89	95 7	2	531.100	3.000	0.0	1.000	0.0	0.0	0.0	0.0	0.0	0.0	0.0	0.0
90	9528	1	521.900	0.0	0.0	1.000	0.0	0.0	0.0	0.0	0.0	0.0	0.0	0.0
91	9590	2	518.100	7.000	0.0	1.000	0.0	0.0	0.0	0.0	0.0	0.0	0.0	0.0
92	9518	1	512.300	0.0	0.0	1.000	0.0	0.0	0.0	0.0	0.0	0.0	0.0	0.0
93	9536	2	510.200	28.000	0.0	1.000	0.0	0.0	0.0	0.0	0.0	0.0	0.0	0.0
94	91 0	1	624.000	0.0	0.0	1.000	1.000	0.0	0.0	0.0	0.0	0.0	0.0	0.0
95	91 3	2	540.000	90.000	0.0	1.000	1.000	0.0	0.0	0.0	0.0	0.0	0.0	0.0
96	91 7	1	532.700	0.0	0.0	0.0	1.000	0.0	0.0	0.0	0.0	0.0	0.0	0.0
97	9128	2	524.100	180.000	0.0	0.0	1.000	0.0	0.0	0.0	0.0	0.0	0.0	0.0
98	9190	1	517.800	0.0	0.0	0.0	1.000	0.0	0.0	0.0	0.0	0.0	0.0	0.0
99	9118	2	514.600	365.000	0.0	0.0	1.000	0.0	0.0	0.0	0.0	0.0	0.0	0.0
100	9136	1	507.700	0.0	0.0	0.0	1.000	0.0	0.0	0.0	0.0	0.0	0.0	0.0
101	70 0	1	564.700	0.0	0.0	0.0	0.0	1.000	0.0	0.0	0.0	0.0	0.0	0.0
102	70 3	2	536.400	3.000	0.0	0.0	0.0	1.000	0.0	0.0	0.0	0.0	0.0	0.0
103	70 7	1	528.400	7.000	0.0	0.0	0.0	1.000	0.0	0.0	0.0	0.0	0.0	0.0
104	7028	1	514.900	28.000	0.0	0.0	0.0	1.000	0.0	0.0	0.0	0.0	0.0	0.0
105	7036	2	502.100	365.000	0.0	0.0	0.0	1.000	0.0	0.0	0.0	0.0	0.0	0.0
106		1	0.0	999.000	0.0	0.0	0.0	0.0	0.0	0.0	0.0	0.0	0.0	0.0

Figure 9A.1 (*continued*)

230

NON-LINEAR LEAST-SQUARES CURVE FITTING PROGRAM

CEMENT HEAT, PASS 1

Z = D' - C' + (A' / (T + F')EXP B)
Z = ACID HEAT OF SOLUTION
B = EXPONENT COMMON TO ALL CEMENTS
D' = INITIAL ACID HEAT OF SOLUTION = D1V1 + D2V2 + --- + D14V14
A' = CONSTANT = A1V1 + A2V2 + --- + A14V14
C' = ULTIMATE CUMULATIVE HEAT OF HARDENING = C1V1 + C2V2 + -- + C14V14
F' = TIME CORRECTION TERM = (A'/C')EXP(1/B)
T = TIME, MEASURED IN DAYS, AFTER WATER IS ADDED TO THE CEMENT
     V1 --- V14 = INDICATOR VARIABLES FOR EACH SAMPLE OF CEMENT
NO OBSERVATIONS OMITTED.

					95% CONFIDENCE LIMITS	
IND.VAR(I)	NAME	COEF.B(I)	S.E. COEF.	T-VALUE	LOWER	UPPER
1	B	2.19917E-01	3.66E-02	6.0	1.47E-01	2.93E-01
2	D1 22	5.88674E 02	2.20E 00	267.9	5.84E 02	5.93E 02
3	D2 23	5.75654E 02	2.20E 00	262.0	5.71E 02	5.80E 02
4	D3 92	6.23848E 02	1.55E 00	401.5	6.21E 02	6.27E 02
5	D4 88	5.95490E 02	2.20E 00	271.0	5.91E 02	6.00E 02
6	D5 96	6.03849E 02	1.55E 00	388.6	6.01E 02	6.07E 02
7	D6 85	6.18813E 02	2.20E 00	281.6	6.14E 02	6.23E 02
8	D7 94	6.10850E 02	1.55E 00	393.1	6.08E 02	6.14E 02
9	D8 24	5.72993E 02	2.20E 00	260.8	5.69E 02	5.77E 02
10	D9 89	5.99008E 02	2.20E 00	272.6	5.95E 02	6.03E 02
11	D10 90	6.33399E 02	2.20E 00	288.3	6.29E 02	6.38E 02
12	D11 25	5.85896E 02	2.20E 00	266.6	5.82E 02	5.90E 02
13	D12 95	6.25601E 02	2.20E 00	284.7	6.21E 02	6.30E 02
14	D13 91	6.23997E 02	2.20E 00	284.0	6.20E 02	6.28E 02
15	D14 70	5.64729E 02	2.20E 00	257.2	5.60E 02	5.69E 02
16	A1 22	7.25552E 01	6.25E 00	11.6	6.01E 01	8.51E 01
17	A2 23	7.80953E 01	6.50E 00	12.0	6.51E 01	9.11E 01
18	A3 92	6.54756E 01	6.00E 00	10.9	5.35E 01	7.75E 01
19	A4 88	7.12716E 01	6.20E 00	11.5	5.89E 01	8.37E 01
20	A5 96	5.67378E 01	5.75E 00	9.9	4.52E 01	6.82E 01
21	A6 85	6.20204E 01	4.74E 00	13.1	5.25E 01	7.15E 01
22	A7 94	5.53831E 01	5.72E 00	9.7	4.40E 01	6.68E 01
23	A8 24	6.66299E 01	6.14E 00	10.8	5.43E 01	7.89E 01
24	A9 89	5.80704E 01	5.79E 00	10.0	4.65E 01	6.97E 01
25	A10 90	6.21030E 01	5.89E 00	10.5	5.03E 01	7.39E 01
26	A11 25	5.18421E 01	5.65E 00	9.2	4.05E 01	6.31E 01
27	A12 95	5.78616E 01	5.78E 00	10.0	4.63E 01	6.94E 01
28	A13 91	5.89554E 01	5.81E 00	10.2	4.73E 01	7.06E 01
29	A14 70	7.14802E 01	7.37E 00	9.7	5.67E 01	8.62E 01
30	C1 22	1.04970E 02	6.47E 00	16.2	9.20E 01	1.18E 02
31	C2 23	9.86860E 01	6.77E 00	14.6	8.52E 01	1.12E 02
32	C3 92	1.26151E 02	5.88E 00	21.5	1.14E 02	1.38E 02
33	C4 88	1.08394E 02	6.41E 00	16.9	9.56E 01	1.21E 02
34	C5 96	1.15153E 02	5.31E 00	21.7	1.05E 02	1.26E 02
35	C6 85	1.28164E 02	5.59E 00	22.9	1.17E 02	1.39E 02
36	C7 94	1.22164E 02	5.23E 00	23.4	1.12E 02	1.33E 02
37	C8 24	9.25444E 01	6.07E 00	15.2	8.04E 01	1.05E 02
38	C9 89	1.12629E 02	5.61E 00	20.1	1.01E 02	1.24E 02
39	C10 90	1.37739E 02	5.88E 00	23.4	1.26E 02	1.49E 02
40	C11 25	9.83243E 01	5.23E 00	18.8	8.79E 01	1.09E 02
41	C12 95	1.30974E 02	5.61E 00	23.3	1.20E 02	1.42E 02
42	C13 91	1.29475E 02	5.67E 00	22.8	1.18E 02	1.41E 02
43	C14 70	8.27440E 01	6.64E 00	12.5	6.95E 01	9.60E 01

NO. OF OBSERVATIONS   105
NO. OF COEFFICIENTS   43
RESIDUAL DEGREES OF FREEDOM   62
RESIDUAL ROOT MEAN SQUARE   2.19731045
RESIDUAL MEAN SQUARE   4.82817554
RESIDUAL SUM OF SQUARES   299.34692383

Figure 9A.2

231

CEMENT HEAT, PASS 1

--------ORDERED BY COMPUTER INPUT--------				---------------ORDERED BY RESIDUALS---------------				
OBS. NO.	OBS. Z	FITTED Z	RESIDUAL	OBS. NO.	OBS. Z	FITTED Z	ORDERED RESID.	SEQ.
22 0	588.700	588.674	0.026	85 3	544.800	539.227	5.573	1
22 3	537.500	539.935	-2.435	22 7	535.600	530.726	4.874	2
22 7	535.600	530.726	4.874	8890	517.800	513.580	4.220	3
2228	516.200	518.520	-2.320	2218	510.200	506.856	3.344	4
2290	507.500	510.662	-3.162	2436	501.400	498.650	2.750	5
2218	510.200	506.856	3.344	23 7	530.000	527.339	2.661	6
2236	503.200	503.525	-0.325	24 7	526.200	523.581	2.619	7
23 0	575.700	575.654	0.046	89 3	534.300	531.823	2.477	8
23 3	535.100	536.851	-1.751	9428	517.600	515.295	2.305	9
23 7	530.000	527.339	2.661	9228	531.400	529.149	2.251	10
2328	514.700	514.396	0.303	9690	511.900	509.785	2.115	11
2390	503.800	505.974	-2.174	9018	517.500	515.481	2.019	12
2318	501.400	501.884	-0.484	9590	518.100	516.135	1.965	13
2336	499.700	498.301	1.399	9618	508.600	506.804	1.796	14
92 0	623.600	623.848	-0.248	25 7	523.100	521.308	1.792	15
92 0	624.100	623.848	0.252	8536	509.000	507.594	1.406	16
92 3	548.800	548.930	-0.130	2336	499.700	498.301	1.399	17
92 7	538.900	540.310	-1.410	8590	515.100	513.702	1.398	18
9228	531.400	529.149	2.251	9190	517.800	516.437	1.363	19
9290	523.000	522.033	0.967	8518	511.800	510.444	1.356	20
9218	519.300	518.594	0.706	9118	514.600	513.339	1.261	21
9236	513.200	515.585	-2.385	9128	524.100	522.848	1.252	22
88 0	595.500	595.490	0.010	2536	502.800	501.736	1.064	23
88 3	542.300	542.478	-0.178	8918	505.900	504.913	0.987	24
88 7	532.400	533.340	-0.940	9290	523.000	522.033	0.967	25
8828	521.800	521.306	0.494	2590	507.800	506.841	0.959	26
8890	517.800	513.580	4.220	9418	507.300	506.362	0.938	27
8818	507.900	509.839	-1.939	8990	508.900	507.963	0.937	28
8836	504.900	506.566	-1.666	8536	508.500	507.594	0.906	29
96 0	604.500	603.849	0.651	94 0	611.700	610.850	0.850	30
96 0	603.200	603.849	-0.649	95 3	540.800	539.989	0.811	31
96 3	532.600	533.126	-0.526	9218	519.300	518.594	0.706	32
96 7	525.700	525.634	0.066	90 7	536.800	536.108	0.691	33
9628	515.700	515.953	-0.253	96 0	604.500	603.849	0.651	34
9690	511.900	509.785	2.115	94 3	532.700	532.095	0.605	35
9618	508.600	506.804	1.796	8590	514.300	513.702	0.598	36
9636	501.000	504.197	-3.197	7036	502.100	501.509	0.591	37
85 0	618.800	618.813	-0.013	9090	519.300	518.744	0.555	38
85 3	538.200	539.227	-1.027	70 7	528.400	527.859	0.541	39
85 3	544.800	539.227	5.573	8828	521.800	521.306	0.494	40
85 7	528.500	531.031	-2.531	2328	514.700	514.396	0.303	41
85 7	530.200	531.031	-0.831	92 0	624.100	623.848	0.252	42
8528	517.800	520.445	-2.645	9490	509.500	509.272	0.228	43
8528	517.100	520.445	-3.345	70 3	536.400	536.204	0.196	44
8590	515.100	513.702	1.398	96 7	525.700	525.634	0.066	45
8590	514.300	513.702	0.598	23 0	575.700	575.654	0.046	46
8518	509.600	510.444	-0.844	22 0	588.700	588.674	0.026	47
8518	511.800	510.444	1.356	88 0	595.500	595.490	0.010	48
8536	509.000	507.594	1.406	24 0	573.000	572.992	0.008	49
8536	508.500	507.594	0.906	25 0	585.900	585.896	0.004	50

Figure 9A.2 (*continued*)

94 0	610.000	610.850	-0.850	91 0	624.000	623.997	0.003	51
94 0	611.700	610.850	0.850	90 0	633.400	633.399	0.001	52
94 3	532.700	532.095	0.605	95 0	625.600	625.601	-0.001	53
94 7	522.500	524.757	-2.257	89 0	599.000	599.008	-0.008	54
9428	517.600	515.295	2.305	85 0	618.800	618.813	-0.013	55
9490	509.500	509.272	0.228	70 0	564.700	564.729	-0.029	56
9418	507.300	506.362	0.938	92 3	548.800	548.930	-0.130	57
9436	502.000	503.817	-1.817	88 3	542.300	542.478	-0.178	58
24 0	573.000	572.992	0.008	91 7	532.700	532.919	-0.219	59
24 3	530.900	531.953	-1.053	9536	510.200	510.436	-0.236	60
24 7	526.200	523.581	2.619	8936	502.000	502.244	-0.244	61
2428	511.500	512.411	-0.911	92 0	623.600	623.848	-0.248	62
2490	503.000	505.203	-2.203	9628	515.700	515.953	-0.253	63
2418	500.500	501.709	-1.209	2236	503.200	503.525	-0.325	64
2436	501.400	498.650	2.750	90 3	543.900	544.339	-0.439	65
89 0	599.000	599.008	-0.008	2318	501.400	501.884	-0.484	66
89 3	534.300	531.823	2.477	96 3	532.600	533.126	-0.526	67
89 7	521.100	524.174	-3.074	9528	521.900	522.428	-0.528	68
8928	513.200	514.274	-1.074	96 0	603.200	603.849	-0.649	69
8990	508.900	507.963	0.937	91 3	540.000	540.730	-0.730	70
8918	505.900	504.913	0.987	2528	511.700	512.474	-0.774	71
8936	502.000	502.244	-0.244	9518	512.300	513.094	-0.795	72
90 0	633.400	633.399	0.001	85 7	530.200	531.031	-0.831	73
90 3	543.900	544.339	-0.439	8518	509.600	510.444	-0.844	74
90 7	536.800	536.108	0.691	94 0	610.000	610.850	-0.850	75
9028	524.600	525.498	-0.898	9028	524.600	525.498	-0.898	76
9090	519.300	518.744	0.555	2428	511.500	512.411	-0.911	77
9018	517.500	515.481	2.019	88 7	532.400	533.340	-0.940	78
9036	510.700	512.628	-1.928	25 3	527.100	528.126	-1.026	79
25 0	585.900	585.896	0.004	85 3	538.200	539.227	-1.027	80
25 3	527.100	528.126	-1.026	24 3	530.900	531.953	-1.053	81
25 7	523.100	521.308	1.792	8928	513.200	514.274	-1.074	82
2528	511.700	512.474	-0.774	2418	500.500	501.709	-1.209	83
2590	507.800	506.841	0.959	95 7	531.100	532.315	-1.216	84
2518	502.100	504.117	-2.018	7028	514.900	516.198	-1.298	85
2536	502.800	501.736	1.064	92 7	538.900	540.310	-1.410	86
95 0	625.600	625.601	-0.001	8836	504.900	506.566	-1.666	87
95 3	540.800	539.989	0.811	23 3	535.100	536.851	-1.751	88
95 7	531.100	532.315	-1.216	9436	502.000	503.817	-1.817	89
9528	521.900	522.428	-0.528	9036	510.700	512.628	-1.928	90
9590	518.100	516.135	1.965	8818	507.900	509.839	-1.939	91
9518	512.300	513.094	-0.795	2518	502.100	504.117	-2.018	92
9536	510.200	510.436	-0.236	2390	503.800	505.974	-2.174	93
91 0	624.000	623.997	0.003	2490	503.000	505.203	-2.203	94
91 3	540.000	540.730	-0.730	94 7	522.500	524.757	-2.257	95
91 7	532.700	532.919	-0.219	2228	516.200	518.520	-2.320	96
9128	524.100	522.848	1.252	9236	513.200	515.585	-2.385	97
9190	517.800	516.437	1.363	22 3	537.500	539.935	-2.435	98
9118	514.600	513.339	1.261	85 7	528.500	531.031	-2.531	99
9136	507.700	510.630	-2.930	8528	517.800	520.445	-2.645	100
70 0	564.700	564.729	-0.029	9136	507.700	510.630	-2.930	101
70 3	536.400	536.204	0.196	89 7	521.100	524.174	-3.074	102
70 7	528.400	527.859	0.541	2290	507.500	510.662	-3.162	103
7028	514.900	516.198	-1.298	9636	501.000	504.197	-3.197	104
7036	502.100	501.509	0.591	8528	517.100	520.445	-3.345	105

Figure 9A.2 (*continued*)

CEMENT HEAT, PASS 1

CUMULATIVE DISTRIBUTION OF RESIDUALS

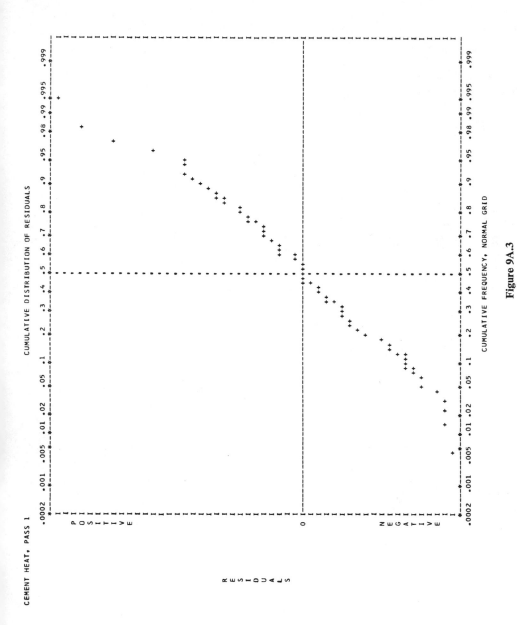

Figure 9A.3

234

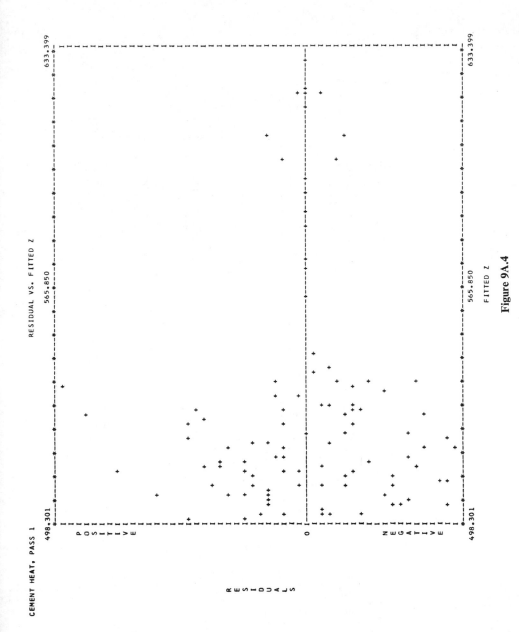

Figure 9A.4

CEMENT HEAT, PASS 2

$Z = D' - C' + (\ A' / (\ T + F'\ )EXP\ B\ )$

Z = ACID HEAT OF SOLUTION
B = EXPONENT COMMON TO ALL CEMENTS
D' = INITIAL ACID HEAT OF SOLUTION = $D1V1 + D2V2 + --- + D14V14$
A' = CONSTANT = $A1V1 + A2V2 + --- + A14V14$
C' = ULTIMATE CUMULATIVE HEAT OF HARDENING = $C1V1 + C2V2 + -- + C14V14$
F' = TIME CORRECTION TERM = $(A'/C')EXP(1/B)$
T = TIME, MEASURED IN DAYS, AFTER WATER IS ADDED TO THE CEMENT
$V1 --- V14$ = INDICATOR VARIABLES FOR EACH SAMPLE OF CEMENT
OBSERVATION 85-3-2 OMITTED.

IND.VAR(I)	NAME		COEF.B(I)	S.E. COEF.	T-VALUE	95% CONFIDENCE LIMITS	
						LOWER	UPPER
1	B		1.99732E-01	3.40E-02	5.9	1.32E-01	2.68E-01
2	D1	22	5.88682E 02	2.04E 00	289.1	5.85E 02	5.93E 02
3	D2	23	5.75667E 02	2.04E 00	282.7	5.72E 02	5.80E 02
4	D3	92	6.23849E 02	1.44E 00	433.1	6.21E 02	6.27E 02
5	D4	88	5.95493E 02	2.04E 00	292.4	5.91E 02	6.00E 02
6	D5	96	6.03849E 02	1.44E 00	419.2	6.01E 02	6.07E 02
7	D6	85	6.18803E 02	2.04E 00	303.8	6.15E 02	6.23E 02
8	D7	94	6.10850E 02	1.44E 00	424.1	6.08E 02	6.14E 02
9	D8	24	5.72999E 02	2.04E 00	281.3	5.69E 02	5.77E 02
10	D9	89	5.99007E 02	2.04E 00	294.1	5.95E 02	6.03E 02
11	D10	90	6.33400E 02	2.04E 00	311.0	6.29E 02	6.37E 02
12	D11	25	5.85898E 02	2.04E 00	287.6	5.82E 02	5.90E 02
13	D12	95	6.25601E 02	2.04E 00	307.1	6.22E 02	6.30E 02
14	D13	91	6.23999E 02	2.04E 00	306.3	6.20E 02	6.28E 02
15	D14	70	5.64736E 02	2.04E 00	277.3	5.61E 02	5.69E 02
16	A1	22	7.47356E 01	6.63E 00	11.3	6.15E 01	8.80E 01
17	A2	23	8.02845E 01	6.87E 00	11.7	6.65E 01	9.40E 01
18	A3	92	6.77074E 01	6.34E 00	10.7	5.50E 01	8.04E 01
19	A4	88	7.35552E 01	6.58E 00	11.2	6.04E 01	8.67E 01
20	A5	96	5.86780E 01	5.99E 00	9.8	4.67E 01	7.07E 01
21	A6	85	5.75986E 01	5.10E 00	11.3	4.74E 01	6.78E 01
22	A7	94	5.72813E 01	5.94E 00	9.6	4.54E 01	6.92E 01
23	A8	24	6.86008E 01	6.42E 00	10.7	5.58E 01	8.14E 01
24	A9	89	5.99506E 01	6.04E 00	9.9	4.79E 01	7.20E 01
25	A10	90	6.42191E 01	6.21E 00	10.3	5.18E 01	7.66E 01
26	A11	25	5.35725E 01	5.81E 00	9.2	4.20E 01	6.52E 01
27	A12	95	5.98080E 01	6.04E 00	9.9	4.77E 01	7.19E 01
28	A13	91	6.10152E 01	6.09E 00	10.0	4.88E 01	7.32E 01
29	A14	70	7.32536E 01	7.43E 00	9.9	5.84E 01	8.81E 01
30	C1	22	1.08332E 02	7.07E 00	15.3	9.42E 01	1.22E 02
31	C2	23	1.02250E 02	7.41E 00	13.8	8.74E 01	1.17E 02
32	C3	92	1.29293E 02	6.46E 00	20.0	1.16E 02	1.42E 02
33	C4	88	1.11759E 02	7.00E 00	16.0	9.78E 01	1.26E 02
34	C5	96	1.17876E 02	5.79E 00	20.4	1.06E 02	1.29E 02
35	C6	85	1.28488E 02	5.57E 00	23.1	1.17E 02	1.40E 02
36	C7	94	1.24822E 02	5.69E 00	21.9	1.13E 02	1.36E 02
37	C8	24	9.56193E 01	6.60E 00	14.5	8.24E 01	1.09E 02
38	C9	89	1.15363E 02	6.05E 00	19.1	1.03E 02	1.27E 02
39	C10	90	1.40714E 02	6.37E 00	22.1	1.28E 02	1.53E 02
40	C11	25	1.00795E 02	5.60E 00	18.0	8.96E 01	1.12E 02
41	C12	95	1.33731E 02	6.05E 00	22.1	1.22E 02	1.46E 02
42	C13	91	1.32326E 02	6.14E 00	21.5	1.20E 02	1.45E 02
43	C14	70	8.58738E 01	7.14E 00	12.0	7.16E 01	1.00E 02

NO. OF OBSERVATIONS   104
NO. OF COEFFICIENTS   43
RESIDUAL DEGREES OF FREEDOM   61
RESIDUAL ROOT MEAN SQUARE   2.03694439
RESIDUAL MEAN SQUARE   4.14914322
RESIDUAL SUM OF SQUARES   253.09777832

Figure 9A.5

CEMENT HEAT, PASS 2

--------ORDERED BY COMPUTER INPUT----------					---------------ORDERED BY RESIDUALS----------------				
OBS. NO.	OBS. Z	FITTED Z	RESIDUAL		OBS. NO.	OBS. Z	FITTED Z	ORDERED RESID.	SEQ.
22 0	588.700	588.681	0.019		22 7	535.600	530.795	4.805	1
22 3	537.500	539.756	-2.256		8890	517.800	513.669	4.131	2
22 7	535.600	530.795	4.805		2218	510.200	506.835	3.365	3
2228	516.200	518.720	-2.521		2436	501.400	498.490	2.909	4
2290	507.500	510.762	-3.262		89 3	534.300	531.663	2.637	5
2218	510.200	506.835	3.365		23 7	530.000	527.396	2.604	6
2236	503.200	503.349	-0.149		24 7	526.200	523.641	2.559	7
23 0	575.700	575.667	0.033		9428	517.600	515.466	2.134	8
23 3	535.100	536.676	-1.576		9228	531.400	529.347	2.052	9
23 7	530.000	527.396	2.604		9018	517.500	515.448	2.052	10
2328	514.700	514.596	0.104		9690	511.900	509.857	2.042	11
2390	503.800	506.077	-2.278		9590	518.100	516.214	1.885	12
2318	501.400	501.864	-0.464		9618	508.600	506.770	1.830	13
2336	499.700	498.122	1.578		25 7	523.100	521.380	1.720	14
92 0	623.600	623.849	-0.249		85 3	538.200	536.510	1.690	15
92 0	624.100	623.849	0.251		2336	499.700	498.122	1.578	16
92 3	548.800	548.783	0.017		8590	515.100	513.761	1.339	17
92 7	538.900	540.408	-1.508		9118	514.600	513.299	1.301	18
9228	531.400	529.347	2.052		9190	517.800	516.509	1.291	19
9290	523.000	522.116	0.884		2536	502.800	501.591	1.209	20
9218	519.300	518.554	0.746		9128	524.100	523.029	1.071	21
9236	513.200	515.394	-2.194		8518	511.800	510.730	1.070	22
88 0	595.500	595.493	0.007		8918	505.900	504.893	1.007	23
88 3	542.300	542.325	-0.025		9418	507.300	506.330	0.969	24
88 7	532.400	533.429	-1.029		95 3	540.800	539.837	0.963	25
8828	521.800	521.509	0.291		8536	509.000	508.042	0.958	26
8890	517.800	513.669	4.131		2590	507.800	506.909	0.891	27
8818	507.900	509.802	-1.903		9290	523.000	522.116	0.884	28
8836	504.900	506.371	-1.471		85 7	530.200	529.344	0.855	29
96 0	604.500	603.849	0.651		8990	508.900	508.047	0.853	30
96 0	603.200	603.849	-0.649		94 0	611.700	610.850	0.850	31
96 3	532.600	532.995	-0.395		9218	519.300	518.554	0.746	32
96 7	525.700	525.720	-0.020		94 3	532.700	531.962	0.738	33
9628	515.700	516.126	-0.426		7036	502.100	501.402	0.698	34
9690	511.900	509.857	2.042		96 0	604.500	603.849	0.651	35
9618	508.600	506.770	1.830		90 7	536.800	536.199	0.600	36
9636	501.000	504.032	-3.032		8590	514.300	513.761	0.539	37
85 0	618.800	618.803	-0.003		70 7	528.400	527.909	0.491	38
85 3	538.200	536.510	1.690		9090	519.300	518.827	0.473	39
85 7	528.500	529.344	-0.844		8536	508.500	508.042	0.458	40
85 7	530.200	529.344	0.855		70 3	536.400	536.060	0.340	41
8528	517.800	519.916	-2.116		8828	521.800	521.509	0.291	42
8528	517.100	519.916	-2.816		92 0	624.100	623.849	0.251	43
8590	515.100	513.761	1.339		9490	509.500	509.344	0.156	44
8590	514.300	513.761	0.539		2328	514.700	514.596	0.104	45
8518	509.600	510.730	-1.130		23 0	575.700	575.667	0.033	46
8518	511.800	510.730	1.070		22 0	588.700	588.681	0.019	47
8536	509.000	508.042	0.958		92 3	548.800	548.783	0.017	48
8536	508.500	508.042	0.458		88 0	595.500	595.493	0.007	49

Figure 9A.5 (*continued*)

237

94 0	610.000	610.850	-0.850	25 0	585.900	585.898	0.002	50
94 0	611.700	610.850	0.850	91 0	624.000	623.998	0.002	51
94 3	532.700	531.962	0.738	24 0	573.000	572.999	0.001	52
94 7	522.500	524.840	-2.340	90 0	633.400	633.399	0.000	53
9428	517.600	515.466	2.134	95 0	625.600	625.601	-0.001	54
9490	509.500	509.344	0.156	85 0	618.800	618.803	-0.003	55
9418	507.300	506.330	0.969	89 0	599.000	599.007	-0.007	56
9436	502.000	503.657	-1.657	96 7	525.700	525.720	-0.020	57
24 0	573.000	572.999	0.001	88 3	542.300	542.325	-0.025	58
24 3	530.900	531.794	-0.894	70 0	564.700	564.736	-0.036	59
24 7	526.200	523.641	2.559	9536	510.200	510.276	-0.076	60
2428	511.500	512.592	-1.092	8936	502.000	502.095	-0.095	61
2490	503.000	505.293	-2.293	2236	503.200	503.349	-0.149	62
2418	500.500	501.689	-1.189	92 0	623.600	623.849	-0.249	63
2436	501.400	498.490	2.909	90 3	543.900	544.185	-0.285	64
89 0	599.000	599.007	-0.007	91 7	532.700	533.014	-0.314	65
89 3	534.300	531.663	2.637	96 3	532.600	532.995	-0.395	66
89 7	521.100	524.245	-3.145	9628	515.700	516.126	-0.426	67
8928	513.200	514.450	-1.250	2318	501.400	501.864	-0.464	68
8990	508.900	508.047	0.853	91 3	540.000	540.599	-0.599	69
8918	505.900	504.893	1.007	96 0	603.200	603.849	-0.649	70
8936	502.000	502.095	-0.095	9528	521.900	522.606	-0.706	71
90 0	633.400	633.399	0.000	9518	512.300	513.067	-0.768	72
90 3	543.900	544.185	-0.285	85 7	528.500	529.344	-0.844	73
90 7	536.800	536.199	0.600	94 0	610.000	610.850	-0.850	74
9028	524.600	525.689	-1.089	24 3	530.900	531.794	-0.894	75
9090	519.300	518.827	0.473	25 3	527.100	528.000	-0.901	76
9018	517.500	515.448	2.052	2528	511.700	512.630	-0.930	77
9036	510.700	512.450	-1.750	88 7	532.400	533.429	-1.029	78
25 0	585.900	585.898	0.002	9028	524.600	525.689	-1.089	79
25 3	527.100	528.000	-0.901	2428	511.500	512.592	-1.092	80
25 7	523.100	521.380	1.720	8518	509.600	510.730	-1.130	81
2528	511.700	512.630	-0.930	2418	500.500	501.689	-1.189	82
2590	507.800	506.909	0.891	8928	513.200	514.450	-1.250	83
2518	502.100	504.091	-1.991	95 7	531.100	532.396	-1.296	84
2536	502.800	501.591	1.209	8836	504.900	506.371	-1.471	85
95 0	625.600	625.601	-0.001	7028	514.900	516.394	-1.494	86
95 3	540.800	539.837	0.963	92 7	538.900	540.408	-1.508	87
95 7	531.100	532.396	-1.296	23 3	535.100	536.676	-1.576	88
9528	521.900	522.606	-0.706	9436	502.000	503.657	-1.657	89
9590	518.100	516.214	1.885	9036	510.700	512.450	-1.750	90
9518	512.300	513.067	-0.768	8818	507.900	509.802	-1.903	91
9536	510.200	510.276	-0.076	2518	502.100	504.091	-1.991	92
91 0	624.000	623.998	0.002	8528	517.800	519.916	-2.116	93
91 3	540.000	540.599	-0.599	9236	513.200	515.394	-2.194	94
91 7	532.700	533.014	-0.314	22 3	537.500	539.756	-2.256	95
9128	524.100	523.029	1.071	2390	503.800	506.077	-2.278	96
9190	517.800	516.509	1.291	2490	503.000	505.293	-2.293	97
9118	514.600	513.299	1.301	94 7	522.500	524.840	-2.340	98
9136	507.700	510.451	-2.751	2228	516.200	518.720	-2.521	99
70 0	564.700	564.736	-0.036	9136	507.700	510.451	-2.751	100
70 3	536.400	536.060	0.340	8528	517.100	519.916	-2.816	101
70 7	528.400	527.909	0.491	9636	501.000	504.032	-3.032	102
7028	514.900	516.394	-1.494	89 7	521.100	524.245	-3.145	103
7036	502.100	501.402	0.698	2290	507.500	510.762	-3.262	104

Figure 9A.5 (*continued*)

CEMENT HEAT, PASS 2

CUMULATIVE DISTRIBUTION OF RESIDUALS

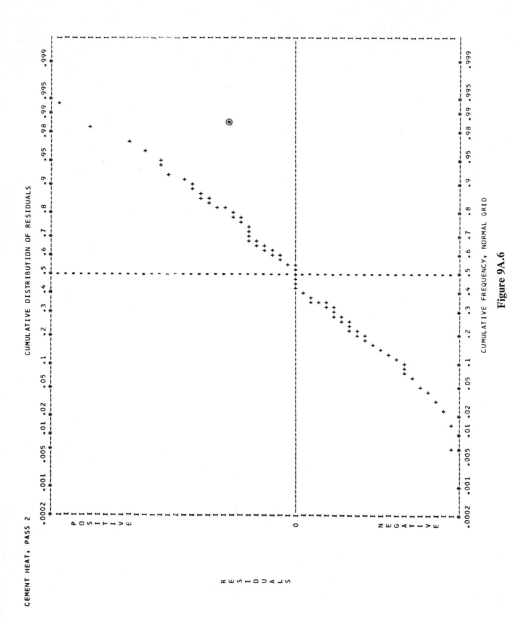

Figure 9A.6

239

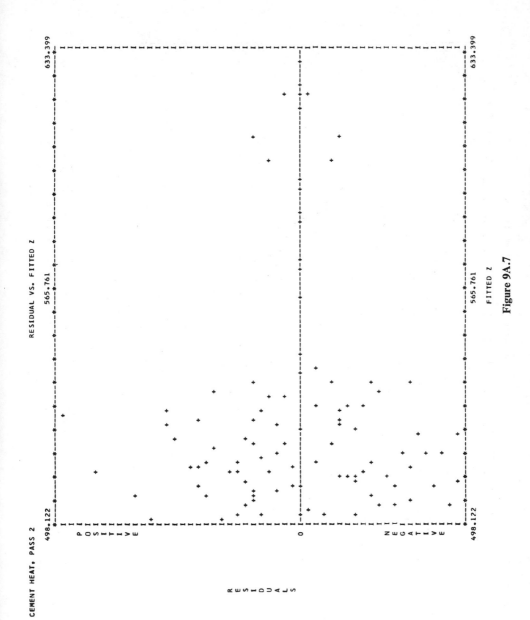

Figure 9A.7

LINEAR LEAST-SQUARES CURVE FITTING PROGRAM

CEMENT HEAT PASS 3   DEP VAR 1: C   MIN Y = 8.587D 01   MAX Y = 1.407D 02   RANGE Y = 5.484D 01

Y = B(0) + B(1)X(1);   C = H + GD;   C = -324. + 0.73D
Y = C, THE ULTIMATE CUMULATIVE HEAT OF HARDENING, CALORIES/GRAM
B(0) = H, A CONSTANT
X = D, THE INITIAL ACID HEAT OF SOLUTION, CALORIES/GRAM
    C AND D VALUES FROM NON-LINEAR FIT WITH B COMMON, PASS 2.

IND-VAR(I)	NAME	COEF-B(I)	S.E. COEF.	T-VALUE	R(I)SQRD	MIN X(I)	MAX X(I)	RANGE X(I)	REL.INF.X(I)
0		-3.235630 02	3.070-02	23.8	0.0				
1	G	7.31004D-01				5.647D 02	6.334D 02	6.860D 01	0.92

NO. OF OBSERVATIONS   14
NO. OF IND. VARIABLES   1
RESIDUAL DEGREES OF FREEDOM   12
F-VALUE   567.9
RESIDUAL ROOT MEAN SQUARE   2.43035632
RESIDUAL MEAN SQUARE   5.90663185
RESIDUAL SUM OF SQUARES   70.89958217
TOTAL SUM OF SQUARES   3425.35599575
MULT. CORREL. COEF. SQUARED   .9793

--------ORDERED BY COMPUTER INPUT--------

IDENT.	OBSV.	WS DISTANCE	OBS. Y	FITTED Y	RESIDUAL
D1	22	15.	108.332	106.766	1.566
D2	23	61.	102.250	97.252	4.998
D3	92	45.	129.293	132.473	-3.180
D4	88	3.	111.759	111.745	0.013
D5	96	0.	117.876	117.853	0.023
D6	85	2.	128.488	128.784	-0.296
D7	94	8.	124.822	122.971	1.851
D8	24	74.	95.619	95.301	0.318
D9	89	1.	115.363	114.314	1.049
D10	90	91.	140.714	139.455	1.259
D11	25	22.	100.795	104.731	-3.936
D12	95	52.	133.731	133.754	-0.022
D13	91	45.	132.326	132.582	-0.257
D14	70	123.	85.874	89.261	-3.387

--------ORDERED BY RESIDUALS--------

OBSV.	OBS. Y	FITTED Y	ORDERED RESID.	SEQ
23	102.250	97.252	4.998	1
94	124.822	122.971	1.851	2
22	108.332	106.766	1.566	3
90	140.714	139.455	1.259	4
89	115.363	114.314	1.049	5
24	95.619	95.301	0.318	6
96	117.876	117.853	0.023	7
88	111.759	111.745	0.013	8
95	133.731	133.754	-0.022	9
91	132.326	132.582	-0.257	10
85	128.488	128.784	-0.296	11
92	129.293	132.473	-3.180	12
70	85.874	89.261	-3.387	13
25	100.795	104.731	-3.936	14

Figure 9A.8

241

CEMENT HEAT, PASS 4

$$Z = H + (G - 1)D' + (A' / (T + F')EXP B)$$

Z = ACID HEAT OF SOLUTION
B, G AND H ARE COMMON TO ALL CEMENTS
D' = INITIAL ACID HEAT OF SOLUTION = D1V1 + D2V2 + --- + D14V14
A' = CONSTANT = A1V1 + A2V2 + --- + A14V14
F' = TIME CORRECTION TERM = (A' / (GD' - H))EXP(1/B)
T = TIME, MEASURED IN DAYS, AFTER WATER IS ADDED TO THE CEMENT
V1 --- V14 = INDICATOR VARIABLES FOR EACH SAMPLE OF CEMENT
OBSERVATION 85-3-2 OMITTED.

					95% CONFIDENCE LIMITS	
IND.VAR(I)	NAME	COEF.B(I)	S.E. COEF.	T-VALUE	LOWER	UPPER
1	B	1.92961E-01	2.78E-02	6.9	1.37E-01	2.49E-01
2	D1 22	5.88475E 02	1.96E 00	300.0	5.85E 02	5.92E 02
3	D2 23	5.75044E 02	1.97E 00	292.1	5.71E 02	5.79E 02
4	D3 92	6.24083E 02	1.40E 00	446.0	6.21E 02	6.27E 02
5	D4 88	5.95496E 02	1.96E 00	303.7	5.92E 02	5.99E 02
6	D5 96	6.03860E 02	1.40E 00	432.6	6.01E 02	6.07E 02
7	D6 85	6.18885E 02	1.94E 00	319.8	6.15E 02	6.23E 02
8	D7 94	6.10704E 02	1.40E 00	437.2	6.08E 02	6.13E 02
9	D8 24	5.73016E 02	1.97E 00	291.2	5.69E 02	5.77E 02
10	D9 89	5.98872E 02	1.96E 00	306.3	5.95E 02	6.03E 02
11	D10 90	6.33143E 02	1.96E 00	323.1	6.29E 02	6.37E 02
12	D11 25	5.86587E 02	1.96E 00	299.5	5.83E 02	5.91E 02
13	D12 95	6.25581E 02	1.96E 00	319.7	6.22E 02	6.29E 02
14	D13 91	6.24014E 02	1.96E 00	319.0	6.20E 02	6.28E 02
15	D14 70	5.64969E 02	1.98E 00	284.7	5.61E 02	5.69E 02
16	A1 22	7.31728E 01	3.89E 00	18.8	6.54E 01	8.09E 01
17	A2 23	7.29480E 01	4.27E 00	17.1	6.44E 01	8.15E 01
18	A3 92	7.35550E 01	3.42E 00	21.5	6.67E 01	8.04E 01
19	A4 88	7.44962E 01	3.66E 00	20.4	6.72E 01	8.18E 01
20	A5 96	5.96849E 01	4.36E 00	13.7	5.10E 01	6.84E 01
21	A6 85	5.88155E 01	4.16E 00	14.1	5.05E 01	6.71E 01
22	A7 94	5.50717E 01	4.55E 00	12.1	4.60E 01	6.42E 01
23	A8 24	6.95947E 01	4.54E 00	15.3	6.05E 01	7.87E 01
24	A9 89	5.93339E 01	4.50E 00	13.2	5.03E 01	6.83E 01
25	A10 90	6.24585E 01	4.10E 00	15.2	5.43E 01	7.07E 01
26	A11 25	6.16220E 01	4.64E 00	13.3	5.23E 01	7.09E 01
27	A12 95	6.03975E 01	4.19E 00	14.4	5.20E 01	6.88E 01
28	A13 91	6.20098E 01	4.10E 00	15.1	5.38E 01	7.02E 01
29	A14 70	8.03994E 01	4.10E 00	19.6	7.22E 01	8.86E 01
30	G	7.17966E-01	3.46E-02	20.8	6.49E-01	7.87E-01
31	H	3.14493E 02	2.29E 01	13.7	2.69E 02	3.60E 02

NO. OF OBSERVATIONS   104
NO. OF COEFFICIENTS   31
RESIDUAL DEGREES OF FREEDOM   73
RESIDUAL ROOT MEAN SQUARE   1.99544048
RESIDUAL MEAN SQUARE   3.98178577
RESIDUAL SUM OF SQUARES   290.67041016

Figure 9A.9

CEMENT HEAT, PASS 4

OBS. NO.	OBS. Z	FITTED Z	RESIDUAL
22 0	588.700	588.475	0.225
22 3	537.500	539.165	-1.665
22 7	535.600	530.548	5.052
2228	516.200	518.897	-2.697
2290	507.500	511.163	-3.663
2218	510.200	507.323	2.876
2236	503.200	503.900	-0.700
23 0	575.700	575.044	0.656
23 3	535.100	534.915	0.185
23 7	530.000	526.499	3.501
2328	514.700	514.970	-0.270
2390	503.800	507.275	-3.476
2318	501.400	503.451	-2.051
2336	499.700	500.039	-0.339
92 0	623.600	624.083	-0.483
92 0	624.100	624.083	0.017
92 3	548.800	549.838	-1.038
92 7	538.900	540.972	-2.072
9228	531.400	529.163	2.237
9290	523.000	521.372	1.628
9218	519.300	517.509	1.791
9236	513.200	514.066	-0.866
88 0	595.500	595.496	0.004
88 3	542.300	542.272	0.027
88 7	532.400	533.458	-1.058
8828	521.800	521.577	0.223
8890	517.800	513.699	4.100
8818	507.900	509.790	-1.890
8836	504.900	506.304	-1.404
96 0	604.500	603.860	0.640
96 0	603.200	603.860	-0.660
96 3	532.600	533.000	-0.400
96 7	525.700	525.772	-0.072
9628	515.700	516.174	-0.474
9690	511.900	509.849	2.051
9618	508.600	506.714	1.886
9636	501.000	503.920	-2.920
85 0	618.800	618.885	-0.085
85 3	538.200	536.570	1.630
85 7	528.500	529.425	-0.925
85 7	530.200	529.425	0.775
8528	517.800	519.957	-2.157
8528	517.100	519.957	-2.857
8590	515.100	513.722	1.378
8590	514.300	513.722	0.578
8518	509.600	510.633	-1.033
8518	511.800	510.633	1.167
8536	509.000	507.879	1.121
8536	508.500	507.879	0.621

OBS. NO.	OBS. Z	FITTED Z	ORDERED RESID.	SEQ.
22 7	535.600	530.548	5.052	1
8890	517.800	513.699	4.100	2
23 7	530.000	526.499	3.501	3
2536	502.800	499.669	3.131	4
2436	501.400	498.394	3.006	5
89 3	534.300	531.298	3.002	6
2218	510.200	507.323	2.876	7
24 7	526.200	523.678	2.521	8
7036	502.100	499.580	2.520	9
9228	531.400	529.163	2.237	10
9690	511.900	509.849	2.051	11
2590	507.800	505.789	2.011	12
9428	517.600	515.682	1.918	13
9618	508.600	506.714	1.886	14
9590	518.100	516.274	1.825	15
9218	519.300	517.509	1.791	16
85 3	538.200	536.570	1.630	17
9290	523.000	521.372	1.628	18
9018	517.500	515.991	1.509	19
94 3	532.700	531.241	1.458	20
8590	515.100	513.722	1.378	21
9118	514.600	513.252	1.348	22
9190	517.800	516.509	1.291	23
8518	511.800	510.633	1.167	24
8536	509.000	507.879	1.121	25
95 3	540.800	539.739	1.061	26
9128	524.100	523.082	1.018	27
94 0	611.700	610.704	0.996	28
25 7	523.100	522.195	0.905	29
90 7	536.800	535.949	0.850	30
85 7	530.200	529.425	0.775	31
8918	505.900	505.178	0.722	32
23 0	575.700	575.044	0.656	33
96 0	604.500	603.860	0.640	34
8536	508.500	507.879	0.621	35
8990	508.900	508.294	0.605	36
8590	514.300	513.722	0.578	37
90 3	543.900	543.539	0.361	38
9418	507.300	506.951	0.349	39
90 0	633.400	633.143	0.257	40
22 0	588.700	588.475	0.225	41
8828	521.800	521.577	0.223	42
23 3	535.100	534.915	0.185	43
89 0	599.000	598.872	0.128	44
70 7	528.400	528.303	0.097	45
9090	519.300	519.272	0.028	46
88 3	542.300	542.272	0.027	47
95 0	625.600	625.581	0.019	48
92 0	624.100	624.083	0.017	49

Figure 9A.9 (*continued*)

94 0	610.000	610.704	-0.704		88 0	595.500	595.496	0.004	50
94 0	611.700	610.704	0.996		91 0	624.000	624.014	-0.014	51
94 3	532.700	531.241	1.458		24 0	573.000	573.015	-0.015	52
94 7	522.500	524.549	-2.049		96 7	525.700	525.772	-0.072	53
9428	517.600	515.682	1.918		9536	510.200	510.274	-0.074	54
9490	509.500	509.844	-0.344		85 0	618.800	618.885	-0.085	55
9418	507.300	506.951	0.349		70 0	564.700	564.968	-0.268	56
9436	502.000	504.373	-2.373		2328	514.700	514.970	-0.270	57
24 0	573.000	573.015	-0.015		2336	499.700	500.039	-0.339	58
24 3	530.900	531.775	-0.875		9490	509.500	509.844	-0.344	59
24 7	526.200	523.678	2.521		91 7	532.700	533.063	-0.363	60
2428	511.500	512.646	-1.146		96 3	532.600	533.000	-0.400	61
2490	503.000	505.299	-2.299		8936	502.000	502.401	-0.401	62
2418	500.500	501.649	-1.149		2518	502.100	502.553	-0.453	63
2436	501.400	498.394	3.006		9628	515.700	516.174	-0.474	64
89 0	599.000	598.872	0.128		92 0	623.600	624.083	-0.483	65
89 3	534.300	531.298	3.002		70 3	536.400	536.891	-0.491	66
89 7	521.100	524.120	-3.020		91 3	540.000	540.591	-0.591	67
8928	513.200	514.582	-1.382		2528	511.700	512.314	-0.614	68
8990	508.900	508.294	0.605		96 0	603.200	603.860	-0.660	69
8918	505.900	505.178	0.722		25 0	585.900	586.587	-0.687	70
8936	502.000	502.401	-0.401		2236	503.200	503.900	-0.700	71
90 0	633.400	633.143	0.257		94 0	610.000	610.704	-0.704	72
90 3	543.900	543.539	0.361		9528	521.900	522.677	-0.777	73
90 7	536.800	535.949	0.850		9518	512.300	513.102	-0.802	74
9028	524.600	525.893	-1.293		9236	513.200	514.066	-0.866	75
9090	519.300	519.272	0.028		24 3	530.900	531.775	-0.875	76
9018	517.500	515.991	1.509		85 7	528.500	529.425	-0.925	77
9036	510.700	513.067	-2.367		8518	509.600	510.633	-1.033	78
25 0	585.900	586.587	-0.687		92 3	548.800	549.838	-1.038	79
25 3	527.100	529.597	-2.497		7028	514.900	515.951	-1.051	80
25 7	523.100	522.195	0.905		88 7	532.400	533.458	-1.058	81
2528	511.700	512.314	-0.614		2428	511.500	512.646	-1.146	82
2590	507.800	505.789	2.011		2418	500.500	501.649	-1.149	83
2518	502.100	502.553	-0.453		9028	524.600	525.893	-1.293	84
2536	502.800	499.669	3.131		95 7	531.100	532.401	-1.301	85
95 0	625.600	625.581	0.019		8928	513.200	514.582	-1.382	86
95 3	540.800	539.739	1.061		8836	504.900	506.304	-1.404	87
95 7	531.100	532.401	-1.301		22 3	537.500	539.165	-1.665	88
9528	521.900	522.677	-0.777		8818	507.900	509.790	-1.890	89
9590	518.100	516.274	1.825		94 7	522.500	524.549	-2.049	90
9518	512.300	513.102	-0.802		2318	501.400	503.451	-2.051	91
9536	510.200	510.274	-0.074		92 7	538.900	540.972	-2.072	92
91 0	624.000	624.014	-0.014		8528	517.800	519.957	-2.157	93
91 3	540.000	540.591	-0.591		2490	503.000	505.299	-2.299	94
91 7	532.700	533.063	-0.363		9036	510.700	513.067	-2.367	95
9128	524.100	523.082	1.018		9436	502.000	504.373	-2.373	96
9190	517.800	516.509	1.291		25 3	527.100	529.597	-2.497	97
9118	514.600	513.252	1.348		9136	507.700	510.349	-2.649	98
9136	507.700	510.349	-2.649		2228	516.200	518.897	-2.697	99
70 0	564.700	564.968	-0.268		8528	517.100	519.957	-2.857	100
70 3	536.400	536.891	-0.491		9636	501.000	503.920	-2.920	101
70 7	528.400	528.303	0.097		89 7	521.100	524.120	-3.020	102
7028	514.900	515.951	-1.051		2390	503.800	507.275	-3.476	103
7036	502.100	499.580	2.520		2290	507.500	511.163	-3.663	104

Figure 9A.9 (*continued*)

LINEAR LEAST-SQUARES CURVE FITTING PROGRAM

CEMENT HEAT PASS 5

EQUATION 1 OF A MULTI-EQUATION PROBLEM

DATA READ WITH SPECIAL FORMAT

FORMAT CARD 1 (A6,I4,I2,4F 5.2,F6.2, 3X , 11F1.0/10X,3F10.5 / 12X,20F1.0)

BZERO = 0

DATA INPUT   36 INDEPENDENT VARIABLES   2 DEPENDENT VARIABLES

OBSV.	SEQ.	1-11-21	2-12-22	3-13-23	4-14-24	5-15-25	6-16-26	7-17-27	8-18-28	9-19-29	10-20-30
22	1	27.680	3.760	1.980	0.0	64.970	0.0	0.0	0.0	0.0	0.0
	2	0.0	0.0	0.0	0.0	0.0	0.0	588.475	73.173	0.718	0.0
	3	0.0	0.0	0.0	0.0	0.0	0.0	0.0	0.0	0.0	0.0
	4	0.0	0.0	0.0	0.0	0.0	0.0	0.0	0.0	0.0	0.0

DATA TRANSFORMATIONS

POSITION	CODE	OPERATION	CONSTANT	LOCATION	OMIT	VARIABLE
1		NONE		1	1	
2		NONE		2	1	
3		NONE		3	1	
4		NONE		4	1	
5		NONE		5	1	
6		NONE		6	1	
7	11	ADD POSITIONS,	1+ 2	7	1	
8	11	ADD POSITIONS,	3+ 7	8	1	
9	11	ADD POSITIONS,	4+ 8	9	1	
10	11	ADD POSITIONS,	5+ 9	11	1	
11	7	DIVIDE BY CONSTANT	100.	11	1	
12	10	DIVIDE POSITIONS,	1/11	20	1	
13	10	DIVIDE POSITIONS,	2/11	21	1	
14	10	DIVIDE POSITIONS,	3/11	22	1	
15	10	DIVIDE POSITIONS,	4/11	23	1	
16	10	DIVIDE POSITIONS,	5/11	24	1	
17		NONE		37	1	
18		NONE		38	1	
19		NONE		39	1	
20	8	ADD CONSTANT	-21.0	20	1	
21	8	ADD CONSTANT	-3.48	21	1	
22	8	ADD CONSTANT	-1.18	22	1	
23	8	ADD CONSTANT	-2.08	23	1	
24	8	ADD CONSTANT	-61.6	24	1	
25	8	ADD CONSTANT	10.66	25	0	
26	10	DIVIDE POSITIONS,	20/25	26	0	1
27	10	DIVIDE POSITIONS,	21/25	27	0	2
28	10	DIVIDE POSITIONS,	22/25	28	0	3
29	10	DIVIDE POSITIONS,	24/25	29	0	4
30	10	DIVIDE POSITIONS,	23/25	30	0	5
31	9	MULTIPLY POSITIONS,	26X27	31	0	6
32	9	MULTIPLY POSITIONS,	26X28	32	0	7
33	9	MULTIPLY POSITIONS,	26X29	33	0	8
34	9	MULTIPLY POSITIONS,	27X28	34	0	9
35	9	MULTIPLY POSITIONS,	27X29	35	0	10
36	9	MULTIPLY POSITIONS,	28X29	37	0	11
37		NONE		37	0	12
38		NONE		38	0	13

Figure 9A.10

DATA AFTER TRANSFORMATIONS   THE FITTED EQUATION HAS 11 INDEPENDENT VARIABLES, 2 DEPENDENT VARIABLES

```
          X1          X2          A           X3          X4          X5          X1X2        X1X3        X1X4        X2X3        X2X4
          X3X4        D
OBSV.   1-11-21     2-12-22     3-13-23     4-14-24     5-15-25     6-16-26     7-17-27     8-18-28     9-19-29     10-20-30
 22   6.04240-01  2.32242D-02  7.34449D-02  3.64120D-01  2.63568D-01  3.55169D-02  1.40331D-02  4.43788D-02  1.59260D-01  1.70570D-03  6.12116D-03
      1.93572D-02  5.88472D-02                           2.94106D-01  4.56541D-05  1.50971D-02  1.69688D-01  6.85896D-02  3.56712D-05  1.44187D-05
 23   4.46022D-02  9.79656D-05  7.29480D 02  3.64120D-01  1.47181D-01  2.25794D-02  4.56541D-05  1.06117D-02  2.10850D-02  1.70570D-03  1.44187D-05
      5.35916D-02  5.75044D 02                           2.94106D-01  2.75604D-01  1.50971D-02  2.10850D-02  3.11699D-02  3.11699D-02  6.19335D-02
 92   7.16921D-02  2.10583D-01  7.35500D 01  1.48017D-01  2.94106D-01  2.75604D-01  1.50971D-02  1.06117D-02  2.10850D-02  3.40288D-02  6.19335D-02
      4.35327D-02  6.24083D 02                           2.52279D-01  3.04022D-02  7.62751D-02  8.61870D-02  3.40288D-02  3.40288D-02  5.63240D-02
 88   3.41641D-01  2.23261D-01  7.35500D 01  1.52417D-01  2.52279D-01  3.04022D-02  7.62751D-02  5.20720D-02  8.61870D-02  3.11317D-03  5.63240D-02
      3.84516D-02  6.24083D 02                           4.69978D-01  2.63993D-02  1.36428D-02  3.32318D-02  1.78778D-01  1.13317D-03  1.68556D-02
 96   3.80397D-01  3.58647D-01  7.44962D 02  8.73609D-02  4.69978D-01  2.63993D-02  1.36428D-02  3.23180D-02  1.78778D-01  1.13317D-03  1.68556D-02
      4.10577D-02  5.95496D 02                           4.41919D-01  3.22872D-02  2.98176D-02  1.85374D-02  3.91548D-02  3.91548D-02  1.09040D-01
 85   1.18814D-01  2.50959D-01  5.96849D 01  5.60020D-01  4.41919D-01  3.22872D-02  2.98176D-02  1.85374D-02  5.25063D-02  3.91548D-02  1.09040D-01
      6.89483D-02  6.03860D 02                           4.58980D-01  6.85325D-04  3.63574D-05  1.41565D-04  5.25063D-02  4.74115D-02  5.06470D-02
 94   3.29483D-04  1.10347D-01  5.88155D 01  4.29659D-01  4.58980D-01  6.85325D-04  3.63574D-05  1.41565D-04  4.74115D-02  4.74115D-02  5.06470D-02
      1.97205D-01  6.10704D-01                           4.90792D-02  1.53251D-02  3.07600D-02  1.36905D-01  7.21513D-02  7.21513D-02  6.24880D-03
 24   2.41531D-01  1.27322D-01  5.50717D 01  5.66882D-01  4.90792D-02  1.53251D-02  3.07600D-02  1.36905D-01  1.18571D-02  7.21513D-02  6.24880D-03
      2.78123D-02  5.73016D 02                           2.88740D-01  2.45793D-02  1.24834D-02  5.07666D-02  5.16722D-03  5.27541D-02  3.28862D-02
 89   1.09604D-01  1.13895D-01  6.95947D 01  4.63181D-01  2.88740D-01  2.45793D-02  1.24834D-02  5.07666D-02  3.16472D-02  5.27541D-02  3.28862D-02
      1.33739D-01  5.98872D 01                           4.33041D-01  4.35010D-02  1.40931D-02  6.49994D-06  1.23354D-02  1.12892D-02  2.14245D-01
 90   2.84856D-01  4.94744D-01  5.93339D 01  2.28183D-04  4.33041D-01  4.35010D-02  1.40931D-02  6.49994D-06  8.43780D-02  1.12892D-02  2.14245D-01
      9.88126D-05  6.33143D 02                           1.10955D-01  1.52197D-02  2.01638D-02  8.31336D-02  8.43780D-02  8.43780D-02  1.58731D-02
 25   1.40948D-01  1.43059D-01  6.24585D 02  5.89819D-01  1.10955D-01  1.52197D-02  2.01638D-02  8.31336D-02  1.56389D-02  1.58731D-02  1.58731D-02
      6.54435D-02  5.86587D 02                           5.13786D-01  4.51964D-01  8.10332D-03  5.47265D-03  1.71076D-02  3.99986D-02  1.25037D-01
 95   3.32972D-02  2.43363D-01  6.16220D 01  1.64358D-01  5.13786D-01  4.51964D-01  8.10332D-03  5.47265D-03  3.99986D-02  3.99986D-02  1.25037D-01
      8.44440D-02  6.25581D 02                           5.51684D-01  3.29955D-02  1.57775D-02  1.13576D-02  4.79720D-02  2.71735D-02  1.07179D-01
 91   8.11261D-02  1.94290D-01  6.16220D 01  1.39861D-01  5.51684D-01  3.29955D-02  1.57775D-02  1.13576D-02  4.79720D-02  2.71735D-02  1.07179D-01
      1.71541D-01  1.24209D 02                           4.50145D-03  1.91122D-01  2.97678D-02  2.16113D-01  2.09952D-03  2.95728D-02  2.87297D-04
 70   4.66410D-01  6.38232D-02  6.03994D 01  4.63554D-01  4.50145D-03  1.91122D-01  2.97678D-02  2.16113D-01  2.09952D-03  2.95728D-02  2.87297D-04
      2.08577D-03  5.64969D 02  8.03994D 01
```

MEANS OF VARIABLES
```
2.20335D-01  1.59631D-01  2.71323D-01  3.05697D-01  3.63956D-01  3.44687D-01  2.00069D-02  5.94583D-02  3.30558D-02  5.74682D-02
6.09230D-02  6.01623D-02  6.59686D 01
```

ROOT MEAN SQUARES OF VARIABLES
```
3.01185D-01  2.08777D-01  3.44687D-01  3.63956D-01  8.15665D-02  2.80465D-02  9.20413D-02  7.67126D-02  4.31853D-02  8.61972D-02
8.22661D-02  6.24720D 02  6.88977D 01
```

Figure 9A.10 (*continued*)

LINEAR LEAST-SQUARES CURVE FITTING PROGRAM

CEMENT HEAT PASS 5   DEP VAR 1: D   MIN Y = 5.650D 02   MAX Y = 6.331D 02   RANGE Y = 6.817D 01

```
FULL EQUATION - - - SEARCH FOR INFLUENTIAL VARIABLES
D = B(1)X1 + B(2)X2 + B(3)X3 + B(4)X4 + B(5)X5 + B(6)X1X2 + B(7)X1X3
  + B(8)X1X4 + B(9)X2X3 + B(10)X2X4 + B(11)X3X4

    D  = INITIAL ACID HEAT OF SOLUTION, CALORIES/GRAM
    X1 = (PERCENT SILICA IN THE CLINKER - 21.0) / 10.66
    X2 = (PERCENT ALUMINA - 3.48) / 10.66
    X3 = (PERCENT FERRIC OXIDE - 1.18) / 10.66
    X4 = (PERCENT LIME - 61.6) / 10.66
    X5 = (PERCENT MAGNESIA - 2.08) / 10.66
    D VALUES FROM NON-LINEAR FIT WITH B, G AND H COMMON.
```

IND.VAR(I)	NAME	COEF.B(I)	S.E. COEF.	T-VALUE	R(I)SQRD	MIN X(I)	MAX X(I)	RANGE X(I)	REL.INF.X(I)
0		-3.49587D-11							
1	X1	5.65902D 02	1.28D 01	44.4	0.9741	3.295D-04	6.042D-01	6.039D-01	5.01
2	X2	5.65201D 02	1.07D 02	5.3	0.9992	9.797D-05	4.947D-01	4.946D-01	4.14
3	X3	5.82379D 02	6.05D 01	9.6	0.9991	2.282D-04	5.898D-01	5.896D-01	5.04
4	X4	6.27214D 02	2.17D 01	28.9	0.9939	4.501D-03	5.516D-01	5.471D-01	5.03
5	X5	6.61219D 02	2.89D 01	22.9	0.9311	6.853D-04	2.756D-01	2.749D-01	2.67
6	X1X2	-1.25588D 00	2.02D 02	0.1	0.9881	3.636D-05	7.628D-02	7.624D-02	0.01
7	X1X3	-5.19942D 01	1.11D 02	0.5	0.9963	6.500D-06	2.161D-01	2.161D-01	0.16
8	X1X4	2.11143D 01	5.27D 01	0.4	0.9766	1.512D-04	1.788D-01	1.786D-01	0.06
9	X2X3	2.20681D 01	1.83D 02	0.1	0.9939	3.456D-05	8.338D-02	8.434D-02	0.03
10	X2X4	1.57850D 02	2.58D 02	0.6	0.9992	4.442D-05	2.142D-01	2.142D-01	0.50
11	X3X4	7.43544D-01	8.18D 01	0.0	0.9916	9.881D-05	1.972D-01	1.971D-01	0.00

```
NO. OF OBSERVATIONS              14
NO. OF IND. VARIABLES            11
RESIDUAL DEGREES OF FREEDOM       3
RESIDUAL ROOT MEAN SQUARE         2.22854150
RESIDUAL MEAN SQUARE              4.96639721
RESIDUAL SUM OF SQUARES          14.89919163
TOTAL SUM OF SQUARES           6271.14395981
MULT. CORREL. COEF. SQUARED       .9976
```

--------ORDERED BY COMPUTER INPUT--------

IDENT.	OBSV.	WS DISTANCE	OBS.Y	FITTED Y	RESIDUAL
1	22	13560.	588.475	588.677	-0.202
2	23	8203.	575.044	575.746	-0.702
3	92	7408.	624.083	623.959	0.124
4	88	7142.	595.496	595.787	-0.291
5	96	3616.	603.860	603.499	0.361
6	85	7014.	618.885	618.167	0.718
7	94	11356.	610.704	610.955	-0.252
8	24	3498.	573.016	575.729	-2.713
9	89	16180.	598.872	598.510	0.362
10	90	10433.	633.143	633.129	0.014
11	25	6953.	586.587	584.803	1.785
12	95	7321.	625.581	626.366	-0.785
13	91	14391.	624.014	623.936	0.078
14	70		564.969	563.466	1.502

--------ORDERED BY RESIDUALS--------

OBS.	FITTED Y	ORDERED RESID.	SEQ
25	584.803	1.785	1
70	563.466	1.502	2
85	618.167	0.718	3
89	598.510	0.362	4
96	603.499	0.361	5
92	623.959	0.124	6
91	623.936	0.078	7
90	633.129	0.014	8
22	588.677	-0.202	9
94	610.955	-0.252	10
88	595.787	-0.291	11
23	575.746	-0.702	12
95	626.366	-0.785	13
24	575.729	-2.713	14

Figure 9A.11

CP VALUES FOR THE SELECTION OF VARIABLES

CEMENT HEAT PASS 5   DEP VAR 1: D

NUMBER OF OBSERVATIONS   14
NUMBER OF VARIABLES IN FULL EQUATION   11
NUMBER OF VARIABLES IN BASIC EQUATION   5
REMAINDER OF VARIABLES TO BE CONSIDERED   6

P	CP	X1	X2	X3	X4	X5	X1X2	X1X3	X1X4	X2X3	X2X4	X3X4
7	3.4	X1	X2	X3	X4	X5		X1X3			X2X4	
7	4.2	X1	X2	X3	X4	X5	X1X2	X1X3				
5	4.3	X1	X2	X3	X4	X5						
6	4.4	X1	X2	X3	X4	X5	X1X2					
7	4.8	X1	X2	X3	X4	X5	X1X2			X2X3		
8	5.1	X1	X2	X3	X4	X5		X1X3	X1X4		X2X4	
6	5.1	X1	X2	X3	X4	X5		X1X3				
8	5.2	X1	X2	X3	X4	X5		X1X3			X2X4	X3X4
8	5.3	X1	X2	X3	X4	X5	X1X2	X1X3			X2X4	
8	5.4	X1	X2	X3	X4	X5		X1X3		X2X3	X2X4	
8	5.4	X1	X2	X3	X4	X5	X1X2			X2X3		X3X4
6	5.8	X1	X2	X3	X4	X5					X2X4	
7	5.9	X1	X2	X3	X4	X5	X1X2					X3X4
8	5.9	X1	X2	X3	X4	X5	X1X2	X1X3		X2X3		
7	6.0	X1	X2	X3	X4	X5	X1X2		X1X4			
6	6.1	X1	X2	X3	X4	X5						X3X4
8	6.1	X1	X2	X3	X4	X5	X1X2	X1X3	X1X4			
7	6.2	X1	X2	X3	X4	X5	X1X2				X2X4	
9	7.0	X1	X2	X3	X4	X5		X1X3	X1X4	X2X3	X2X4	
10	9.0	X1	X2	X3	X4	X5		X1X3	X1X4	X2X3	X2X4	X3X4
11	11.0	X1	X2	X3	X4	X5	X1X2	X1X3	X1X4	X2X3	X2X4	X3X4

Figure 9A.12

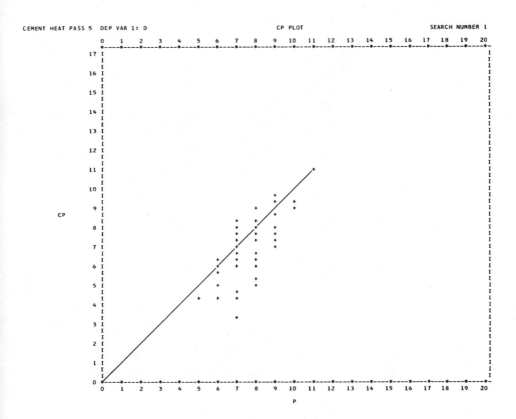

Figure 9A.13

LINEAR LEAST-SQUARES CURVE FITTING PROGRAM

CEMENT HEAT PASS 6   DEP VAR 1: A   MIN Y = 5.507D 01   MAX Y = 8.040D 01   RANGE Y = 2.533D 01

FULL EQUATION -- - SEARCH FOR INFLUENTIAL VARIABLES
A = B(1)X1 + B(2)X2 + B(3)X3 + B(4)X4 + B(5)X5 + B(6)X1X2 + B(7)X1X3
    + B(8)X1X4 + B(9)X2X3 + B(10)X2X4 + B(11)X3X4

A = COEFFICIENT IN NON-LINEAR EQUATION
X1 = (PERCENT SILICA IN THE CLINKER - 21.0) / 10.66
X2 = (PERCENT ALUMINA - 3.48) / 10.66
X3 = (PERCENT FERRIC OXIDE - 1.18) / 10.66
X4 = (PERCENT LIME - 61.6) / 10.66
X5 = (PERCENT MAGNESIA - 2.08) / 10.66

A VALUES FROM NON-LINEAR FIT WITH B, G AND H COMMON.

IND.VAR(I)	NAME	COEF.B(I)	S.E. COEF.	T-VALUE	R(I)SQRD	MIN X(I)	MAX X(I)	RANGE X(I)	REL.INF.X(I)
1	X1	-3.65930D-12	9.44D 00	11.7	0.9741	3.29D-04	6.042D-01	6.039D-01	2.63
2	X2	1.10375D 01	7.96D 01	0.5	0.9992	9.79D-05	4.947D-01	4.946D-01	0.75
3	X3	-3.83723D 01	4.48D 01	2.7	0.9991	2.28D-04	5.898D-01	5.896D-01	2.85
4	X4	1.22400D 01	1.61D 01	3.9	0.9939	4.50D-03	5.516D-01	5.471D-01	1.34
5	X5	6.20427D 01	2.14D 01	6.1	0.9311	2.75D-01	2.749D-01	2.749D-01	1.42
6	X1X2	1.31025D 02	1.49D 02	2.1	0.9881	3.63D-05	7.628D-02	7.624D-02	0.97
7	X1X3	3.20975D 02	8.19D 01	1.5	0.9963	6.50D-06	2.161D-01	2.161D-01	1.08
8	X1X4	-1.26387D 02	3.90D 01	3.3	0.9766	1.51D-04	1.788D-01	1.786D-01	0.89
9	X2X3	-1.26724D 02	1.35D 02	1.0	0.9999	3.56D-07	8.438D-02	8.434D-02	0.77
10	X2X4	-2.31286D 02	1.91D 02	1.0	0.9992	1.54D-05	2.142D-01	2.142D-01	1.69
11	X3X4	-1.99748D 02	6.06D 01	1.7	0.9916	9.88D-05	1.972D-01	1.971D-01	0.82
	X3X4	-1.05430D 02							

NO. OF OBSERVATIONS   14
NO. OF IND. VARIABLES   11
RESIDUAL DEGREES OF FREEDOM   3
RESIDUAL ROOT MEAN SQUARE   1.65014279
RESIDUAL MEAN SQUARE   2.72297122
RESIDUAL SUM OF SQUARES   8.16891367
TOTAL SUM OF SQUARES   783.62088279
MULT. CORREL. COEF. SQUARED   .9896

ORDERED BY COMPUTER INPUT

IDENT.	OBSV.	WS DISTANCE	OBSV.	OBS-Y	FITTED Y	RESIDUAL
1	22	1026.	95	73.173	73.298	-0.126
2	23	528.	88	72.948	72.312	0.636
3	92	547.	24	73.555	73.712	-0.157
4	88	281.	23	74.496	73.738	0.758
5	96	496.	90	59.685	59.757	-0.072
6	85	207.	89	58.815	60.630	-1.815
7	94	531.	96	55.072	55.291	-0.220
8	24	702.	91	69.595	68.890	0.704
9	89	304.	22	59.334	59.402	-0.068
10	90	1080.	92	62.458	62.472	-0.014
11	25	332.	94	61.622	61.878	-0.256
12	95	398.	25	60.398	58.815	1.583
13	91	321.	70	62.010	62.117	-0.107
14	70	841.	85	80.399	81.248	-0.848

ORDERED BY RESIDUALS

OBS-Y	FITTED Y	ORDERED RESID.	SEQ
60.398	58.815	1.583	1
74.496	73.738	0.758	2
69.595	68.890	0.704	3
72.948	72.312	0.636	4
62.458	62.472	-0.014	5
59.334	59.402	-0.068	6
59.685	59.757	-0.072	7
62.010	62.117	-0.107	8
73.173	73.298	-0.126	9
73.555	73.712	-0.157	10
55.072	55.291	-0.220	11
61.622	61.878	-0.256	12
80.399	81.248	-0.848	13
58.815	60.630	-1.815	14

Figure 9A.14

CP VALUES FOR THE SELECTION OF VARIABLES

CEMENT HEAT PASS 6   DEP VAR 2: A

NUMBER OF OBSERVATIONS   14
NUMBER OF VARIABLES IN FULL EQUATION   11
NUMBER OF VARIABLES IN BASIC EQUATION   5
REMAINDER OF VARIABLES TO BE CONSIDERED   6

P	CP	VARIABLES IN EQUATION							
7	6.4	X1	X2	X3	X4	X5	X1X2	X1X4	
8	8.0	X1	X2	X3	X4	X5	X1X2	X1X4	X3X4
8	8.1	X1	X2	X3	X4	X5	X1X2	X1X4 X2X3	
8	8.3	X1	X2	X3	X4	X5	X1X2 X1X3 X1X4		
8	8.4	X1	X2	X3	X4	X5	X1X2	X1X4	X2X4
9	9.5	X1	X2	X3	X4	X5	X1X2	X1X4 X2X3	X3X4
9	9.9	X1	X2	X3	X4	X5	X1X2	X1X4	X2X4 X3X4
10	10.1	X1	X2	X3	X4	X5	X1X2 X1X3 X1X4 X2X3	X3X4	
11	11.0	X1	X2	X3	X4	X5	X1X2 X1X3 X1X4 X2X3 X2X4 X3X4		

Figure 9A.15

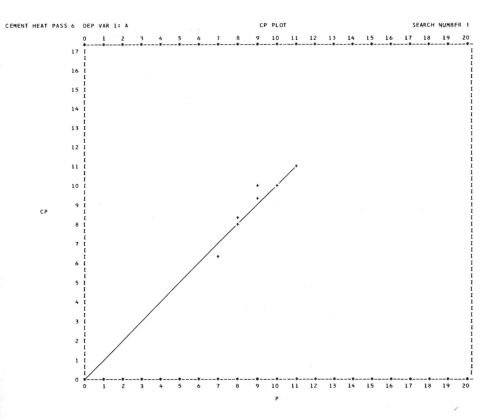

Figure 9A.16

251

LINEAR LEAST-SQUARES CURVE FITTING PROGRAM

CEMENT HEAT PASS 7   DEP VAR 1: D   MIN Y = 5.650D 02   MAX Y = 6.331D 02   RANGE Y = 6.817D 01

SELECTED EQUATION
D = B(1)X1 + B(2)X2 + B(3)X3 + B(4)X4 + B(5)X5 + B(6)X1X3 + B(7)X2X4
D = INITIAL ACID HEAT OF SOLUTION, CALORIES/GRAM
X1 = (PERCENT SILICA IN THE CLINKER - 21.01) / 10.66
X2 = (PERCENT ALUMINA - 3.48) / 10.66
X3 = (PERCENT FERRIC OXIDE - 1.18) / 10.66
X4 = (PERCENT LIME - 61.6) / 10.66
X5 = (PERCENT MAGNESIA - 2.08) / 10.66
D VALUES FROM NON-LINEAR FIT WITH B, G AND H COMMON.

IND.VAR(I)	NAME	COEF.B(I)	S.E. COEF.	T-VALUE	R(I)SQRD	MIN X(I)	MAX X(I)	RANGE X(I)	REL.INF.X(I)
		9.094950-13							
1	X1	5.70172D 02	5.33D 00	107.0	0.9278	3.2950-04	6.0420-01	6.0390-01	5.05
2	X2	5.77082D 02	1.70D 01	34.0	0.9852	9.7970-05	4.9470-01	4.9460-01	4.19
3	X3	5.83677D 02	5.51D 00	105.9	0.9485	2.2820-04	5.8980-01	5.8960-01	5.05
4	X4	6.59703D 02	7.70D 00	85.7	0.5289	6.8530-04	2.7560-01	2.7490-01	2.66
5	X5	6.31448D 02	6.33D 00	99.7	0.9650	4.5010-03	5.5160-01	5.4710-01	5.07
6	X1X3	-6.20689D 01	2.07D 01	3.0	0.9487	6.5000-06	2.1610-01	2.1610-01	5.07
7	X2X4	1.35580D 02	4.94D 01	2.7	0.9897	1.4420-05	2.1420-01	2.1420-01	0.43

NO. OF OBSERVATIONS   14
NO. OF IND. VARIABLES   7
RESIDUAL DEGREES OF FREEDOM   7
RESIDUAL ROOT MEAN SQUARE   1.55473912
RESIDUAL MEAN SQUARE   2.41721372
RESIDUAL SUM OF SQUARES   16.92049605
TOTAL SUM OF SQUARES   627I.14395981
MULT. CORREL. COEF. SQUARED   .9973

----ORDERED BY COMPUTER INPUT----

IDENT.	OBSV.	WS DISTANCE	OBS.Y	FITTED Y	RESIDUAL
1	22	28228.	588.475	588.730	-0.255
2	92	15233.	575.044	575.600	-0.556
3	93	15233.	624.083	624.061	0.022
4	88	5029.	595.496	596.358	-0.862
5	96	14841.	603.860	602.984	0.875
6	85	7515.	618.885	617.869	1.016
7	94	14582.	610.704	611.781	-1.077
8	24	23530.	573.016	575.434	-2.418
9	89	7238.	598.872	598.415	0.456
10	90	33647.	633.143	633.071	0.072
11	25	21591.	586.587	584.280	2.308
12	95	14466.	625.581	626.215	-0.635
13	91	2223.	624.014	623.987	0.027
14	70	29938.	564.969	563.943	1.026

----ORDERED BY RESIDUALS----

OBSV.	OBS.Y	FITTED Y	ORDERED RESID.	SEQ
25	586.587	584.280	2.308	1
70	564.969	563.943	1.026	2
85	618.885	617.869	1.016	3
96	603.860	602.984	0.875	4
89	598.872	598.415	0.456	5
90	633.143	633.071	0.072	6
91	624.014	623.987	0.027	7
93	624.083	624.061	0.022	8
22	588.475	588.730	-0.255	9
92	575.044	575.600	-0.556	10
95	625.581	626.215	-0.635	11
88	595.496	596.358	-0.862	12
94	610.704	611.781	-1.077	13
24	573.016	575.434	-2.418	14

Figure 9A.17

LINEAR LEAST-SQUARES CURVE FITTING PROGRAM

CEMENT HEAT PASS 8   DEP VAR 1: A   MIN Y = 5.507D 01   MAX Y = 8.040D 01   RANGE Y = 2.533D 01

```
SELECTED EQUATION
A = B(1)X1 + B(2)X2 + B(3)X3 + B(4)X4 + B(5)X5 + B(6)X1X2 + B(7)X1X4
A  = COEFFICIENT IN NON-LINEAR EQUATION
X1 = (PERCENT SILICA IN THE CLINKER - 21.01) / 10.66
X2 = (PERCENT ALUMINA - 3.481) / 10.66
X3 = (PERCENT FERRIC OXIDE - 1.181) / 10.66
X4 = (PERCENT LIME - 61.61) / 10.66
X5 = (PERCENT MAGNESIA - 2.08) / 10.66

A VALUES FROM NON-LINEAR FIT WITH B, G AND H COMMON.
```

IND.VAR(I)	NAME	COEF.B(I)	S.E. COEF.	T-VALUE	R(I)SQRD	MIN X(I)	MAX X(I)	RANGE X(I)	REL.INF.X(I)
0		8.17124D-14							
1	X1	1.08936D 02	4.14D 00	26.3	0.8776	3.295D-04	6.042D-01	6.039D-01	2.60
2	X2	5.73104D 01	5.56D 00	10.3	0.8585	9.797D-05	4.947D-01	4.946D-01	1.12
3	X3	5.05094D 01	2.35D 00	21.5	0.7100	2.282D-04	5.898D-01	5.896D-01	1.18
4	X4	5.96873D 01	3.75D 00	15.9	0.8979	4.501D-03	5.516D-01	5.471D-01	1.29
5	X5	1.07567D 02	6.50D 00	16.6	0.3206	6.853D-01	2.756D-01	2.749D-01	1.17
6	X1X2	-1.00885D 02	2.88D 01	3.5	0.7072	3.636D-05	7.628D-02	7.624D-02	0.30
7	X1X4	-1.15227D 01	1.83D 01	6.3	0.9031	1.512D-04	1.788D-01	1.786D-01	0.81

```
NO. OF OBSERVATIONS            14
NO. OF IND. VARIABLES           7
RESIDUAL DEGREES OF FREEDOM     7
RESIDUAL ROOT MEAN SQUARE      1.57455660
RESIDUAL MEAN SQUARE           2.47922848
RESIDUAL SUM OF SQUARES       17.35459938
TOTAL SUM OF SQUARES         783.62088279
MULT. CORREL. COEF. SQUARED     .9779
```

	----ORDERED BY COMPUTER INPUT----						----ORDERED BY RESIDUALS----			
IDENT.	OBSV.	WS DISTANCE	OBS. Y	FITTED Y	RESIDUAL	OBSV.	OBS. Y	FITTED Y	ORDERED RESID.	SEQ
1	22	837.	73.173	73.481	-0.309	91	62.010	59.951	2.059	1
2	23	373.	72.948	72.478	0.470	24	69.595	68.552	1.042	2
3	92	373.	73.555	73.648	-0.093	88	74.496	73.802	0.694	3
4	88	115.	74.496	73.802	0.694	23	72.948	72.478	0.470	4
5	96	307.	59.685	59.574	0.110	90	62.458	61.995	0.463	5
6	85	102.	58.815	62.014	-3.199	95	60.398	60.250	0.147	6
7	94	318.	55.072	55.517	-0.445	96	59.685	59.574	0.110	7
8	24	200.	69.595	68.552	1.042	89	59.334	59.353	-0.019	8
9	89	103.	59.334	59.353	-0.019	92	73.555	73.648	-0.093	9
10	90	432.	62.458	61.995	0.463	25	61.622	61.836	-0.214	10
11	25	199.	61.622	61.836	-0.214	22	73.173	73.481	-0.309	11
12	95	257.	60.398	60.250	0.147	94	55.072	55.517	-0.445	12
13	91	200.	62.010	59.951	2.059	70	80.399	81.106	-0.706	13
14	70	491.	80.399	81.106	-0.706	85	58.815	62.014	-3.199	14

Figure 9A.18

CEMENT HEAT, PASS 9

Z = H + $(G - 1)D'$ + $(A' / (T + F')$EXP B $)$
Z = ACID HEAT OF SOLUTION AT TIME "T"
B, G AND H ARE COMMON TO ALL CEMENTS
D' = $D1X1 + D2X2 + D3X3 + D4X4 + D5X5 - D13X1X3 + D24X2X4$
A' = $A1X1 + A2X2 + A3X3 + A4X4 + A5X5 + A12X1X2 - A14X1X4$
F' = TIME CORRECTION TERM = $(A' / (GD' - H))$EXP$(1/B)$
    $X1$ = (PERCENT SILICA IN THE CLINKER - 21.0) / 10.66
    $X2$ = (PERCENT ALUMINA - 3.48) / 10.66
    $X3$ = (PERCENT FERRIC OXIDE - 1.18) / 10.66
    $X4$ = (PERCENT LIME - 61.6) / 10.66
    $X5$ = (PERCENT MAGNESIA - 2.08) / 10.66
OBSERVATION 85-3-2 OMITTED.

					95% CONFIDENCE LIMITS	
IND.VAR(I)	NAME	COEF.B(I)	S.E. COEF.	T-VALUE	LOWER	UPPER
1	B		1.98204E-01	2.80E-02	7.1	1.42E-01	2.54E-01
2	D1	S	5.68108E 02	5.84E 00	97.3	5.56E 02	5.80E 02
3	D2	A	5.84257E 02	1.90E 01	30.7	5.46E 02	6.22E 02
4	D3	F	5.80589E 02	5.58E 00	104.0	5.69E 02	5.92E 02
5	D4	L	6.34258E 02	6.40E 00	99.1	6.21E 02	6.47E 02
6	D5	M	6.58238E 02	7.66E 00	85.9	6.43E 02	6.74E 02
7	D13	SF	5.21298C 01	2.14E 01	2.4	9.36E 00	9.49E 01
8	D24	AL	1.13566E 02	5.34E 01	2.1	6.73E 00	2.20E 02
9	A1	S	1.07869E 02	5.30E 01	20.4	9.73E 01	1.18E 02
10	A2	A	5.84145E 01	6.92E 00	8.4	4.46E 01	7.23E 01
11	A3	F	5.00791E 01	5.98E 00	8.4	3.81E 01	6.20E 01
12	A4	L	5.83754E 01	5.59E 00	10.4	4.72E 01	6.95E 01
13	A5	M	1.08437E 02	7.37E 00	14.7	9.37E 01	1.23E 02
14	A12	SA	8.71551E 01	3.00E 01	2.9	2.71E 01	1.47E 02
15	A14	SL	1.11729E 02	2.22E 01	5.0	6.73E 01	1.56E 02
16	G		7.25998E-01	3.25E-02	22.3	6.61E-01	7.91E-01
17	H		3.20155E 02	2.14E 01	15.0	2.77E 02	3.63E 02

NO. OF OBSERVATIONS   104
NO. OF COEFFICIENTS   17
RESIDUAL DEGREES OF FREEDOM   87
RESIDUAL ROOT MEAN SQUARE   1.97992229
RESIDUAL MEAN SQUARE   3.92009544
RESIDUAL SUM OF SQUARES   341.04833984

Figure 9A.19

CEMENT HEAT, PASS 9

	--------ORDERED BY COMPUTER INPUT----------				---------------ORDERED BY RESIDUALS----------------			
OBS. NO.	OBS. Z	FITTED Z	RESIDUAL	OBS. NO.	OBS. Z	FITTED Z	ORDERED RESID.	SEQ.
22 0	588.700	588.418	0.281	22 7	535.600	530.741	4.859	1
22 3	537.500	539.461	-1.961	8890	517.800	513.397	4.403	2
22 7	535.600	530.741	4.859	2536	502.800	499.178	3.622	3
2228	516.200	518.996	-2.796	23 7	530.000	526.460	3.540	4
2290	507.500	511.246	-3.746	89 3	534.300	531.186	3.114	5
2218	510.200	507.417	2.783	2218	510.200	507.417	2.783	6
2236	503.200	504.015	-0.815	2590	507.800	505.257	2.543	7
23 0	575.700	575.582	0.118	2436	501.400	498.867	2.533	8
23 3	535.100	534.909	0.190	7036	502.100	499.595	2.505	9
23 7	530.000	526.460	3.540	24 7	526.200	523.697	2.503	10
2328	514.700	514.949	-0.250	9190	517.800	515.432	2.367	11
2390	503.800	507.318	-3.519	9128	524.100	521.755	2.345	12
2318	501.400	503.543	-2.144	9228	531.400	529.106	2.294	13
2336	499.700	500.189	-0.489	9118	514.600	512.315	2.285	14
92 0	623.600	624.133	-0.533	9690	511.900	509.683	2.217	15
92 0	624.100	624.133	-0.033	9618	508.600	506.579	2.021	16
92 3	548.800	550.061	-1.261	25 0	585.900	583.960	1.939	17
92 7	538.900	541.050	-2.150	9590	518.100	516.178	1.922	18
9228	531.400	529.106	2.294	9218	519.300	517.412	1.887	19
9290	523.000	521.276	1.724	9018	517.500	515.687	1.813	20
9218	519.300	517.412	1.887	9290	523.000	521.276	1.724	21
9236	513.200	513.982	-0.782	9428	517.600	516.020	1.580	22
88 0	595.500	596.726	-1.226	25 7	523.100	521.719	1.381	23
88 3	542.300	541.625	0.675	91 7	532.700	531.421	1.279	24
88 7	532.400	532.857	-0.457	91 3	540.000	538.756	1.244	25
8828	521.800	521.121	0.679	90 7	536.800	535.622	1.178	26
8890	517.800	513.397	4.403	95 3	540.800	539.685	1.115	27
8818	507.900	509.583	-1.683	96 0	604.500	603.428	1.072	28
8836	504.900	506.195	-1.295	94 3	532.700	531.757	0.943	29
96 0	604.500	603.428	1.072	85 0	618.800	617.880	0.920	30
96 0	603.200	603.428	-0.228	8918	505.900	505.031	0.869	31
96 3	532.600	532.868	-0.268	89 0	599.000	598.133	0.867	32
96 7	525.700	525.590	0.110	8990	508.900	508.121	0.779	33
9628	515.700	515.977	-0.277	8828	521.800	521.121	0.679	34
9690	511.900	509.683	2.217	88 3	542.300	541.625	0.675	35
9618	508.600	506.579	2.021	90 3	543.900	543.275	0.625	36
9636	501.000	503.823	-2.823	70 0	564.700	564.159	0.541	37
85 0	618.800	617.880	0.920	8590	515.100	514.588	0.512	38
85 3	538.200	538.698	-0.498	8536	509.000	508.498	0.502	39
85 7	528.500	531.123	-2.623	8518	511.800	511.362	0.438	40
85 7	530.200	531.123	-0.923	90 0	633.400	632.997	0.403	41
8528	517.800	521.128	-3.328	9090	519.300	518.940	0.360	42
8528	517.100	521.128	-4.028	22 0	588.700	588.418	0.281	43
8590	515.100	514.588	0.512	94 0	611.700	611.421	0.279	44
8590	514.300	514.588	-0.288	23 3	535.100	534.909	0.190	45
8518	509.600	511.362	-1.762	23 0	575.700	575.582	0.118	46
8518	511.800	511.362	0.438	96 7	525.700	525.590	0.110	47
8536	509.000	508.498	0.502	2518	502.100	502.037	0.063	48
8536	508.500	508.498	0.002	70 7	528.400	528.342	0.058	49

Figure 9A.19 (*continued*)

94 0	610.000	611.421	-1.421		9418	507.300	507.282	0.018	50
94 0	611.700	611.421	0.279		8536	508.500	508.498	0.002	51
94 3	532.700	531.757	0.943		92 0	624.100	624.133	-0.033	52
94 7	522.500	524.967	-2.467		91 0	624.000	624.037	-0.037	53
9428	517.600	516.020	1.580		9536	510.200	510.247	-0.047	54
9490	509.500	510.168	-0.668		2528	511.700	511.781	-0.081	55
9418	507.300	507.282	0.018		96 0	603.200	603.428	-0.228	56
9436	502.000	504.720	-2.720		2328	514.700	514.949	-0.250	57
24 0	573.000	575.438	-2.437		96 3	532.600	532.868	-0.268	58
24 3	530.900	531.773	-0.873		9628	515.700	515.977	-0.277	59
24 7	526.200	523.697	2.503		8590	514.300	514.588	-0.288	60
2428	511.500	512.792	-1.292		8936	502.000	502.288	-0.288	61
2490	503.000	505.590	-2.590		88 7	532.400	532.857	-0.457	62
2418	500.500	502.030	-1.530		95 0	625.600	626.066	-0.466	63
2436	501.400	498.867	2.533		2336	499.700	500.189	-0.489	64
89 0	599.000	598.133	0.867		85 3	538.200	538.698	-0.498	65
89 3	534.300	531.186	3.114		92 0	623.600	624.133	-0.533	66
89 7	521.100	523.950	-2.850		70 3	536.400	536.948	-0.548	67
8928	513.200	514.385	-1.185		9528	521.900	522.550	-0.650	68
8990	508.900	508.121	0.779		9490	509.500	510.168	-0.668	69
8918	505.900	505.031	0.869		9518	512.300	513.036	-0.737	70
8936	502.000	502.288	-0.288		9236	513.200	513.982	-0.782	71
90 0	633.400	632.997	0.403		2236	503.200	504.015	-0.815	72
90 3	543.900	543.275	0.625		24 3	530.900	531.773	-0.873	73
90 7	536.800	535.622	1.178		85 7	530.200	531.123	-0.923	74
9028	524.600	525.536	-0.937		9028	524.600	525.536	-0.937	75
9090	519.300	518.940	0.360		7028	514.900	515.937	-1.037	76
9018	517.500	515.687	1.813		8928	513.200	514.385	-1.185	77
9036	510.700	512.799	-2.099		95 7	531.100	532.292	-1.192	78
25 0	585.900	583.960	1.939		88 0	595.500	596.726	-1.226	79
25 3	527.100	529.191	-2.091		92 3	548.800	550.061	-1.261	80
25 7	523.100	521.719	1.381		2428	511.500	512.792	-1.292	81
2528	511.700	511.781	-0.081		8836	504.900	506.195	-1.295	82
2590	507.800	505.257	2.543		94 0	610.000	611.421	-1.421	83
2518	502.100	502.037	0.063		2418	500.500	502.030	-1.530	84
2536	502.800	499.178	3.622		8818	507.900	509.583	-1.683	85
95 0	625.600	626.066	-0.466		8518	509.600	511.362	-1.762	86
95 3	540.800	539.685	1.115		9136	507.700	509.547	-1.847	87
95 7	531.100	532.292	-1.192		22 3	537.500	539.461	-1.961	88
9528	521.900	522.550	-0.650		25 3	527.100	529.191	-2.091	89
9590	518.100	516.178	1.922		9036	510.700	512.799	-2.099	90
9518	512.300	513.036	-0.737		2318	501.400	503.543	-2.144	91
9536	510.200	510.247	-0.047		92 7	538.900	541.050	-2.150	92
91 0	624.000	624.037	-0.037		24 0	573.000	575.438	-2.437	93
91 3	540.000	538.756	1.244		94 7	522.500	524.967	-2.467	94
91 7	532.700	531.421	1.279		2490	503.000	505.590	-2.590	95
9128	524.100	521.755	2.345		85 7	528.500	531.123	-2.623	96
9190	517.800	515.432	2.367		9436	502.000	504.720	-2.720	97
9118	514.600	512.315	2.285		2228	516.200	518.996	-2.796	98
9136	507.700	509.547	-1.847		9636	501.000	503.823	-2.823	99
70 0	564.700	564.159	0.541		89 7	521.100	523.950	-2.850	100
70 3	536.400	536.948	-0.548		8528	517.800	521.128	-3.328	101
70 7	528.400	528.342	0.058		2390	503.800	507.318	-3.519	102
7028	514.900	515.937	-1.037		2290	507.500	511.246	-3.746	103
7036	502.100	499.595	2.505		8528	517.100	521.128	-4.028	104

Figure 9A.19 (*continued*)

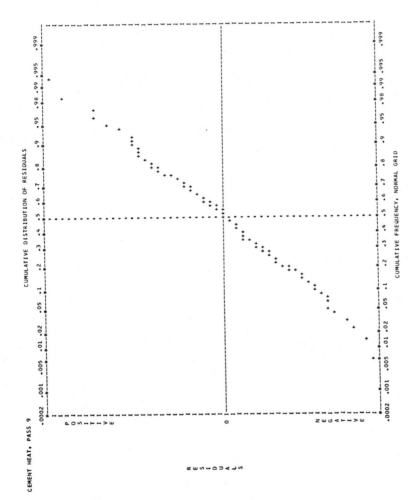

Figure 9A.20

257

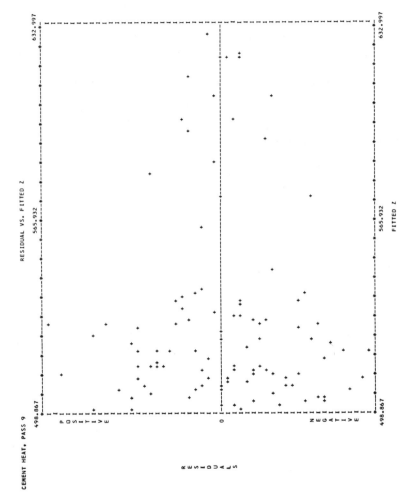

Figure 9A.21

258

NON-LINEAR LEAST-SQUARES CURVE FITTING PROGRAM

CEMENT HEAT, PASS 10

Z = H + (G - 1)D' + (A' / (T + F')EXP B)
Z = ACID HEAT OF SOLUTION AT TIME "T"
   B, G AND H ARE COMMON TO ALL CEMENTS
A' = A1X1 - A2X2 + A3X3 + A4X4 + A5X5 + A12X1X2 - A13X1X3 - A14X1X4
   - A23X2X3 + A24X2X4 - A34X3X4
D' = D1X1 + D2X2 + D3X3 + D4X4 + D5X5
F' = TIME CORRECTION TERM = (A' / (GD' - H))EXP(1/B)
X1 = (MOL PERCENT SILICA IN THE CLINKER - 20.9) / 7.43
X2 = (MOL PERCENT ALUMINA - 2.03) / 7.43
X3 = (MOL PERCENT FERRIC OXIDE - 0.44) / 7.43
X4 = (MOL PERCENT LIME - 66.1) / 7.43
X5 = (MOL PERCENT MAGNESIA - 3.10) / 7.43 ; OBSV. 85-3-2 OMITTED.

IND.VAR(I)	NAME	COEF.B(I)	S.E. COEF.	T-VALUE	95% CONFIDENCE LIMITS	
					LOWER	UPPER
1	B	1.95268E-01	2.82E-02	6.9	1.39E-01	2.52E-01
2	G	7.18791E-01	3.43E-02	21.0	6.50E-01	7.87E-01
3	H	3.15396E 02	2.26E 01	13.9	2.70E 02	3.61E 02
4	A1	8.51041E 01	4.80E 00	17.7	7.55E 01	9.47E 01
5	A2	4.35535E 01	6.36E 01	0.7	-8.35E 01	1.71E 02
6	A3	3.22453E 02	1.23E 02	2.6	7.72E 01	5.68E 02
7	A4	6.54490E 01	1.10E 01	5.9	4.34E 01	8.75E 01
8	A5	9.18853E 01	1.05E 01	8.8	7.10E 01	1.13E 02
9	A12	3.27494E 02	1.21E 02	2.7	8.63E 01	5.69E 02
10	A13	3.40659E 02	1.67E 02	2.0	7.75E 00	6.74E 02
11	A14	8.53281E 01	2.04E 01	4.2	4.45E 01	1.26E 02
12	A23	7.97744E 02	3.45E 02	2.3	1.09E 02	1.49E 03
13	A24	1.89868E 02	1.31E 02	1.4	-7.21E 01	4.50E 02
14	A34	3.47132E 02	1.57E 02	2.2	3.27E 01	6.62E 02
15	D1	5.75547E 02	1.62E 00	356.3	5.72E 02	5.79E 02
16	D2	6.22688E 02	5.49E 00	113.5	6.12E 02	6.34E 02
17	D3	5.21692E 02	4.23E 00	123.5	5.13E 02	5.30E 02
18	D4	6.40787E 02	2.08E 00	308.4	6.37E 02	6.45E 02
19	D5	6.36470E 02	2.95E 00	215.6	6.31E 02	6.42E 02

NO. OF OBSERVATIONS   104
NO. OF COEFFICIENTS   19
RESIDUAL DEGREES OF FREEDOM   85
RESIDUAL ROOT MEAN SQUARE   1.99987602
RESIDUAL MEAN SQUARE   3.99950790
RESIDUAL SUM OF SQUARES   339.95825195

Figure 9A.22

CEMENT HEAT, PASS 10

--------ORDERED BY COMPUTER INPUT---------				----------------ORDERED BY RESIDUALS----------------				
OBS. NO.	OBS. Z	FITTED Z	RESIDUAL	OBS. NO.	OBS. Z	FITTED Z	ORDERED RESID.	SEQ.
22 0	588.700	585.464	3.236	22 7	535.600	530.924	4.676	1
22 3	537.500	539.712	-2.212	8890	517.800	513.628	4.172	2
22 7	535.600	530.924	4.676	2536	502.800	499.007	3.793	3
2228	516.200	518.993	-2.793	23 7	530.000	526.287	3.713	4
2290	507.500	511.075	-3.575	22 0	588.700	585.464	3.236	5
2218	510.200	507.151	3.049	2218	510.200	507.151	3.049	6
2236	503.200	503.657	-0.457	89 3	534.300	531.448	2.852	7
23 0	575.700	575.279	0.421	2590	507.800	505.150	2.649	8
23 3	535.100	534.697	0.403	7036	502.100	499.541	2.559	9
23 7	530.000	526.287	3.713	25 0	585.900	583.413	2.487	10
2328	514.700	514.799	-0.099	2436	501.400	499.142	2.258	11
2390	503.800	507.157	-3.357	9228	531.400	529.165	2.235	12
2318	501.400	503.367	-1.967	9590	518.100	515.875	2.224	13
2336	499.700	499.592	-0.292	9690	511.900	509.863	2.037	14
92 0	623.600	624.184	-0.584	24 7	526.200	524.256	1.944	15
92 0	624.100	624.184	-0.084	9618	508.600	506.774	1.826	16
92 3	548.800	549.914	-1.114	95 3	540.800	538.989	1.811	17
92 7	538.900	541.005	-2.105	9218	519.300	517.521	1.779	18
9228	531.400	529.165	2.235	9428	517.600	515.830	1.770	19
9290	523.000	521.375	1.625	9290	523.000	521.375	1.625	20
9218	519.300	517.521	1.779	9018	517.500	515.941	1.559	21
9236	513.200	514.092	-0.892	25 7	523.100	521.687	1.413	22
88 0	595.500	598.128	-2.628	9118	514.600	513.319	1.281	23
88 3	542.300	541.602	0.698	94 3	532.700	531.436	1.264	24
88 7	532.400	532.927	-0.527	9190	517.800	516.561	1.239	25
8828	521.800	521.302	0.498	89 0	599.000	597.783	1.217	26
8890	517.800	513.628	4.172	8590	515.100	513.936	1.164	27
8818	507.900	509.829	-1.929	8518	511.800	510.803	0.996	28
8836	504.900	506.447	-1.547	85 3	538.200	537.212	0.988	29
96 0	604.500	604.740	-0.240	8536	509.000	508.017	0.983	30
96 0	603.200	604.740	-1.541	9128	524.100	523.117	0.983	31
96 3	532.600	532.797	-0.198	90 0	633.400	632.460	0.939	32
96 7	525.700	525.617	0.083	8918	505.900	505.103	0.797	33
9628	515.700	516.110	-0.410	90 7	536.800	536.010	0.790	34
9690	511.900	509.863	2.037	85 0	618.800	618.017	0.783	35
9618	508.600	506.774	1.826	94 0	611.700	610.997	0.703	36
9636	501.000	504.025	-3.025	88 3	542.300	541.602	0.698	37
85 0	618.800	618.017	0.783	8990	508.900	508.233	0.667	38
85 3	538.200	537.212	0.988	8828	521.800	521.302	0.498	39
85 7	528.500	529.917	-1.417	8536	508.500	508.017	0.483	40
85 7	530.200	529.917	0.282	23 0	575.700	575.279	0.421	41
8528	517.800	520.271	-2.471	23 3	535.100	534.697	0.403	42
8528	517.100	520.271	-3.171	8590	514.300	513.936	0.364	43
8590	515.100	513.936	1.164	85 7	530.200	529.917	0.282	44
8590	514.300	513.936	0.364	90 3	543.900	543.673	0.227	45
8518	509.600	510.803	-1.204	70 7	528.400	528.195	0.205	46
8518	511.800	510.803	0.996	2518	502.100	501.899	0.200	47
8536	509.000	508.017	0.983	9536	510.200	510.000	0.200	48
8536	508.500	508.017	0.483	9418	507.300	507.114	0.186	49

Figure 9A.23

94 0	610.000	610.997	-0.997	96 7	525.700	525.617	0.083	50
94 0	611.700	610.997	0.703	9090	519.300	519.230	0.070	51
94 3	532.700	531.436	1.264	95 0	625.600	625.573	0.026	52
94 7	522.500	524.714	-2.214	2528	511.700	511.719	-0.019	53
9428	517.600	515.83C	1.770	92 0	624.100	624.184	-0.084	54
9490	509.500	509.998	-0.498	2328	514.700	514.799	-0.099	55
9418	507.300	507.114	0.186	96 3	532.600	532.797	-0.198	56
9436	502.000	504.548	-2.548	96 0	604.500	604.740	-0.240	57
24 0	573.000	576.099	-3.099	9528	521.900	522.163	-0.263	58
24 3	530.900	532.375	-1.475	91 0	624.000	624.280	-0.280	59
24 7	526.200	524.256	1.944	2336	499.700	499.992	-0.292	60
2428	511.500	513.260	-1.760	8936	502.000	502.317	-0.317	61
2490	503.000	505.971	-2.971	70 0	564.700	565.033	-0.333	62
2418	500.500	502.358	-1.858	91 7	532.700	533.100	-0.400	63
2436	501.400	499.142	2.258	70 3	536.400	536.803	-0.404	64
89 0	599.000	597.783	1.217	9628	515.700	516.110	-0.410	65
89 3	534.300	531.448	2.852	2236	503.200	503.657	-0.457	66
89 7	521.100	524.189	-3.089	9518	512.300	512.766	-0.467	67
8928	513.200	514.562	-1.362	9490	509.500	509.998	-0.498	68
8990	508.900	508.233	0.667	88 7	532.400	532.927	-0.527	69
8918	505.900	505.103	0.797	92 0	623.600	624.184	-0.584	70
8936	502.000	502.317	-0.317	95 7	531.100	531.742	-0.642	71
90 0	633.400	632.460	0.939	91 3	540.000	540.649	-0.649	72
90 3	543.900	543.673	0.227	9236	513.200	514.092	-0.892	73
90 7	536.800	536.010	0.790	7028	514.900	515.843	-0.943	74
9028	524.600	525.88C	-1.281	94 0	610.000	610.997	-0.997	75
9090	519.300	519.230	0.070	92 3	548.800	549.914	-1.114	76
9018	517.500	515.941	1.559	8518	509.600	510.803	-1.204	77
9036	510.700	513.016	-2.316	9028	524.600	525.880	-1.281	78
25 0	585.900	583.413	2.487	8928	513.200	514.562	-1.362	79
25 3	527.100	529.158	-2.058	85 7	528.500	529.917	-1.417	80
25 7	523.100	521.687	1.413	24 3	530.900	532.375	-1.475	81
2528	511.700	511.719	-0.019	96 0	603.200	604.740	-1.541	82
2590	507.800	505.150	2.649	8836	504.900	506.447	-1.547	83
2518	502.100	501.899	0.200	2428	511.500	513.260	-1.760	84
2536	502.800	499.007	3.793	2418	500.500	502.358	-1.858	85
95 0	625.600	625.573	0.026	8818	507.900	509.829	-1.929	86
95 3	540.800	538.989	1.811	2318	501.400	503.367	-1.967	87
95 7	531.100	531.742	-0.642	25 3	527.100	529.158	-2.058	88
9528	521.900	522.163	-0.263	92 7	538.900	541.005	-2.105	89
9590	518.100	515.875	2.224	22 3	537.500	539.712	-2.212	90
9518	512.300	512.766	-0.467	94 7	522.500	524.714	-2.214	91
9536	510.200	510.000	0.200	9036	510.700	513.016	-2.316	92
91 0	624.000	624.280	-0.280	8528	517.800	520.271	-2.471	93
91 3	540.000	540.649	-0.649	9436	502.000	504.548	-2.548	94
91 7	532.700	533.100	-0.400	88 0	595.500	598.128	-2.628	95
9128	524.100	523.117	0.983	9136	507.700	510.435	-2.735	96
9190	517.800	516.561	1.239	2228	516.200	518.993	-2.793	97
9118	514.600	513.319	1.281	2490	503.000	505.971	-2.971	98
9136	507.700	510.435	-2.735	9636	501.000	504.025	-3.025	99
70 0	564.700	565.033	-0.333	89 7	521.100	524.189	-3.089	100
70 3	536.400	536.803	-0.404	24 0	573.000	576.099	-3.099	101
70 7	528.400	528.195	0.205	8528	517.100	520.271	-3.171	102
7028	514.900	515.843	-0.943	2390	503.800	507.157	-3.357	103
7036	502.100	499.541	2.559	2290	507.500	511.075	-3.575	104

Figure 9A.23 (*continued*)

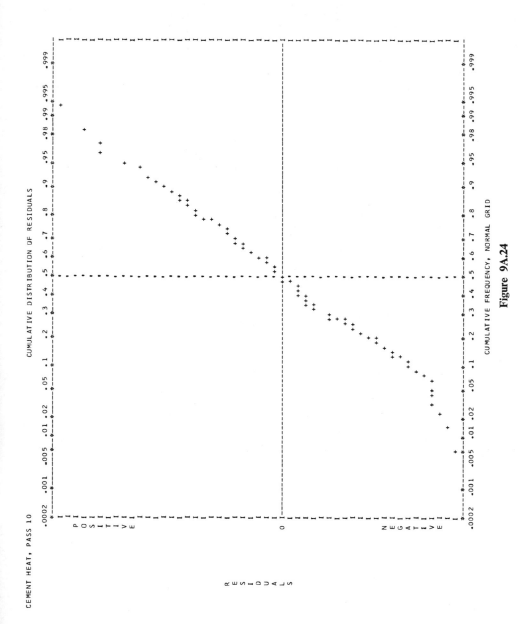

CEMENT HEAT, PASS 10

CUMULATIVE DISTRIBUTION OF RESIDUALS

CUMULATIVE FREQUENCY, NORMAL GRID

Figure 9A.24

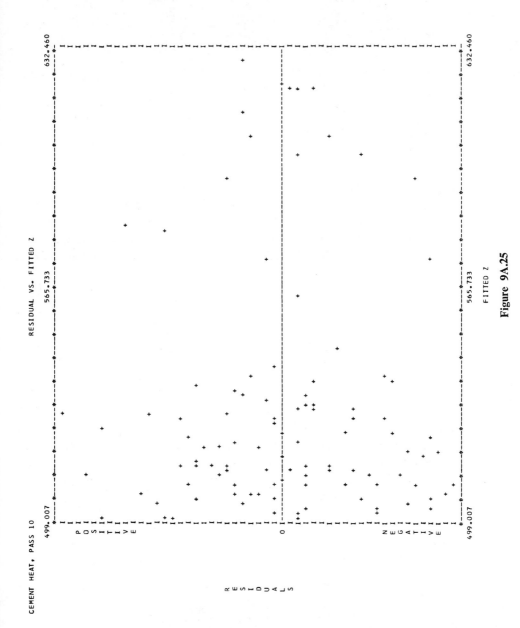

Figure 9A.25

Glossary

Conventions

A clear distinction is drawn between parameters and estimates, that is, between quantities which characterize the "true" situation and estimates of these quantities calculated from observations. When possible, a Greek letter is used to represent the parameter and the corresponding Roman letter to represent the estimate. Thus σ represents the standard deviation of the total population of values, and s the estimate of σ. Likewise, $\eta = \beta_0 + \beta_1 x_1$ represents the true equation, and $Y = b_0 + b_1 x_1$ its estimate. Notice that Y is used for the estimated or fitted value of the dependent variable and y for the observed value. Since computers to date print only capital letters, Y is listed on our printout as FITTED VALUE and y as OBS. VALUE.

Independent variables and their coefficients are indexed by the subscript i from 1 through K, that is, $b_1, b_2, \ldots, b_i, \ldots, b_K$. Dependent variables are indexed by the subscript m from 1 through Q: $y_1, y_2, \ldots, y_m, \ldots, y_Q$. Observations are indexed by the second subscript, j, from 1 through N; that is, the ith independent variable's observations are $x_{i1}, x_{i2}, \ldots, x_{ij}, \ldots, x_{iN}$, and the mth dependent variable's observations are $y_{m1}, y_{m2}, \ldots, y_{mj}, \ldots, y_{mN}$. When there is only one dependent variable, however, the subscript m is usually dropped and only the observation subscript j is used; $y_1, y_2, \ldots, y_j, \ldots, y_N$ instead of $y_{11}, y_{12}, \ldots, y_{1j}, \ldots, y_{1N}$. The computer printout lists variables numerically under IND. VAR. (I) and coefficients under COEF. B(I). Observations are listed in the order in which they are given to the computer.

Symbols

$\beta_1, \beta_2, \beta_3$	"Beta (1), beta (2), beta (3)," true values of the coefficients of three independent variables.
b_1, b_2, b_3	Estimates of the coefficients of three independent variables.
$c_{ii'}$	Elements of the inverse matrix: i indexes rows; i', columns.

C(I, I PRIME)	Computer designation for $c_{ii'}$.
C_p	An estimate of the standard total squared error.
$D_j{}^2$	The squared standardized distance of the jth observation from the centroid of all the observations:

$$D_j{}^2 = \sum_{i=1}^{K} \left[\frac{b_i(x_{ij} - \bar{x}_i)}{s_{ij}} \right]^2.$$

df	Degrees of freedom.
e_j	Random error associated with the jth observation.
$E\{\ \}$	The expected value of the quantity in braces.
Est. Var. ( )	Estimated variance of the quantity in parentheses.
F	F-value, Fisher's variance ratio: the ratio of two independent estimates of variance:

$$F = \frac{s_1{}^2}{s_2{}^2}.$$

Γ_p	"Gamma sub p," the standard total squared error:

$$\Gamma_p = E\{C_p\}.$$

i	Subscript i, index of the independent variable x, its parameter β, and its coefficient b.
(I)	Computer index of the independent variable X.
j	Subscript j, index of the observation.
K	The total number of independent variables, the terminal value of index i.
m	Subscript m, index for the dependent variables, y_m: $m = 1, \ldots, m, \ldots, Q$.
N	The total number of observations, the terminal value of index j.
η	"Eta," the "true" value of a response, or dependent variable.
p	The number of constants to be estimated. Used in plots of p and C_p; $p_{\max} = k + 1$ with b_0 present, $p_{\max} = k$ without b_0.
Q	The total number of dependent variables, the terminal value of index m.
$r_{12}{}^2$	The degree of nonorthogonality between x_1 and x_2.

R_i^2	When the x_i are correlated, the overall linear dependence of x_i on all the $x_{i'}$ ($i' \neq i$) may be expressed by $X_i = \sum_{i'} a_{i'} x_{i'}$.

$$R_i^2 = \frac{[X_i X_i]}{[x_i x_i]}.$$

R_y^2	The multiple correlation coefficient squared (sometimes called the coefficient of determination): the fraction of the total variation accounted for by the fitted equation:

$$R_y^2 = \frac{[YY]}{[yy]}.$$

RMS	The residual mean square, the residual sum of squares divided by $(N - p)$ degrees of freedom:

$$\text{RMS} = \frac{\text{RSS}}{N - p} \equiv \frac{[rr]}{N - p}.$$

RSS	The residual sum of squares, that part of the total sum of squares not accounted for by the fitted equation:

$$\text{RSS} \equiv [rr].$$

$\sigma(y)$	"Sigma," the standard deviation of y.
$\sigma^2(y)$	"Sigma squared," the variance of y.
Σ	The summation of the values which follow; for example

$$\sum_{j=1}^{N} x_{ij}$$

is the summation of the values of the x_i variable from 1 through N observations.

s_i	The root-mean-square value of the x_{ij}.
$s(b_1)$	The estimated standard error of the coefficient, b_1.
$s(y)$	The estimate of the standard deviation of y, $\sigma(y)$.
$s^2(y)$	The estimate of the variance of y, $\sigma^2(y)$.
SSB	The sum of squares due to bias; one of the components of C_p.
SSFE	The sum of squares accounted for by the fitted equation: $\text{SSFE} \equiv [YY]$.
TSS	The total sum of squares: $\text{TSS} \equiv [yy]$.

Var. ( )	The variance of the quantity in parentheses.
w_i	The range of observations, $x_{i,\max} - x_{i,\min}$.
x_{ij}	The independent variable indexed by i from 1 to K with observations indexed by j from 1 to N.
$\bar{x}_i$	The arithmetic mean of N observations of the ith independent variable:

$$\bar{x}_i \equiv \frac{\sum\limits_{j=1}^{N} x_{ij}}{N}.$$

y_{mj}	The dependent variable indexed by m from 1 to Q with observations indexed by j from 1 to N.
$\bar{y}_m$	The arithmetic mean of N observations of the mth dependent variable:

$$\bar{y}_m \equiv \frac{\sum\limits_{j=1}^{N} y_{mj}}{N}.$$

Y_{mj}	The estimated or fitted value of the dependent variable indexed by the subscript m from 1 to Q with observations indexed by j from 1 to N.
z	A random normal deviate with a standard error of 1 and a mean of 0.
[ ]	Gaussian brackets, indicating summation of products obtained by multiplying deviations from means of factors shown inside the brackets.
$[rr]$	The residual sum of squares:

$$[rr] \equiv \sum_{j=1}^{N} (y_j - Y_j)^2.$$

$[yy]$	The total sum of squares:

$$[yy] \equiv \sum_{j=1}^{N} (y_j - \bar{y})^2.$$

$[YY]$	The sum of squares accounted for by the fitted equation:

$$[YY] \equiv \sum_{j=1}^{N} (Y_j - \bar{y})^2.$$

[1 y]   The corrected (for the mean) sum of the products of the independent and dependent variables, x_{1j} and y_j:

$$[1\ y] \equiv \sum_{j=1}^{N}(x_{1j} - \bar{x}_1)(y_j - \bar{y}).$$

[1 2]   The corrected (for the mean) sum of the products of two independent variables, x_{1j} and x_{2j}:

$$[1\ 2] \equiv \sum_{j=1}^{N}(x_{1j} - \bar{x}_1)(x_{2j} - \bar{x}_2).$$

Computer Terms

This portion of the glossary is a connecting link between the computer Linear and Nonlinear Least-Squares Curve Fitting Programs and the text. Because of the limitations of both space and computer printing symbols, many of the computer listings are necessarily abbreviations and are restricted to capital letters. In the text, on the other hand, it was necessary to use mathematical abbreviations to avoid cumbersome rewriting of certain summations. Thus we have the residual sum of squares in four forms:

$$\sum_{j=1}^{N}(y_j - Y_j)^2 \equiv [rr] \equiv \text{RESIDUAL SUM OF SQUARES} \equiv \text{RSS}$$

(Normal form)   (Gaussian bracket abbreviation)   (Computer listing)   (Abbreviation)

The Glossary of Computer Terms is arranged in the alphabetical order of the computer listings. Here, along with the definition of terms, we have included enough of the mathematics of the curve fitting programs so that the reader can understand what the computer is doing. Those wishing a more complete exposition of the mathematics may refer to Brownlee, Hald, Bennett and Franklin, or Anderson and Bancroft.

B ZERO = CALCULATED VALUE.   This statement on the computer printout indicates that the b_0 term in the fitted equation is a calculated value and has not been set equal to zero.

B ZERO = ZERO.   This statement on the computer printout indicates that the b_0 term in the fitted equation has been set, a priori, equal to zero. (See the User's Manual for instructions on control card options.)

COEF. B(I). Under this heading, the coefficients of the variables are indexed by i and the values calculated for the fitted equation are tabulated starting with b_0. (See FITTED EQUATION.)

CUMULATIVE DISTRIBUTION OF RESIDUALS. The program plots the ranked residuals, OBSERVED Y − FITTED Y, on a CUMULATIVE FREQUENCY NORMAL PROBABILITY GRID. If the residuals are normally distributed, the points should fall approximately on a straight line with mean zero. The distance between the 50% and the 84% points should be approximately equal to the residual root mean square of the fitted equation for $N \geq 3K \geq 16$.

On a normal probability grid, the abscissa is specially ruled so that the cumulative distribution function of a normal distribution can be plotted as a straight line against the residuals as ordinate. Plotting points for the abscissa, P, are calculated from the formula:

$$P = \frac{i - \frac{1}{2}}{N},$$

where i = rank of observation, and N = total number of observations. Examples of plots for samples ranging in size from 8 to 384, drawn from a population of known normal distribution with mean 0 and variance 1, are shown in Appendix 3A (lines on the graphs represent the true population line). In the long run, the larger the sample the better is the approximation to a straight line. (See Blom, pages 71–72, and Hald, Sections 6.6 and 6.8, for additional examples.)

DATA INPUT. This is a listing of the data read into the computer from the data input cards. The format card determines which columns are used for each variable. Instructions for submitting problems to the computer are given in the User's Manual.

DATA TRANSFORMATIONS. This is a listing of the transformed values of each variable as used in the fitting equation. The program can convert the DATA INPUT to several forms such as sums, differences, products, ratios, and logarithms. Also, the input and transformed variables can be put in or left out in any desired pattern. (See the User's Manual for specific instructions.)

FITTED EQUATION

FIRST-ORDER EQUATION. This is an equation with K independent variables, indexed by i. It is generally of the form:

$$Y = b_0 + b_1 x_1 + b_2 x_2 + b_3 x_3 + b_i x_i + \cdots + b_K x_K,$$

where $Y =$ FITTED Y, the dependent variable; $x_1, x_2, x_3, \ldots, x_K =$ IND. VAR. (I), the independent variables indexed by i from 1 to K; $b_0 =$ is listed under COEF. B(I) when I is 0; and $b_1, b_2, \ldots, b_i, \ldots, b_K$ are listed under COEF. B(I) when $I = 1, 2, \ldots, i, \ldots, K$.

SECOND-ORDER EQUATION. This is an equation in K variables which includes K linear terms, the $K(K-1)/2$ cross-product (or interaction) terms, and the K square terms:

$$Y = b_0 + b_1 x_1 + \cdots + b_K x_K + b_{12} x_1 x_2 + \cdots + b_{K-1,K} x_{K-1} x_K \\ + b_{11} x_1{}^2 + \cdots + b_{KK} x_K{}^2.$$

FITTED Y. OBSERVED Y is the observed response, and FITTED Y is the corresponding value calculated from the fitted equation. Both are indexed by the OBSERVATION NUMBER, j from 1 to N.

F-VALUE. The F-VALUE can be compared with tabled values to give a joint test of the hypothesis that all the coefficients are zero against the alternative that the equation as a whole produces a significant reduction in the total sum of squares. The F-value is calculated from the multiple correlation coefficient squared, the residual degrees of freedom, and the degrees of freedom associated with the fitted equation:

$$F\text{-value} = \frac{\text{SSFE}/K}{\text{RSS}/(N-K-1)} = \left(\frac{N-K-1}{K}\right)\left(\frac{R^2}{1-R^2}\right).$$

The ratio is compared to the corresponding value from an F-table with K and $(N-K-1)$ degrees of freedom.

IND. VAR. (I). The independent variables, indexed by i, are tabulated under IND. VAR. (I), starting with 0 (to identify the b_0 term) and proceeding through the K variables, $0, 1, 2, \ldots, i, \ldots, K$. (See FITTED EQUATION.)

INVERSE, C(I, I PRIME).   Elements of the inverse matrix are listed in the following form:

$$
\begin{array}{cccccc}
C_{11} & C_{12} & C_{13} & \cdots & C_{1i'} & \cdots & C_{1K} \\
C_{21} & C_{22} & C_{23} & \cdots & C_{2i'} & \cdots & C_{2K} \\
C_{31} & C_{32} & C_{33} & \cdots & C_{3i'} & \cdots & C_{3K} \\
\cdot & \cdot & \cdot & \cdots & \cdot & \cdots & \cdot \\
\cdot & \cdot & \cdot & \cdots & \cdot & \cdots & \cdot \\
\cdot & \cdot & \cdot & \cdots & \cdot & \cdots & \cdot \\
C_{i1} & C_{i2} & C_{i3} & \cdots & C_{ii'} & \cdots & C_{iK} \\
\cdot & \cdot & \cdot & \cdots & \cdot & \cdots & \cdot \\
\cdot & \cdot & \cdot & \cdots & \cdot & \cdots & \cdot \\
\cdot & \cdot & \cdot & \cdots & \cdot & \cdots & \cdot \\
C_{K1} & C_{K2} & C_{K3} & \cdots & C_{Ki'} & \cdots & C_{KK}
\end{array}
$$

(See Appendix 2B for details on the inverse matrix as used in the computer program.)

LINEAR LEAST-SQUARES ESTIMATES.   See Appendix 2B for a detailed description of linear least-squares and computer computations.

MAX X(I).   The largest value of each independent variable is tabulated under this heading.

MEANS OF VARIABLES.   The arithmetic means of N observations of each independent and dependent variable are listed under this heading in the same order that they appear on the data cards or as dictated by transformation instructions.

MIN X(I).   The minimum value of each independent variable is tabulated under this heading.

MULT. CORREL. COEF. SQUARED.   The multiple correlation coefficient squared (sometimes called the coefficient of determination) is defined as follows:

MULT. CORREL. COEF. SQUARED $\equiv$

$$
\frac{\text{Sum of squares due to the fitted equation}}{\text{Total sum of squares}}.
$$

(See "Partitioning Sums of Squares" in Appendix 2B.)

OBS. NO.   The observation number is read from the computer card, and the numbers are listed in the order in which the cards are given to the computer. If the observations are in the same order in which they were taken, their time independence can be verified by a "run-of-signs" test on the residuals.

OBS. Y.   The observed value of the dependent variable, y, is indexed by j, the observation number from 1 to N.

ORDERED RESID.   Under this heading the residuals are ranked in order from the largest positive to the largest negative value. The observation numbers are listed with the ordered residuals so that they can be easily identified.

RANGE X(I).   The range of each independent variable, w_i, is determined by subtracting its minimum value from its maximum value.

RAW SUMS OF SQUARES + CROSS PRODUCTS.   The raw sum of squares for the first variable over the N observations is the quantity
$$\sum_{j=1}^{N} x_{1j}{}^2.$$
The raw sum of cross products for the first and second variables over the N observations is the quantity $\sum_{j=1}^{N} x_{1j} x_{2j}$.

The computer printout lists the raw sums of squares + cross products for all the independent and dependent variables in the following form:

	Col. 1	Col. 2	Col. 3	$\cdots$	Col. $(K + Q)$
Row 1	$\sum_{j=1}^{N} x_{1j}{}^2$				
Row 2	$\Sigma x_{2j} x_{1j}$				
Row 3	$\Sigma x_{3j} x_{1j}$	$\Sigma x_{3j} x_{2j}$	$\Sigma x_{3j}{}^2$		
	$\cdot$	$\cdot$	$\cdot$		
	$\cdot$	$\cdot$	$\cdot$		
	$\cdot$	$\cdot$	$\cdot$		
Row $(K + Q)$	$\Sigma x_{1j} x_{(K+Q)j}$	$\Sigma x_{2j} x_{(K+Q)j}$		$\cdots$	$\Sigma x_{(K+Q)}^2$

where $K = $ number of independent variables, and $Q = $ number of dependent variables.

REL. INF. X(I).   The relative influence of x_i describes the fraction of the total change in Y that can be accounted for by the accompanying total change in the ith variable:

$$\text{Relative influence } x_i = \frac{b_i w_i}{w_y},$$

where b_i is the coefficient, w_i is the range of the ith independent variable, and w_y is the range of the observed dependent variable.

RESIDUAL.   The residual is the difference between the observed value of the dependent variable, y_j, and its fitted value, Y_j:

$$\text{Residual} = y_j - Y_j.$$

The residuals of the fitted equation are listed twice, first in the order of the observation number, and then by rank, from the largest positive to the largest negative value.

RESIDUAL DEGREES OF FREEDOM.   The residual degrees of freedom of a set of observations is the number of observations minus the number of constants estimated from the same set of observations. For example, the sum of squares of deviations from the mean in a sample of N observations,

$$\Sigma(x - \bar{x})^2 = (x_1 - \bar{x})^2 + (x_2 - \bar{x})^2 + \cdots + (x_j - \bar{x})^2 + \cdots$$
$$+ (x_N - \bar{x})^2,$$

has $(N - 1)$ residual degrees of freedom, because one constant, $\bar{x}$, has been estimated from the N observations. Calculation of the residual mean square involves the sum of squares of the differences between the observed and the fitted values. The fitted values are obtained from the equation which has $(K + 1)$ constants—K coefficients plus the mean. Hence the residual mean square has $[N - (K + 1)]$ residual degrees of freedom.

RESIDUAL MEAN SQUARE.   The residual mean square is calculated by dividing the residual sum of squares by the residual degrees of freedom $(N - K - 1)$:

$$\text{RMS} = \frac{[rr]}{N - K - 1} = \frac{\text{RSS}}{N - K - 1}.$$

RESIDUAL ROOT MEAN SQUARE.   The residual root mean square is the square root of the residual mean square:

$$RRMS = (RMS)^{1/2}$$

RESIDUAL SUM OF SQUARES.   The residual sum of squares is a measure of the squared scatter of the observed values around those calculated by the fitted equation:

$$RSS = \sum_{j=1}^{N}(y_j - Y_j)^2 \equiv [rr] = [yy] - [YY].$$

(See "Partitioning Sums of Squares" in Appendix 2B.)

RESIDUAL SUMS OF SQUARES AND CROSS PRODUCTS.   The residual sum of squares for the first variable about its mean is the quantity

$$\sum_{j=1}^{N}(x_{1j} - \bar{x}_1)^2 \equiv [11].$$

The residual sum of cross products for the first and the second variables about their means is the quantity

$$\sum_{j=1}^{N}(x_{1j} - \bar{x}_1)(x_{2j} - \bar{x}_2) \equiv [12] \equiv [21].$$

The computer printout lists the residual sums for all independent and dependent variables in the following form:

[11]

[21]   [22]

[31]   [32]   [33]

  .   .   .

  .   .   .

  .   .   .

$[K + Q]$ $[(K + Q)^2]$ $\cdots$ $[(K + Q)(K + Q)]$

where K = number of independent variables, and Q = number of dependent variables.

RESIDUAL VS FITTED Y.   An option of the computer program is a plot of each residual versus its corresponding fitted Y-value.

R(I)SQRD. When one x_i is correlated with the other x_i' ($i' \neq i$), we can express this as a linear least-squares equation, showing X_i as a function of the other $x_{i'}$:

$$X_i = \sum_{i'} a_{i'} x_{i'}.$$

The squared multiple correlation coefficient, R_i^2, shows how well this equation fits and hence measures the degree of linear dependence of x_i on the other $x_{i'}$. It is then a measure of the nonorthogonality of the ith independent variable. The simple(!) correlation coefficient between X_{ij} and x_{ij} is R_i.

S. E. COEF. Under this heading is listed the estimated standard error of the coefficient, $\sqrt{\text{Var.}(b_i)}$. (See Appendix 2B.)

SIMPLE CORRELATION COEFFICIENT, R(I, I PRIME). The simple correlation coefficient is a measure of the linear interdependence between two variables. It is a pure number varying between -1 and $+1$; a value of 0 indicates no linear correlation. The limiting values of -1 and $+1$ indicate perfect negative or positive linear correlation. The correlation coefficient for the first and second variables is:

$$r_{12} = \frac{\sum\limits_{j=1}^{N} (x_{1j} - \bar{x}_1)(x_{2j} - \bar{x}_2)}{\sqrt{\left[\sum\limits_{j=1}^{N}(x_{1j} - \bar{x}_1)^2\right]\left[\sum\limits_{j=1}^{N}(x_{2y} - \bar{x}_2)^2\right]}} \equiv \frac{[12]}{\sqrt{[11][22]}}.$$

The computer printout lists the simple correlation coefficient between all independent and dependent variables in the following form:

r_{11}

r_{21} r_{22}

r_{31} r_{32} r_{33}

.   .   .

.   .   .

.   .   .

$r_{(K+Q)1}$ $r_{(K+Q)2}$ $\cdots$ $r_{(K+Q)(K+Q)}$

where K = number of independent variables, and Q = number of dependent variables.

STANDARD DEVIATION OF VARIABLES.   The standard deviation for
the first variable over the N observations is the quantity

$$s(x_i) = \sqrt{\frac{\sum_{j=1}^{N}(x_{ij} - \bar{x}_j)^2}{N - 1}} = \sqrt{\frac{[ii]}{N - 1}}$$

The computer printout lists the standard deviation for each independent
and dependent variable in the order of its appearance on data cards or
as dictated by transformation instructions.

SUMS OF VARIABLES.   The computer lists $\sum_{j=1}^{N} X_{ij}$ for each independent
and dependent variable in the order of its appearance on data cards
or as dictated by transformation instructions.

TAYLOR SERIES EXPANSIONS.
    1. *Single independent variable.*   Any function, $f(x)$, that possesses con-
tinuous derivatives $f^n(x)$ in the neighborhood of a can be expanded in
the form:

$$f(x) = f(a) + (x - a)f^1(a) + \frac{(x - a)^2}{2!}f^2(a)$$

$$+ \cdots + \frac{(x - a)^{n-1}}{(n - 1)!}f^{n-1}(a) + R_n$$

where $R_n = (x - a)^n f^n(u)$, $a < u < x$.
By analogy the following equation form can be viewed as the first three
terms in a Taylor series expansion:

$$Y = b_0 + b_1(x - \bar{x}) + b_2(x - \bar{x})^2.$$

2. *Two independent variables.*   Any function $f(x, z)$ that possesses continuous partial derivatives of all orders can be expanded about the point
$(x = a, z = b)$ in the form:

$$f(x, z) = f(a, b) + \left(\frac{\partial f}{\partial x}\right)_{a,b}(x - a) + \left(\frac{\partial f}{\partial z}\right)_{a,b}(z - b)$$

$$+ \frac{1}{2!}\left[\left(\frac{\partial^2 f}{\partial x^2}\right)_{a,b}(x - a)^2 + 2\left(\frac{\partial^2 f}{\partial x\,\partial z}\right)_{a,b}(x - a)(z - b) + \left(\frac{\partial^2 f}{\partial z^2}\right)_{a,b}(z - b)^2\right]$$

$$+ \cdots.$$

The analogous second order fitting equation is:

$$Y = b_0 + b_1(x - \bar{x}) + b_2(z - \bar{z}) + b_3(x - \bar{x})^2 + b_4(x - \bar{x})(z - \bar{z})$$
$$+ b_5(z - \bar{z})^2.$$

Both the Taylor series and the analogous form may be extended to include any number of independent variables.

TOTAL SUM OF SQUARES.   The total sum of squares is a measure of the total variation in the dependent variable:

$$\text{Total sum of squares} = \sum_{j=1}^{N} (y_j - \bar{y}) .$$

T-VALUE.   The t-value is calculated as follows:

$$t_i = \frac{b_i}{s(b_i)} = \frac{\text{COEF. B(I)}}{\text{S. E. COEF.}} .$$

Computer Linear Least-Squares Curve Fitting Program

Abstract

This computer program allows the user to transform data into an appropriate form, fits specified equations to the transformed data by linear least squares, and provides both statistics and plots to aid in evaluating the fit. A C_p-statistic search technique determines whether smaller sets of the variables will represent the data equally well.

The transformations which are available to the user include reciprocals, sums, differences, products, quotients, logarithms, and exponentials. Such transformations are used to convert the observed data to more convenient or more rational units, to add terms that are functions of the data variables, to stabilize variance, and to omit variables.

In addition to the usual statistics, the program calculates the maximum and the minimum value of each variable, as well as its range, the relative influence of each variable, and the weighted squared standardized distance of each observation from the centroid of all observations. Near neighbors are used to estimate the standard deviation of the dependent variable. A table of component effects shows how each variable contributes to the fitted value of each observation.

Listings are made of the observed and fitted values of the dependent variable—both in the sequence in which observations were given to the computer, and in the order of the magnitude of the differences between the observed and fitted values. Plots are made to indicate (1) whether these differences are normally distributed and (2) how they are distributed over all the fitted values of the dependent variable.

The program, as dimensioned, will handle up to 80 variables and 1000 observations. Multiple dependent variables are fitted one at a time, and multiple forms of specified linear equations can be fitted with one data loading.

Examples are given.

General Information

A. Program written and revised by: R. A. Brehmer, P. E. Piechocki, W. B. Traver, F. M. Jacobsen, F. M. Oliva, R. J. Toman, and F. S. Wood.

B. Instructions written and revised by: P. E. Piechocki, W. M Dudley, and F. S. Wood.

C. Form of equations:

$$Y = b_0 + b_1 x_1 + b_2 x_2 + \cdots + b_i x_i + \cdots + b_K x_K,$$

Computer Listing

where Y = fitted value of the dependent variable,   FITTED Y

$x_1, x_2, \ldots, x_i, \ldots, x_K$ = independent variables indexed by i from 1 to K,   IND. VAR. (I)

b_0 = estimated Y intercept when all $x_i = 0$,   COEF. B(I)

$b_1, b_2, \ldots, b_i, \ldots, b_K$ = estimated coefficients of the independent variables.   COEF. B(I)

Note: The dependent and independent variables may be transformations of the observed values, for example, $x_3 = (x_1)(x_2)$, $x_4 = 1/x_4'$, etc.

D. Restrictions:*
    Number of variables
        Maximum
            Before transformations   105
            After transformations   80
        Minimum
            Independent variable   1
            Dependent variable   1
    Number of observations
        Maximum   1000
        Minimum   One greater than the number of coefficients being estimated

Control Cards and Data Entry

To submit a problem, information is normally entered on two types of special forms from which cards can be keypunched (see Figure 1 and Figure 2). The control card and data cards are always used; one or more transformation cards will be needed only if some variables of the entering data are to be transformed or omitted before calculations begin. Information cards

* These restrictions may be altered by changing the dimension statements of the computer program.

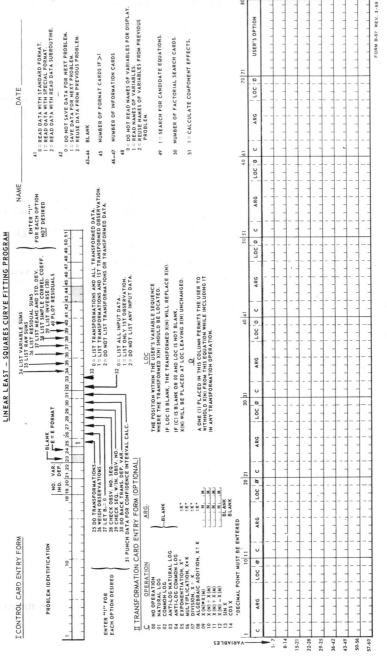

Figure 1

LINEAR LEAST – SQUARES CURVE FITTING PROGRAM

III STANDARD DATA-CARD ENTRY FORM

ENTER WEIGHTING FACTOR, IF ANY, AS LAST ENTRY
"END" CARD MUST BE LAST CARD.

IDENT.	OBSV. NO.	SEQ NO.	x1–11–21	x2–12–22	x3–13–23	x4–14–24	x5–15–25	x6–16–26	x7–17–27	x8–18–28	x9–19–29	x10–20–30	USER'S OPTION

FORM B-56 REV. 3-69

Figure 2

281

may be used to display on the printout statements concerning the form of equation, the purpose of the run, or other pertinent information that will help identify the printout in the future. Variable name cards may be used to display the names of variables in English or alphanumeric code. Such displays help identify both the coefficients of the fitted equation and the variables in each of the candidate C_p-search equations. All possible combinations of the variables can be searched to ascertain whether a smaller set will fit the data equally well. If more than 12 variables are to be searched (4096 equations), fractional factorial searches are required to find all candidate equations. In this case, the variables to be searched are read from factorial search cards.

To avoid obtaining garbled results, particularly on problems in which the transformation option is used, the user is advised to read all of this section before starting to fill out any forms. For routine use, the order of the cards and control card entries are summarized in the "User's Instructions" at the end of this manual.

A.   Control card entry

One control card is necessary for each problem. This card contains information regarding the problem identification, the numbers of independent and dependent variables, and the operating and output options desired by the user. Column entries are as follows:

Cols. 1–18. Problem Identification. These columns may contain any grouping of letters, numbers, and blanks.

Cols. 19 and 20. Number of Independent Variables.

Cols. 21 and 22. Number of Dependent Variables.

Columns 19 and 20 contain either the number of independent variables that are read in as input or the number that will be correlated after transforming the input variables, whichever quantity is greater. Similarly, cols. 21 and 22 contain the larger number of dependent variables. For example, if the user wishes to read in five independent and four dependent variables and perform a transformation to increase the number of independent variables to seven, col. 20 must contain the number 7 and col. 22 the number 4. On the other hand, if seven independent variables and five dependent variables appear on the data cards and the user wishes to correlate only three independent and one dependent variable, while withholding four independent and four dependent variables, then the total number of each, seven and five, are the appropriate quantities to be entered for the number of independent and dependent variables. A subsequent adjustment for the number of variables to be used in correlation is performed within the program after the specified transformations and omissions are completed. Thus the program permits the user to try several different combinations of variables without repunching the data cards; only the control and transformation cards need to be changed in making successive runs.

Col. 23. Blank.

Col. 24. Method of Listing Residuals. When this column is blank, the residuals are printed with a digit field containing three places after the decimal. When the letter E is inserted, the residuals are printed in an E-fixed decimal-point format, for example, 4.5025E-03 is printed instead of 0.004.

Col. 25. Do transformations—Enter 1. When input variables are to be transformed, a one (1) must be placed in this column. This signals the program to read the appropriate number of transformation cards.

Col. 26. Weight Observations—Enter 1. When a user wishes to impute a greater degree of importance to certain observations as opposed to others, these observations may be "weighted." This column entry indicates that weighting is to be done. The actual weighting factor is entered on the data card as an entry following the last dependent variable. When the weighting option is on, every observation must have a nonblank weighting factor, that is, a factor of 1 should be used on unweighted sets. *Note: The weighting factor is not considered a variable and must not be counted in the totals of the number of independent and dependent variables.*

Col. 27. Let $b_0 = 0$—Enter 1. In the fitted equation, $Y = b_0 + b_1x_1 + b_2x_2 + \cdots + b_Kx_K$, the b_0 term will be the calculated value of the intercept on the y-axis. If the user chooses to specify that b_0 be assumed equal to zero, he may indicate this by entering a 1 in col. 27. In most cases the value reported for b_0 under these circumstances will not be exactly zero, because of the nature of the calculations. When this option is used with this program, the means and the standard deviations of the independent and dependent variables require special interpretation, as $x = 0$, $y = 0$ is not included as an observed value but nonetheless $b_0 = 0$ is a restraint imposed upon the fit.

Col. 28. Check Observation Number Sequence—Enter 1. A check is made to ensure that the data card observation numbers are in ascending sequence as they are read into the program. Observations out of sequence are reported on the output listing. If one observation number out of sequence is found, the program bypasses all calculations on the problem but continues checking the observation numbers until either a total of five observations out of sequence have been found and reported, or all observations have been read in. It then proceeds to the next problem, if any.

Col. 29. Check Sequence Numbers within Observation Numbers—Enter 1. The check included here ensures that the sequencing of cards is consecutive and in ascending order, even when the number of variables included in one observation requires more than one data card per observation. The same conditions of operation apply as in checking sequencing of observation numbers (see col. 28 above).

On problems having a large number of observations, checking of observation numbers (and sequence numbers, if used) is desirable to prevent waste of machine time on data which are improperly set up. *Note pitfall:* when there is more than one data card per observation, each card in the sequence must carry the observation number as well as the sequence number. Otherwise, if checking of observation numbers is specified, a card with no observation number will be considered "out of order" and the whole problem will be rejected.

Col. 30. Do Back-Transformation of Dependent Variable—Enter 1. Dependent variables that have been transformed by a log or exponential operation may be back-transformed in the listing of the observed and fitted Y-values and their differences or residuals. Back-transformations are performed only on dependent variables and only for the log and exponential transformations. When back-transformations other than these are requested, the user is notified by a message that the back-transformation has not been effected and that the results listed are as the dependent variables were transformed.

When the fitted values have been back-transformed, the residual plots are not usable to determine the fit of the data—they too will have been transformed. Hence, the residual plots are automatically deleted.

Col. 31. Punch Data for Confidence Interval Calculations—Enter 1. Data required for confidence interval calculations (another computer program) is punched on cards when this option is chosen.

Col. 32. List Input Data.
0 (or blank) List all input data.
1 List only first observation.
2 Do not list any input data.

Col. 33. List Transformations and Transformed Data.
0 (or blank) List transformations and all transformed data.
1 List transformations and first transformed observation.
2 Do not list any transformations or transformed data.

Col. 34. Do not List the Sum of Each Variable—Enter 1.

Col. 35. Do Not List Raw Sums and Cross Products—Enter 1. When b_0 equals zero, the raw sums of squares and cross products of the variables are not listed.

Col. 36. Do Not List Residual Sums and Cross Products—Enter 1. When b_0 is a calculated value, the residual (deviations from the mean) sums of squares and cross products are not listed.

Col. 37. Do Not List the Mean and Standard Deviation of Each Variable— Enter 1.

Col. 38. Do Not List the Simple Correlation Coefficients—Enter 1.

Col. 39. Do Not List the Inverse of the Simple Correlation Coefficients— Enter 1.

Col. 40. 1 Do Not Print Plots of Residuals. The cumulative distribution plot of the residuals and the plot of the residuals versus their corresponding fitted Y-values are not printed. 2 Plot a residuals and b component-effects— plus residuals vs. each independent variable. 3 Plot a only. 4 Plot b only.

Col. 41. Method of Reading Data.

0 (or blank) Read data with standard format—cards keypunched from the standard data entry form.

1 Read data with special format. The format card containing this informa- tion immediately follows the control card. If more than 72 columns are needed, up to 8 additional format cards can be used. The total number of format cards, if more than 1, is entered in col. 45. The standard format, written in FORTRAN, is (A6, I4,I2, 10F6.3). This means there are 6 columns of alphanumeric information for identification, 4 columns of integers for observation numbers, 2 columns of integers for sequence numbers, and 10 fields of 6 columns for fixed-point numbers—each with 3 digits after the decimal point unless the decimal point is punched in some other position. The fields for identification, observation numbers, and sequence numbers are required even though they are not used. If each observation had data on 2 cards with the last variable in floating- point notation, the format card might read (A6, I2, I1, 3F4.1/12x, E 12.5). The / mark indicates that another card should be read and 12 columns omitted before reading the fourth variable. The use of such a format requires that two END cards follow the data cards.

2 Read data, using a read data subroutine. (See printout of the read data subroutine, Figure 3, for available common dimensions and require- ments for returning data to main program.)

                          READ DATA SUBROUTINE

```
       SUBROUTINE REDATA
C
C ---    SUBROUTINE TO READ DATA. CAN BE REVISED TO SUIT SPECIAL NEEDS.
C       CONTROL CARD SHOULD INDICATE THE NUMBER OF INDEPENDENT
C       VARIABLES BEING RETURNED IN ARRAY LABELLED DATA(106).
C       IDENTIFICATION CAN BE RETURNED USING A6 FORMAT IN IDENT.
C       THE OBSERVATION NUMBER CAN BE RETURNED USING I4 FORMAT IN IOBSV.
C       THE SEQUENCE NUMBER CAN BE RETURNED USING I2 FORMAT IN ISEQ IF
C       DESIRED.
C ---    ARRAYS THAT CAN BE USED IN REDATA TO SAVE COMPUTER CORE SPACE
C       ARE COM4, LSORT AND IDZB, E.G.
C       EQUIVALENCE (COM4,NEWARY)  WHERE NEWARY IS USED IN REDATA.
C ---    KSL2 CAN BE USED AS A SWITCH.  KSL2 IS SET = 1 IN MAIN AFTER
C       CONTROL CARD IS READ.  IF KSL2 IS SET = 2 IN REDATA AFTER READING
C       FIRST OBSERVATION. SUBSEQUENT OBSERVATIONS CAN BE READ IN
C       DIFFERENT MANNER USING KSL2 AS A SWITCH.
C
       IMPLICIT REAL*8(A-H,O-Z)
       DOUBLE PRECISION IDENT, IEND
       DIMENSION AVATR(105), DATA(106), FMT(144), IOMIT(105), ITLOC(105)
       COMMON /AAAAAA/ COM1(1000)
       COMMON /BBBBBB/ COM2(186)
       COMMON /CCCCCC/ COM3(240)
       COMMON /DDDDDD/ COM4(3756)
C      COMMON /DDDDDD/ COM4(4096) IN CDC 6400 PROGRAM
       COMMON /EEEEEE/ LSORT(1000)
       COMMON /FFFFFF/ VARTR(105), ITRFM(105)
       COMMON /GGGGGG/ BI(80), EQU(144)
       COMMON /HHHHHH/ BETA(81), C(80,80), Z(81,81), NNNN
       COMMON /JJJJJJ/ IDZB(1000)
       COMMON /KKKKKK/ IQ(40), PR1, PR2, PR3, K, KFM, NODEP, NOIND,
      1                NOOBSV, ICURY
       COMMON /LLLLLL/ XEFMT
       COMMON /MMMMMM/ BZRO, IDENT, IEND, IOBSV, ISEQ, JOBSV, KSL2, M,
      1                NOERR, NVP1
       COMMON /OOOOOO/ KNODEP, KNOIND, KNOVAR, KNVP1, L, NOVAR
       COMMON /QQQQQQ/ KTIN, KTOU, KTPCH, KKK, KONE, KTWO, MNO
       COMMON /RRRRRR/KTBIN1, KTBIN2, KTBIN3, KTBIN4, KTBIN5
C ---        EQUIVALENCE STATEMENTS IN CF 4 ARE THE SAME AS IN CF 1
       EQUIVALENCE (COM1(1),AVATR(1)), (COM1(106),ITLOC(1)),
      1            (COM1(321),IOMIT(1))
       EQUIVALENCE (COM2(1),DATA(1))
       EQUIVALENCE (COM3(1),FMT(1))
     4 IA = IA
       IF(IDENT - IEND ) 4,99,4
    99 RETURN
       END
```
                          Figure 3

Col. 42. Save or Reuse Data.

0 (or blank) Do not save data for next problem.

1 Save data for next problem.

2 Reuse data from previous problem.

Cols. 43 and 44. Blank.

Col. 45. Number of Format Cards If Greater than 1 (8 Max.).

Cols. 46 and 47. Number of Information Cards. If desired, the user can
specify the number of 72-column information cards that are to be read for

display on the computer printout. This information can include the form of equation used, a statement of the purpose of the run, or other pertinent information that will help identify the printout in the future. The identification cards follow the transformation cards.

Col. 48. Names of Variables.

0 (or blank) Do not read names of variables for display on printout.

1 Read names of variables from cards. The names or abbreviations of the names of the variables, independent and dependent, are keypunched in the first six positions of fields in the same locations as the variables in the transformation cards. Which variables are listed and the order in which they are placed depends on the transformations and omission of variables in the problem. If no transformations are made, all variable names are displayed. The variable name cards follow the identification cards. English or alphanumeric code names aid in identifying both the coefficients of the fitted equation and the variables in each of the candidate C_p-search equations.

2 Reuse names of variables from previous problem. Again, which names are displayed depends on the transformations and the omission of variables in the problem.

Col. 49. Search for Candidate Equations via C_p—Enter 1. The program automatically searches for the combination of variables whose C_p-values are less than those of the full equation. If any are found, the candidate equations are listed and their C_p-values plotted versus p (the number of coefficients). A maximum of 12 variables (4096 equations) can be searched at one time. If more than 12 variables remain after a basic set has been chosen, a factorial search is required to find all possible candidate equations.

Col. 50. Number of C_p Fractional Factorial Search Cards Read for Each Dependent Variable.* For a factorial search, the identification numbers of the variables to be searched are listed numerically in ascending order on a factorial search card with a 12(2x, I2) format. All variables not listed are included in the basic equation. The first search card follows the variable name cards. Additional factorial search cards may be placed after the END card(s) of the data set. Each dependent variable requires its own factorial search card(s). An entry in col. 49 is *not* needed to request a factorial search.

Col. 51. Calculate the Component Effect of Each Variable on Each Observation and Estimate the Standard Deviation of y from Near Neighbors —Enter 1. The component effect of each variable on each observation is listed in units of Y. The variables are ordered by their relative influences in the full equation. The observations in turn are ordered by the magnitude of the effect of the most influential variable.

The program also determines the weighted squared standardized distance

*After transformations.

between adjacent observations (ordered by the most influential variable), and if the distance is less than one, squares the difference between the y-values of these two near-neighbor observations. The sum of these squared differences provides an estimate of the standard deviation of y.

B. Standard data card entry

Without a special format card, up to ten variables can be entered on one data card. Therefore, the number of cards read per observation must equal the sum of the number of independent and dependent variables, either before or after transformations (whichever is larger), divided by 10; for example, before transformations 4 ind. $+$ 2 dep. $=$ 6, after transformations 7 ind. $+$ 4 dep. $=$ 11, $11/10 = 1.1$ or 2 cards per observation are required.

The data entry card allows six card columns for each variable. If the decimal point is not entered, it is understood to lie between the third and fourth card columns of each variable field. If more than three digits are required before or after the decimal point, the point must be entered in one of the six spaces. Negative numbers are designated by a minus sign preceding the number as in standard algebraic notation. Plus signs need not be entered. Blanks are interpreted as zeros, whether or not the position is significant.

Normally the data card entry fields are interpreted by the program as numbers of the following magnitude:

-50.423 is entered as	$-$	5	0	4	2	3
121.242 is entered as	1	2	1	2	4	2
50.0 is entered as		5				
-0.4 is entered as			$-$	4		
-0.1424 is entered as	$-$	.	1	4	2	4
$5000.$ is entered as	5	0	0	0	.	
-121.242 is entered as	$-$	1	2	1	.	2

Thus, scaling is accomplished automatically when the decimal point is entered.

Column entries of a data card are as follows:

Columns	Entry
1–6	Card or observation identification
7–10	Observation number
11–12	Sequence number of card within observation
13–18, etc.	Data
73–80	Not used by program. May be used for identification of data sets.

The identification contained in cols. 1–6 may be any combination of letters, numbers, and blanks.

The observation number entered in cols. 7–10 and the sequence number in cols. 11 and 12 *must* be numeric. If no transformations are to be used, independent variables must precede the dependent variables as data card entries. If transformations are being done, there are different restrictions (see below).

Weighting factor, if any, must be the last data entry of an observation sequence.

As an example, suppose that data on four independent and two dependent variables are to be read in, and no transformations or omissions are desired. The independent variables must occupy the fields headed x_1, x_2, x_3, and x_4 on the data card entry sheet; the order of the four variables relative to each other is immaterial. The dependent variables must occupy the x_5 and x_6 fields; x_5 will be identified on the printout as "Dependent Variable 1," and x_6 as "Dependent Variable 2." If the weighting option is used, a weighting factor must appear on *every* data set, in the field to the right of the last dependent variable. In the above example, weighting factors, if used, would appear in field 7. They are not included in the variable count.

The number of observations is limited to 1000. The last card of every problem must be a card with END punched in cols. 1–3.

C. Problems using transformation option

Transformation and data cards

In trying to discover what relationships exist in a body of data, or how to simplify relationships which are already apparent, it is frequently desirable to make simple transformations of the original variables. New forms such as sums, differences, products, ratios, logarithms, or exponentials of variables with other variables or constants can be generated by the program, thus saving time and avoiding arithmetical mistakes.

Whether the program regards a certain field of the data cards as an independent or a dependent variable can be controlled by suitably arranged control and transformation cards. If desired, a "dependent" variable may be combined with one or more of the other independent or dependent variables to develop new variables against which the original dependent variable is to be correlated.

When the mass of data is large, the data need be punched into a set of data cards only once. In running a series of problems using the same data cards (but different control and transformation cards), the individual variables may be worked over and combined or omitted in any desired pattern. *All must use the same format.*

Each transformation card can carry instructions for transformations on as many as seven variables (see transformation card keypunch sheet, shown in Figure 1). The directions for one transformation occupy a field of ten digits. Thus the seven transformations occupy cols. 1–70 of a transformation card; cols. 71–80 are available for sorting or identification.

Each of the operation fields applies to a specific variable whose data entries are in corresponding locations on the data cards. Thus the first transformation card applies to variables 1–7; the second, to 8–14; etc.

Each ten-digit transformation field is divided into four subfields:

Columns of Transformation Field	Column Heading	Subfield Description
1–2	C	Transformation code
3–7	ARG.	Transformation argument
8–9	LOC.	Where to put results
10	ϕ	"Omit" instruction

For a printout of transformed data without a fit of the data, 1 is placed in col. 25, 1 in col. 27 ($b_0 = 0$), and 0 in col. 33, and a blank field is indicated for the independent variable.

The transformations included are as follows:

Code	Transformation
0, blank	None
1	Natural log
2	Common log
3	Antilog of natural log
4	Antilog of common log
5	Exponentiation by indicated entry
6	Multiplication by indicated entry
7	Division by indicated entry
8	Algebraic addition by indicated entry
9	Replacement of variable by product of variables M and N
10	Replacement of variable by ratio of variables M and N
11	Replacement of variable by sum of variables M and N
12	Replacement of variable by difference of variables M and N
13	Replacement of variable by sine of variable, in radians
14	Replacement of variable by cosine of variable, in radians
15	Indicated entry divided by variable
16	Natural log of natural log of variable

17   Common log of common log of variable
18   Common log (indicated entry added to variable)
19   Antilog of antilog of natural log
20   Antilog of antilog of common log

The program is not capable of rounding either input data or transformed data.

The transformation code digits are entered into cols. 1 and 2 of the transformation field, right-justified. Zeros may be omitted.

The transformation argument field occupying cols. 3–7 of the transformation field may contain three different kinds of entries: blanks, constants, or variable subscripts.

Transformation codes 1–4 and 13 and 14 indicate log, antilog, or trigonometric transformations on the appropriate variable. These codes require only the variable itself as an argument and thus require no entry in the argument subfield. For example, to replace x_1 by $\log_e x_1$:

$$x_1$$

C	ARG	LOC	ϕ
1			

Transformation codes 5–8 indicate operations performed on the appropriate variables by a constant entered in the ARG subfield. Entries in the ARG subfield must be entered right-justified. Decimal values must include a decimal point. Some examples of transformation codes using constants as arguments are as follows:

(1) Replace x_1 by x_1^2. (*Note:* If the 2 had been placed in the 5th position a decimal would not have been required, as the format of the field is F5.0.)

$$x_1$$

C	ARG	LOC	ϕ
5	2 .		

(2) Replace x_3 by $x_3/1.5$.

$$x_3$$

C	ARG	LOC	ϕ
7	1 . 5		

(3) Replace x_8 by $x_8 + (-5.2)$.

Second card:

		x_8		
C	ARG		LOC	ϕ
8	$-$	5 . 2		

In any of the preceding cases, the transformed result can be moved to another position in the series of variables by writing the location, right justified, under LOC.

Transformation codes 9–12 indicate an operation performed using two variables to develop a transformed variable. The values inserted in the argument subfields are the subscripts of the variables to be used in the indicated operation; the value inserted in LOC tells where the result should be placed in the user's series of variables.

If one of the two variables entering the transformation is to be omitted from a particular equation, the transformation may be written in that field of the transformation card entry, with a 1 in the ϕ column. As an alternative, the transformation may be written elsewhere; the omission can be accomplished by a no-operation code in the desired field, with a 1 in the ϕ column. An omission code cannot be made to refer to any variable other than the one in whose field it appears. The omissions are not made until after all transformations have been completed; for example, an omit code in field 2 does not prevent the use of x_2 in a transformation appearing in its own or any other field.

Some examples of transformation codes using variable subscripts as arguments are as follows:

(1) Form product x_1 times x_2 and place result at x_4.

	x_1			
C	ARG		LOC	ϕ
9	1	2	4	

(2) Form product x_3 times x_5 and place result at x_5; omit x_3 from the equation. (Note that space is always available for two-digit subscripts. Single-digit subscripts should be right-justified as shown.)

	x_3			
C	ARG		LOC	ϕ
9	3	5	5	1

(3) Form x_4/x_2 and place at x_4. Include x_2 in the equation. (Note that numerator is left entry; divisor is right entry.)

x_2

C		ARG			LOC		ϕ
1	0		4		2		4

(4) Form sum of x_{13} and x_{14} and place at x_{15}.

C		ARG				LOC		ϕ
1	1		1	3	1	4	1	5

(5) Form x_3-x_9 and place at x_{10}. (Note that minuend is left entry, subtrahend right entry.)

x_{10}

C		ARG			LOC		ϕ
1	2		3		9		

LOC may be left blank if the transformation is written in field 10, which is the third field of the second transformation card.

When transformations are to be done in a problem, a 1 must appear in col. 25 of the control card. This tells the program to look for transformation cards following the control card, and to figure out how many such cards to expect. It does this by adding up the total number of independent and dependent variables listed in the control card, dividing by seven operations per card, and taking the next larger whole number. For example, if there are four independent and two dependent variables, there will be one transformation card; but if there are six independent and two dependent variables, there will be two transformation cards. If the number of transformation cards actually placed in the deck is either greater or less than the expected number, the calculated results will be garbled, because the program will be trying to read transformation cards as data cards, or vice versa. Thus, in the second example above, even if only the first three independent variables are to be transformed, two transformation cards must be put in; the first one carries the three transformations, and the second one is blank. When a certain field on a transformation card is blank, the program assumes that no transformation of the corresponding variable is desired.

In problems large enough to approach the capacity of the program (105 input variables, 80 to be used in the equation), the above arrangement limits the number of transformations to an average of about one per variable. In smaller problems, transformations requiring several steps can be done by having the results of intermediate steps listed as "variables" and later omitting them from the equation.

When the transformation option is used, none of the input tells the program directly how many independent or dependent variables are to be included in the calculations. The program develops this information by starting with the number of independent or dependent variables listed in the control card, and subtracting the specified omissions. The various types of situation that may occur are best understood by means of examples.

1. The user has placed data for six variables in fields 1–6 of a set of data cards. He wants to use 1, 2, and 4 as independent variables, and 5 as a single dependent variable, all without any transformations. In the control card he lists 4 independent variables and 1 dependent variable, and enters a 1 in col. 25, calling for transformations. In the transformation card entry, he orders omission of x_3. In the printed output the program reports under "Data Transformations" that "the equation has 3 independent variables and 1 dependent variable"; the transformation instructions will be listed as written, followed by a repetition of the data for x_1, x_2, x_4, and x_5 in the first four columns, x_3 and x_6 having been omitted.

2. With the same raw data as before, the user now wishes to consider x_2 and x_5 as independent variables, and x_3 as the dependent variable. He can use the old data cards by listing "6 independent, 1 dependent" on the control card, and entering the following transformations:

x_1 Mark for "omit"
x_2 No operation (leave blank)
x_3 Move to location 7 and "omit" in 3
x_4 Mark for "omit"
x_5 Leave blank
x_6 Mark for "omit"
x_7 Leave blank

The program, having been told that data fields 1–6 are now reserved for six independent variables, and field 7 for a dependent variable, and that x_1, x_3, x_4, and x_6 are to be omitted, will report the transformed equation as "2 independent and 1 dependent" and proceed, using field 7 (which now contains the data from field 3) as the dependent variable, and cols. 2 and 5 as the independents. The user should note that, when a transformation in field 3 calls for "move to field 7 and omit," it is x_3 that will be omitted, not x_7, which can be omitted only by an omit instruction in field 7.

3. The user has data for five independent variables $x_1 \cdots x_5$ and a dependent variable x_6. He wants to perform a fit of x_6 against x_1, x_2, $x_3 - x_4$, and $x_5 - 100$. On the data sheet, he enters $x_1 \cdots x_6$ in the first six fields. The

control card will list "5 independents, 1 dependent." Transformations will be as follows:

x_1 No operation (leave blank)

x_2 No operation

x_3 Take $x_3 - x_4$

x_4 Mark for "omit"

x_5 Take $x_5 - 100$, leave "location" blank

x_6 No operation

The program reports the transformed equation as "4 independent, 1 dependent." The transformed data x_1, x_2, $x_3 - x_4$, $x_5 - 100$, x_6 will appear in the first five columns.

4. The user has data for $x_1 \cdots x_5$. He wants to perform a fit of x_4/x_5 as dependent variable against $x_3(x_1 - x_2)$ as independent variable. On the data sheet, he enters $x_1 \cdots x_5$ in the first five fields. The control card will list "6 independent, 1 dependent." Transformations will be as follows:

x_1 Take $x_1 - x_2$, place in LOC 6, omit

x_2 Mark for "omit"

x_3 Take x_3 times x_6, place in LOC 6, omit

x_4 Take x_4/x_5, LOC 7, omit

x_5 Mark for "omit"

x_6 No operation

x_7 No operation

The program reports that the fitted equation has one independent and one dependent variable.

The examples illustrate that, when a control card lists seven independent and three dependent variables, the program assumes that:

1. The combination of raw data and transformations is such that, after the transformations are completed, there will be data in the first ten fields, with independent variables in fields 1–7, and dependents in 8–10.

2. The number of independent variables to be used is 7 minus the total omissions listed for fields 1–7.

3. The number of dependent variables is 3 minus the omissions listed for fields 8–10.

Most problems can be handled in more than one way, since a transformed variable can either replace one of the variables from which it came or, if space is available, go to a new field. When a variable is to be omitted, its position in the series identifies it as either independent or dependent; the program adjusts the makeup of the equation accordingly.

Order of cards

First Problem

1. Control card   (0 or blank in col. 42)
2. Format card(s)   (If special format is used, 1 col. 41)
3. Transformation card(s)   (If transformations are used, 1 col. 25)
4. Information card(s)   (If desired, number in cols. 46 and 47)
5. Variable name card(s)   (If desired, 1 col. 48)
6. Factorial search card   (If desired, number in col. 50)
7. Data cards
8. END card(s)   (END in cols. 1–3 of data card identification. The number of END cards *must* equal the number of data cards per observation.)

9. Additional factorial
search card(s)   (If desired, number in col. 50)

Second Problem (Different Data)

1. Control card   (1 in col. 42)
2. Format card(s)   (If special format is used, 1 col. 41)
3. Transformation card(s)   (If transformations are used, 1 col. 25)
4. Information card(s)   (If desired, number in cols. 46 and 47)
5. Variable name card(s)   (If desired, 1 col. 48)
6. Factorial search card   (If desired, number in col. 50)
7. Data cards
8 END card(s)   (END in cols. 1–3 of data card identification. The number of END cards *must* equal the number of data cards per observation.)

Third Problem (Same Data as in Second Problem)

1. Control card   (2 in col. 42)
2. Transformation card(s)   (If transformations are used, 1 col. 25)
3. Information card(s)   (If desired, number in cols. 46 and 47) (Enter 2 in col. 48 if variable names are to be reused.)
4. Factorial search card   (If desired, number in col. 50)

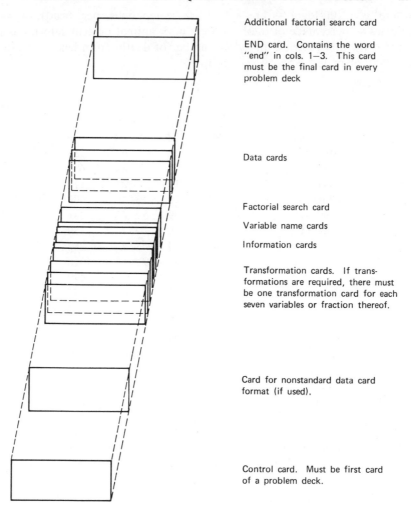

Additional factorial search card

END card. Contains the word "end" in cols. 1–3. This card must be the final card in every problem deck

Data cards

Factorial search card

Variable name cards

Information cards

Transformation cards. If transformations are required, there must be one transformation card for each seven variables or fraction thereof.

Card for nonstandard data card format (if used).

Control card. Must be first card of a problem deck.

Figure 4   Order of cards

Sample problem*

The accompanying table gives, for various countries, data on death rate due to heart disease in males in the 55–59 age group, along with the proportionate number of telephones and the calories of fat and protein in the diet. The fitting equation is

$$Y = b_0 + b_1 x_1 + b_2 x_2 + b_3 x_3,$$

* From K. A. Brownlee (pages 462–464).

where $y = 100(\log y' - 2)$, $x_1 = 1000$ (telephones per head), $x_2 = $ fat calories as percentage of total calories, $x_3 = $ animal protein calories as percentage of total calories, and $y' = $ number of deaths from heart disease per 100,000 for males in 55–59 age group.

	x_1	x_2	x_3	y'
Australia	124	33	8	646
Austria	49	31	6	355
Canada	181	38	8	631
Ceylon	4	17	2	174
Chile	22	20	4	603
Denmark	152	39	6	331
Finland	75	30	7	759
France	54	29	7	282
Germany	43	35	6	316
Ireland	41	31	5	490
Israel	17	23	4	457
Italy	22	21	3	282
Japan	16	8	3	174
Mexico	10	23	3	269
Netherlands	63	37	6	240
New Zealand	170	40	8	525
Norway	125	38	6	257
Portugal	15	25	4	240
Sweden	221	39	7	331
Switzerland	171	33	7	331
United Kingdom	97	38	6	457
United States	254	39	8	776

The control and transformation card entry form and the data card entry form used for this problem are shown in Figure 5 and Figure 6. Following Brownlee, the death rate data were transformed by the formula $100(\log y' - 2)$, before fitting the equation. The three transformations required are shown in fields 4, 5, and 6. The resulting dependent variable is left in field 6, while the intermediate results in 4 and 5 are omitted. The independent variables x_1, x_2, and x_3 did not require any transformations or omissions. In the control card, the numbers of independent and dependent variables were listed as 3 and 3.

Computer printout from this problem are shown in Figures 7–12.

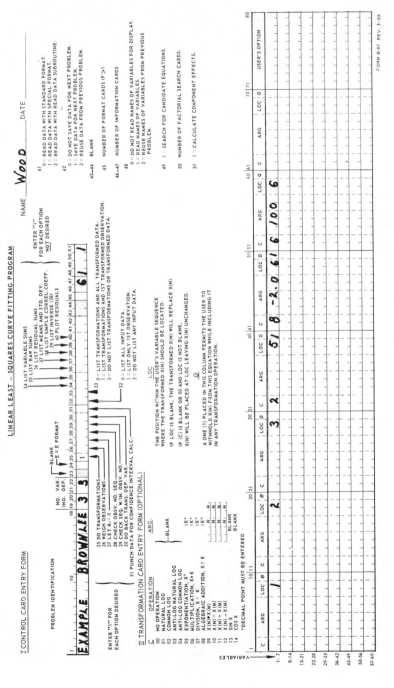

Figure 5

LINEAR LEAST – SQUARES CURVE FITTING PROGRAM

III STANDARD DATA-CARD ENTRY FORM

ENTER WEIGHTING FACTOR, IF ANY, AS LAST ENTRY
"END" CARD MUST BE LAST CARD.

IDENT.	OBSV. NO.	SEQ. NO.	X1-11-21	X2-12-22	X3-13-23	X4-14-24	X5-15-25	X6-16-26	X7-17-27	X8-18-28	X9-19-29	X10-20-30	USER'S OPTION
AUSTRL	1		1.24	.33	.8	.646							
ASTRIA	2		.49	.31	.6	.355							
CANADA	3		1.81	.38	.8	.631							
CEYLON	4		.4	.17	.2	.174							
CHILE	5		.22	.20	.4	.603							
DENMK	6		1.52	.39	.6	.331							
FINLND	7		.75	.30	.7	.759							
FRANCE	8		.54	.29	.7	.282							
GMANY	9		.43	.35	.6	.316							
IRELND	10		.41	.31	.5	.490							
ISRAEL	11		.17	.23	.4	.457							
ITALY	12		.22	.21	.3	.282							
JAPAN	13		.16	.8	.3	.174							
MEXICO	14		.10	.23	.3	.269							
NTHLDS	15		.63	.37	.6	.240							
NZEAL	16		1.70	.40	.8	.525							
NORWAY	17		1.25	.38	.6	.257							
PORTGL	18		.15	.25	.4	.240							
SWEDEN	19		2.21	.39	.7	.331							
SNITED	20		1.71	.33	.7	.331							
UNKING	21		.97	.38	.6	.457							
U.S.A.	22		2.54	.39	.8	.776							
END													

FORM B-55 REV. 3-69

Figure 6

300

EXAMPLE   BROWNLEE

EQUATION 1 OF A MULTI-EQUATION PROBLEM

DATA READ WITH STANDARD FORMAT (A6, I4, I2, 10F6.3)

BZERO = CALCULATED VALUE

DATA INPUT   3 INDEPENDENT VARIABLES   3 DEPENDENT VARIABLES

OBSV.	SEQ.	1-11-21	2-12-22	3-13-23	4-14-24	5-15-25	6-16-26	7-17-27	8-18-28	9-19-29	10-20-30
1	1	124.000	33.000	8.000	646.000	0.0	0.0				
2	1	49.000	31.000	6.000	355.000	0.0	0.0				
3	1	181.000	38.000	8.000	631.000	0.0	0.0				
4	1	4.000	17.000	2.000	174.000	0.0	0.0				
5	1	22.000	20.000	4.000	603.000	0.0	0.0				
6	1	152.000	39.000	6.000	331.000	0.0	0.0				
7	1	75.000	30.000	7.000	759.000	0.0	0.0				
8	1	54.000	29.000	7.000	282.000	0.0	0.0				
9	1	43.000	35.000	6.000	316.000	0.0	0.0				
10	1	41.000	31.000	5.000	490.000	0.0	0.0				
11	1	17.000	23.000	4.000	457.000	0.0	0.0				
12	1	22.000	21.000	3.000	282.000	0.0	0.0				
13	1	16.000	8.000	3.000	174.000	0.0	0.0				
14	1	10.000	23.000	3.000	269.000	0.0	0.0				
15	1	63.000	37.000	6.000	240.000	0.0	0.0				
16	1	170.000	40.000	8.000	525.000	0.0	0.0				
17	1	125.000	38.000	6.000	257.000	0.0	0.0				
18	1	15.000	25.000	4.000	240.000	0.0	0.0				
19	1	221.000	39.000	7.000	331.000	0.0	0.0				
20	1	171.000	33.000	7.000	331.000	0.0	0.0				
21	1	97.000	38.000	6.000	457.000	0.0	0.0				
22	1	254.000	39.000	8.000	776.000	0.0	0.0				

DATA TRANSFORMATIONS

POSITION	CODE	OPERATION	CONSTANT	LOCATION	OMIT	VARIABLE
1		NONE		1	0	1
2		NONE		2	0	2
3		NONE		3	0	3
4	2	COMMON LOG		5	1	
5	8	ADD CONSTANT	-2.	6	1	
6	6	MULTIPLY BY CONSTANT	100.	6	0	4

Figure 7

DATA AFTER TRANSFORMATIONS   THE FITTED EQUATION HAS 3 INDEPENDENT VARIABLES, 1 DEPENDENT VARIABLES

	TELEPH	F. CAL	AP.CAL	DEATHS						
OBSV.	1-11-21	2-12-22	3-13-23	4-14-24	5-15-25	6-16-26	7-17-27	8-18-28	9-19-29	10-20-30
1	1.240000 02	3.300000 01	8.000000 00	8.10233D 01						
2	4.900000 01	3.100000 01	6.000000 00	5.50228D 01						
3	1.810000 02	3.800000 01	8.000000 00	8.00029D 01						
4	4.000000 00	1.700000 01	2.000000 00	2.40549D 01						
5	2.200000 01	2.000000 01	4.000000 00	7.80317D 01						
6	1.520000 02	3.900000 01	6.000000 00	5.19828D 01						
7	7.500000 01	3.000000 01	7.000000 00	8.80242D 01						
8	5.400000 01	2.900000 01	7.000000 00	4.50249D 01						
9	4.300000 01	3.500000 01	6.000000 00	4.99687D 01						
10	4.100000 01	3.100000 01	5.000000 00	6.90196D 01						
11	1.700000 01	2.300000 01	4.000000 00	6.59916D 01						
12	2.200000 01	2.100000 01	3.000000 00	4.50249D 01						
13	1.600000 01	8.000000 00	3.000000 00	2.40549D 01						
14	1.000000 01	2.300000 01	3.000000 00	4.29752D 01						
15	6.300000 01	3.700000 01	6.000000 00	3.80211D 01						
16	1.700000 02	4.000000 01	8.000000 00	7.20159D 01						
17	1.250000 02	3.800000 01	6.000000 00	4.09933D 01						
18	1.500000 01	2.500000 01	4.000000 00	3.80211D 01						
19	2.210000 02	3.900000 01	7.000000 00	5.19828D 01						
20	1.710000 02	3.300000 01	7.000000 00	5.19828D 01						
21	9.700000 01	3.800000 01	6.000000 00	6.59916D 01						
22	2.540000 02	3.900000 01	8.000000 00	8.89862D 01						

SUMS OF VARIABLES
    1.92600D 03 6.67000D 02 1.24000D 02 1.24820D 03

RESIDUAL SUMS OF SQUARES + CROSS PRODUCTS
    1.19455D 05
    1.04452D 04 1.58477D 03
    2.37036D 03 2.82545D 02 7.30909D 01
    1.43159D 04 1.56916D 03 4.69468D 02 7.82405D 03

MEANS OF VARIABLES
    8.75455D 01 3.03182D 01 5.63636D 00 5.67362D 01

ROOT MEAN SQUARES OF VARIABLES
    7.54212D 01 8.68708D 00 1.86562D 00 1.93022D 01

SIMPLE CORRELATION COEFFICIENTS, R(I,I PRIME)

1	1.000	0.759	0.802	0.468
2	0.759	1.000	0.830	0.446
3	0.802	0.830	1.000	0.621
4	0.468	0.446	0.621	1.000

INVERSE, C(I,I PRIME)

3.044	-0.913	-1.684
-0.913	3.491	-2.166
-1.684	-2.166	4.149

Figure 8

301

LINEAR LEAST-SQUARES CURVE FITTING PROGRAM

EXAMPLE   BROWNLEE DEP VAR 1: DEATHS   MIN Y = 2.405D 01   MAX Y = 8.899D 01   RANGE Y = 6.493D 01

        Y = B(0) + B(1)X1 + B(2)X2 + B(3)X3
        Y = 100(LOG(NUMBER OF DEATHS FROM HEART DISEASE PER 100,000 FOR MALES
                   IN THE 55 TO 59 AGE GROUP) - 2)
        X1 = 1000(TELEPHONES PER PERSON)
        X2 = FAT CALORIES AS PER CENT OF TOTAL CALORIES
        X3 = ANIMAL PROTEIN CALORIES AS PER CENT TOTAL CALORIES

IND.VAR(I)	NAME	COEF.B(I)	S.E. COEF.	T-VALUE	R(I)SQRD	MIN X(I)	MAX X(I)	RANGE X(I)	REL.INF.X(I)
0		2.39806D 01							
1	TELEPH	-6.87054D-03	8.14D-02	0.1	0.6715	4.000D 00	2.540D 02	2.500D 02	0.03
2	F. CAL	-4.80840D-01	7.57D-01	0.6	0.7136	8.000D 00	4.000D 01	3.200D 01	0.24
3	AP.CAL	8.50465D 00	3.84D 00	2.2	0.7590	2.000D 00	8.000D 00	6.000D 00	0.79

NO. OF OBSERVATIONS	22
NO. OF IND. VARIABLES	3
RESIDUAL DEGREES OF FREEDOM	18
F-VALUE	4.0
RESIDUAL ROOT MEAN SQUARE	16.13185806
RESIDUAL MEAN SQUARE	260.23684450
RESIDUAL SUM OF SQUARES	4684.26320100
TOTAL SUM OF SQUARES	7824.04773566
MULT. CORREL. COEF. SQUARED	.4013

--------------------ORDERED BY COMPUTER INPUT--------------------						---------------ORDERED BY RESIDUALS---------------				
IDENT.	OBSV.	WS DISTANCE	OBS. Y	FITTED Y	RESIDUAL	OBSV.	OBS. Y	FITTED Y	ORDERED RESID.	SEQ
AUSTRL	1	2.	81.023	75.298	5.725	5	78.032	48.231	29.800	1
ASTRIA	2	0.	55.023	59.766	-4.743	7	88.024	68.573	19.451	2
CANADA	3	2.	80.003	72.502	7.501	11	65.992	46.823	19.169	3
CEYLON	4	4.	24.055	32.788	-8.733	10	69.020	51.316	17.703	4
CHILE	5	1.	78.032	48.231	29.800	22	88.986	71.520	17.466	5
DENMK	6	0.	51.983	55.211	-3.229	21	65.992	56.070	9.921	6
FINLND	7	1.	88.024	68.573	19.451	3	80.003	72.502	7.501	7
FRANCE	8	1.	45.025	69.198	-24.173	12	45.025	39.246	5.779	8
GMANY	9	0.	49.969	57.884	-7.915	1	81.023	75.298	5.725	9
IRELND	10	0.	69.020	51.316	17.703	14	42.975	38.367	4.609	10
ISRAEL	11	1.	65.992	46.823	19.169	16	72.016	71.616	0.400	11
ITALY	12	2.	45.025	39.246	5.779	6	51.983	55.211	-3.229	12
JAPAN	13	2.	24.055	45.538	-21.483	2	55.023	59.766	-4.743	13
MEXICO	14	2.	42.975	38.367	4.609	18	38.021	45.875	-7.854	14
NTHLDS	15	0.	38.021	56.785	-18.763	9	49.969	57.884	-7.915	15
NWZEAL	16	2.	72.016	71.616	0.400	4	24.055	32.788	-8.733	16
NORWAY	17	0.	40.993	55.878	-14.884	19	51.983	63.242	-11.259	17
PORTGL	18	1.	38.021	45.875	-7.854	20	51.983	66.471	-14.488	18
SWEDEN	19	1.	51.983	63.242	-11.259	17	40.993	55.878	-14.884	19
SWITZD	20	1.	51.983	66.471	-14.488	15	38.021	56.785	-18.763	20
UNKING	21	0.	65.992	56.070	9.921	13	24.055	45.538	-21.483	21
U.S.A.	22	2.	88.986	71.520	17.466	8	45.025	69.198	-24.173	22

Figure 9

302

LINEAR LEAST-SQUARES CURVE FITTING PROGRAM

EXAMPLE   BROWNLEE   DEP VAR 1: DEATHS

COMPONENT EFFECT OF EACH VARIABLE ON EACH OBSERVATION (IN UNITS OF Y)
(VARIABLES ORDERED BY THEIR RELATIVE INFLUENCE --- OBSERVATIONS ORDERED BY INFLUENCE OF MOST INFLUENTIAL VARIABLE)

| | | VARIABLES | | |
| | | 3 | 2 | 1 |
SEQ.	OBSV.	AP.CAL	F. CAL	TELEPH
1	1	20.10	-1.29	-0.25
2	3	20.10	-3.69	-0.64
3	16	20.10	-4.66	-0.57
4	22	20.10	-4.17	-1.14
5	7	11.60	-0.15	0.09
6	8	11.60	0.63	0.23
7	19	11.60	-4.17	-0.92
8	20	11.60	-1.29	-0.57
9	2	3.09	-0.33	0.26
10	6	3.09	-4.17	-0.44
11	9	3.09	-2.25	0.31
12	15	3.09	-3.21	0.17
13	17	3.09	-3.69	-0.26
14	21	3.09	-3.69	-0.06
15	10	-5.41	-0.33	0.32
16	5	-13.92	4.96	0.45
17	11	-13.92	3.52	0.48
18	18	-13.92	2.56	0.50
19	12	-22.42	4.48	0.45
20	13	-22.42	10.73	0.49
21	14	-22.42	3.52	0.53
22	4	-30.93	6.40	0.57

STANDARD DEVIATION ESTIMATED FROM NEIGHBORING OBSERVATIONS (OBSERVATIONS 1 TO 4 APART IN FITTED Y ORDER).

| | CUMULATIVE | ORDERED BY WSSD | | ----------ORDERED BY FITTED Y---------- | | | | |
NO.	STD DEV	WSSD	DEL OBS.Y	WSSD	DEL OBS.Y	FITTED Y	OBSV.	SEQ.
1	22.16	0.00	25.00	0.31	18.92	32.79	4	1
2	30.14	0.00	43.00	0.00	2.05	38.37	14	2
3	23.34	0.00	10.99	0.15	20.97	39.25	12	3
4	23.71	0.00	27.97	0.53	13.97	45.54	13	4
5	21.45	0.00	14.01	0.00	27.97	45.88	18	5
6	18.31	0.00	2.97	0.01	12.04	46.82	11	6
7	16.83	0.00	8.98	0.39	9.01	48.23	5	7
8	16.61	0.00	16.97	0.34	17.04	51.32	10	8
9	17.52	0.00	27.97	0.00	10.99	55.21	6	9
10	16.48	0.00	7.99	0.00	25.00	55.88	17	10
11	15.14	0.00	2.05	0.00	27.97	56.07	21	11
12	14.76	0.00	11.95	0.00	11.95	56.78	15	12
13	14.58	0.00	13.96	0.01	5.05	57.88	9	13
14	14.30	0.01	12.04	0.34	3.04	59.77	2	14
15	14.30	0.01	16.02	0.03	0.0	63.24	19	15
16	13.90	0.01	8.98	0.01	36.04	66.47	20	16
17	14.96	0.01	36.04	0.37	43.00	68.57	7	17
18	14.38	0.01	5.05	0.00	43.96	69.20	8	18
19	13.72	0.02	2.01	0.00	16.97	71.52	22	19
20	13.34	0.02	6.96	0.00	7.99	71.62	16	20
21	14.39	0.02	40.01	0.02	1.02	72.50	3	21
22	13.78	0.02	1.02	0.29	20.97	75.30	1	22

Figure 10

303

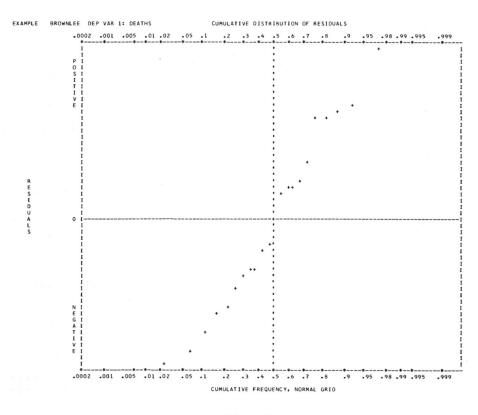

EXAMPLE   BROWNLEE DEP VAR 1: DEATHS   CUMULATIVE DISTRIBUTION OF RESIDUALS

CUMULATIVE FREQUENCY, NORMAL GRID

Figure 11

304

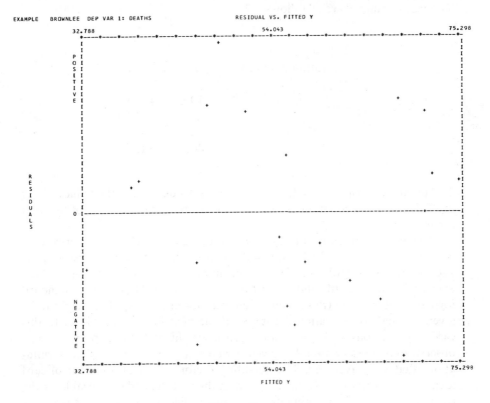

Figure 12

Example of weighted observations*

In this relatively simple problem, four samples have been titrated. Single determinations have been obtained on the first, third, and fourth sample; four determinations have been averaged to give a value for the second sample. The measurements are as follows:

Sample and Observation Number	x	y
1	10	33.8
2	20	62.2
3	30	92.0
4	40	122.4

The fitted equation is $Y = b_0 + bx$. We wish to determine the values of the coefficients of the equation, as well as a measure of the variance of titration error.

The control card entry form, with a 1 in col. 26 for weighted observations, is shown in Figure 13. In Figure 14 the samples have been identified in cols. 1–6 of each data card, the observation numbers in cols. 7–10, the x-values entered in cols. 13–18, and the y-values in cols. 19–24 so that a standard format could be used (this is convenient but not necessary). The weight to be given to each y-observation is entered in the next field, cols. 25–30. In this case, observations 1, 3, and 4 are given a weight of 1.0 to represent single observations. Observation 2 is given a weight of 4.0 for the four determinations that were averaged. The weighting factor switch on the control card identifies the values in the field following the last dependent variable as the weighting factor. These weights are then applied to all of the dependent variables that are present.

The printout is shown in Figures 15 and 16.

The residual mean square is a measure of fit, but because of the weighting factor sample 2 contributed only $\frac{1}{4}$ to the variance of titration error. Whenever possible, more is learned by fitting all of the values at once (include all determinations of sample 2 rather than its average). The plot of the residuals will show the distribution of replicates about the fitted equation. If one value influences the mean, and hence the fit, excessively, it can be seen. In addition, a better overall measure of the residual mean square of the fit is obtained.

* From John Mandel, Section 7.6.

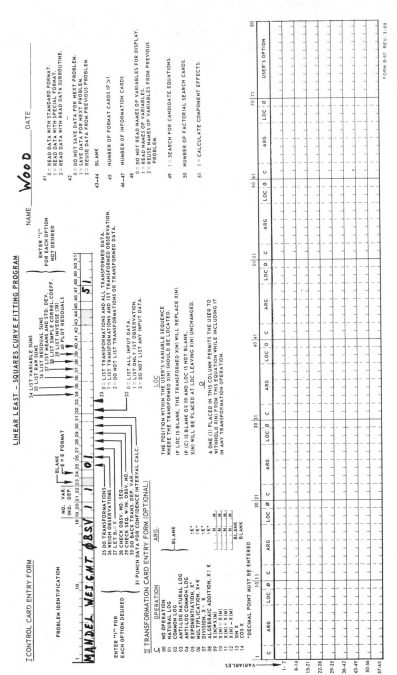

Figure 13

307

III STANDARD DATA-CARD ENTRY FORM

ENTER WEIGHTING FACTOR, IF ANY, AS LAST ENTRY
"END" CARD MUST BE LAST CARD.

IDENT.	OBSV. NO.	SEQ. NO.	X_1-11-21	X_2-12-22	X_3-13-23	X_4-14-24	X_5-15-25	X_6-16-26	X_7-17-27	X_8-18-28	X_9-19-29	X_{10}-20-30	USER'S OPTION
SAMPLE			LEVEL OF X_1		WEIGHT OF OBSERVATION								
1			.10			1.							
2			2.0			4.							
3			3.0			1.							
4			4.0			1.							
LEVEL													
1			10.	33.8	1.								
2			2.0.	62.2	4.								
3			3.0.	92.0	1.								
H			40.	1.22.4	1.								
END													

FORM B-56 REV. 3-69

Figure 14

MANDEL WEIGHT OBSV

PROBLEM HAS ONE EQUATION

DATA READ WITH STANDARD FORMAT (A6, I4, I2, 10F6.3)

BZERO = CALCULATED VALUE

DATA INPUT   1 INDEPENDENT VARIABLES   1 DEPENDENT VARIABLES   WEIGHT OF OBSERVATION

OBSV.	SEQ.	1-11-21	2-12-22	3-13-23	4-14-24	5-15-25	6-16-26	7-17-27	8-18-28	9-19-29	10-20-30
1	1	10.000	33.800	1.000							
2	1	20.000	62.200	4.000							
3	1	30.000	92.000	1.000							
4	1	40.000	122.400	1.000							

SUMS OF VARIABLES
    1.60000D 02 4.97000D 02

RESIDUAL SUMS OF SQUARES + CROSS PRODUCTS
    5.42857D 02
    1.61000D 03 4.77656D 03

MEANS OF VARIABLES
    2.28571D 01 7.10000D 01

ROOT MEAN SQUARES OF VARIABLES
    9.51190D 00 2.82151D 01

SIMPLE CORRELATION COEFFICIENTS, R(I,I PRIME)
    1   1.000   1.000
    2   1.000   1.000

INVERSE, C(I,I PRIME)
    1.000

Figure 15

MANDEL WEIGHT OBSV   DEP VAR 1:   MIN Y = 3.380D 01 MAX Y = 1.224D 02 RANGE Y = 8.860D 01

	SAMPLE	LEVEL OF X	WEIGHT OF OBSERVATION
	1	10	1
	2	20	4
	3	30	1
	4	40	1

IND.VAR(I)	NAME	COEF.B(I)	S.E. COEF.	T-VALUE	R(I)SQRD	MIN X(I)	MAX X(I)	RANGE X(I)	REL.INF.X(I)
0		3.21053D 00							
1	LEVEL	2.96579D 00	3.89D-02	76.3	0.0	1.000D 01	4.000D 01	3.000D 01	1.00

NO. OF OBSERVATIONS   4
NO. OF IND. VARIABLES   1
RESIDUAL DEGREES OF FREEDOM   2
F-VALUE   5826.8
RESIDUAL ROOT MEAN SQUARE 0.90524786
RESIDUAL MEAN SQUARE   0.81947368
RESIDUAL SUM OF SQUARES   1.63894737
TOTAL SUM OF SQUARES   4776.56000000
MULT. CORREL. COEF. SQUARED   .9997

IDENT.	OBSV.	WS DISTANCE	OBS. Y	FITTED Y	RESIDUAL	OBSV.	OBS. Y	FITTED Y	ORDERED RESID.	SEQ
	1	1774.	33.800	32.868	0.932	1	33.800	32.868	0.932	1
	2	88.	62.200	62.526	-0.326	4	122.400	121.842	0.558	2
	3	548.	92.000	92.184	-0.184	3	92.000	92.184	-0.184	3
	4	3154.	122.400	121.842	0.558	2	62.200	62.526	-0.326	4

--------------ORDERED BY COMPUTER INPUT--------------------   --------------ORDERED BY RESIDUALS---------------

Figure 16

309

Example to measure the precision of calculations

Background of problem

In the September 1967 issue of the *Journal of the American Statistical Association* (Vol. 62, No. 319), James W. Longley of the Bureau of Labor Statistics presented the results of a study of the numerical accuracy of various least-squares programs run on a wide range of electronic computers. It had been observed that the results and conclusions drawn by various companies and government agencies using different computers and computer least-squares programs were not consistent. In similar stepwise regression programs different sets of variables had been chosen as important, and even the co-efficients of some important variables had been reported as being opposite in sign.

Thus, the objective of this study was to determine the combined effect of computer precision and program technique on the coefficients of the resulting fitted equation. Other government agencies and some large companies with government contracts were asked to run the same specified problem. This problem, chosen by the Bureau of Labor Statistics, was economic in nature, involving price index, gross national product, unemployment, size of the armed forces, noninstitutional population, and time as the "independent" variables. Eight measures of employment, including total, agricultural, self-employed, family workers, domestic, industrial, and federal and state government workers, were used as the dependent variables.

Comparison of widely used computer programs

In order to check the precision of a computer program, it is necessary to know the correct answer to a sufficient number of digits. In this case a tremendous amount of effort was required to calculate the coefficients of the equations on desk calculators to 12 digits. As examples of the magnitude of the numbers involved, one determinant had 25 digits to the left of the decimal and another 18 digits to the right of the decimal.

As a basis for comparing the accuracy of the calculated coefficients we have chosen the coefficient of the first independent variable, price index, as being representative. This variable is used to measure the effect of inflation on total employment. The "correct" hand-calculated value of this coefficient is +15.061 872 271 373.

In the Bureau of Labor Statistics Study, the percentage of error in the price index coefficient ranged from 0.03% to 375%. As seen in Table 1, four of the nine computer programs did not agree even in the first digit, giving values of −36 and −41 instead of +15. Two were accurate to 1 digit, two to 3 digits, and one to 4 digits. Since accuracy is a prime concern in the

TABLE 1

COMPARISON OF LEAST-SQUARES PROGRAMS AND COMPUTERS
ACCURACY OF PRICE INDEX COEFFICIENT IN TOTAL EMPLOYMENT EQUATION

Program	Computer	Price Index Coefficient	Accuracy, Number of Digits	Per Cent Error
Bureau of Labor Statistics Study				
Hand calculation		15.061 872		
NIPD	IBM 7094	−41.	0	375
Stepwise A	IBM 7094	−36.	0	340
Stepwise B	IBM 7074	2.	0	82
UCLA, BIMD	IBM 7094	27.	0	80
Johns Hopkins	IBM 1401	11.	1	23
Dartmouth (Time-Sharing)	GE 235	13.	1	7
CORRE(SP)	IBM 360	15.00	3	4
CORRE(DP)	IBM 360	15.04	3	1
ORTHO	IBM 7074	15.065	4	0.03
American Oil				
Curve fit(SP)	IBM 7044	15.066	4	0.03
Curve fit(DP)	IBM 360–75	15.061 872	12	10^{-10}
Curve fit(SP)	CDC 6400	15.061 872	10	10^{-8}

use of these techniques, this problem provides a good basis for comparing computer precision and program technique.

The curve fitting single-precision program written for the IBM 7044 computer is accurate on this problem to 4 digits with an error of 0.03%. The curve fitting double-precision program written for the IBM 360–75 computer is accurate to 12 digits. This program adapted for the CDC 6400 computer by Dr. Neil Timm of the University of California, Berkeley, is accurate to 10 digits, using only single precision. To obtain a printout of this precision, alternate formats are provided in the subroutine FIT (CF3), statement numbers 950 and 3012. A printout of this problem is shown in Figures 17–23.

LINEAR LEAST-SQUARES CURVE FITTING PROGRAM

PRECISION - EMPLOY

EQUATION 1 OF A MULTI-EQUATION PROBLEM

DATA READ WITH SPECIAL FORMAT

FORMAT CARD 1 (A6,I4,I2, F6.1,9F6.0/ 12X, 4F6.0)

BZERO = CALCULATED VALUE

DATA INPUT   6 INDEPENDENT VARIABLES   8 DEPENDENT VARIABLES

OBSV.	SEQ.	1-11-21	2-12-22	3-13-23	4-14-24	5-15-25	6-16-26	7-17-27	8-18-28	9-19-29	10-20-30
1947	1	83.000	234289.000	2356.000	1590.000	107608.000	1947.000	60323.000	8256.000	6045.000	427.000
	2	1714.000	38407.000	1892.000	3582.000						
1948	1	88.500	259426.000	2325.000	1456.000	108632.000	1948.000	61122.000	7960.000	6139.000	401.000
	2	1731.000	39241.000	1863.000	3787.000						
1949	1	88.200	258054.000	3682.000	1616.000	109773.000	1949.000	60171.000	8017.000	6208.000	396.000
	2	1772.000	37922.000	1908.000	3948.000						
1950	1	89.500	284599.000	3351.000	1650.000	110929.000	1950.000	61187.000	7497.000	6069.000	404.000
	2	1995.000	39196.000	1928.000	4098.000						
1951	1	96.200	328975.000	2099.000	3099.000	112075.000	1951.000	63221.000	7048.000	5869.000	400.000
	2	2055.000	41460.000	2302.000	4087.000						
1952	1	98.100	346999.000	1932.000	3594.000	113270.000	1952.000	63639.000	6792.000	5670.000	431.000
	2	1922.000	42216.000	2420.000	4188.000						
1953	1	99.000	365385.000	1870.000	3547.000	115094.000	1953.000	64989.000	6555.000	5794.000	423.000
	2	1985.000	43587.000	2305.000	4340.000						
1954	1	100.000	363112.000	3578.000	3350.000	116219.000	1954.000	63761.000	6495.000	5880.000	445.000
	2	1919.000	42271.000	2188.000	4563.000						
1955	1	101.200	397469.000	2904.000	3048.000	117388.000	1955.000	66019.000	6718.000	5886.000	524.000
	2	2216.000	43761.000	2187.000	4727.000						
1956	1	104.600	419180.000	2822.000	2857.000	118734.000	1956.000	67857.000	6572.000	5936.000	581.000
	2	2359.000	45131.000	2209.000	5069.000						
1957	1	108.400	442769.000	2936.000	2798.000	120445.000	1957.000	68169.000	6222.000	6089.000	626.000
	2	2328.000	45278.000	2217.000	5409.000						
1958	1	110.800	444546.000	4681.000	2637.000	121950.000	1958.000	66513.000	5844.000	6185.000	605.000
	2	2456.000	43530.000	2191.000	5702.000						
1959	1	112.600	482704.000	3813.000	2552.000	123366.000	1959.000	68655.000	5836.000	6298.000	597.000
	2	2520.000	45214.000	2233.000	5957.000						
1960	1	114.200	502601.000	3931.000	2514.000	125368.000	1960.000	69564.000	5723.000	6367.000	615.000
	2	2489.000	45850.000	2270.000	6250.000						
1961	1	115.700	518173.000	4806.000	2572.000	127852.000	1961.000	69331.000	5463.000	6388.000	662.000
	2	2594.000	45397.000	2279.000	6548.000						
1962	1	116.900	554894.000	4007.000	2827.000	130081.000	1962.000	70551.000	5190.000	6271.000	623.000
	2	2626.000	46652.000	2340.000	6849.000						

Figure 17

DATA TRANSFORMATIONS

POSITION	CODE	OPERATION	CONSTANT	LOCATION	OMIT	VARIABLE
1		NONE		1	0	1
2		NONE		2	0	2
3		NONE		3	0	3
4		NONE		4	0	4
5		NONE		5	0	5
6		NONE		6	0	6
7		NONE		7	0	7
8		NONE		8	0	8
9		NONE		9	1	
10		NONE		10	1	
11		NONE		11	1	
12		NONE		12	1	
13		NONE		13	1	
14		NONE		14	1	

FIRST OBSERVATION AFTER TRANSFORMATIONS   THE FITTED EQUATION HAS 6 INDEPENDENT VARIABLES, 2 DEPENDENT VARIABLES

OBSV.	PRICE 1-11-21	GNP 2-12-22	UNEMPY 3-13-23	ARMY 4-14-24	NONIST 5-15-25	TIME 6-16-26	TOTAL 7-17-27	FARM 8-18-28	9-19-29	10-20-30
1947	8.30000D 01	2.34289D 05	2.35600D 03	1.59000D 03	1.07608D 05	1.94700D 03	6.03230D 04	8.25600D 03		

MEANS OF VARIABLES
     1.01681D 02 3.87698D 05 3.19331D 03 2.60669D 03 1.17424D 05 1.95450D 03 6.53170D 04 6.63675D 03

ROOT MEAN SQUARES OF VARIABLES
     1.07916D 01 9.93949D 04 9.34464D 02 6.95920D 02 6.95610D 03 4.76095D 00 3.51197D 03 9.30816D 02

SIMPLE CORRELATION COEFFICIENTS, R(I,I PRIME)

1	1.000	0.992	0.621	0.465	0.979	0.991	0.971	-0.982
2	0.992	1.000	0.604	0.446	0.991	0.995	0.984	-0.977
3	0.621	0.604	1.000	-0.177	0.687	0.668	0.502	-0.582
4	0.465	0.446	-0.177	1.000	0.364	0.417	0.457	-0.558
5	0.979	0.991	0.687	0.364	1.000	0.994	0.960	-0.964
6	0.991	0.995	0.668	0.417	0.994	1.000	0.971	-0.975
7	0.971	0.984	0.502	0.457	0.960	0.971	1.000	-0.943
8	-0.982	-0.977	-0.582	-0.558	-0.964	-0.975	-0.943	1.000

Figure 18

LINEAR LEAST-SQUARES CURVE FITTING PROGRAM

PRECISION - EMPLOY DEP VAR 1: TOTAL   MIN Y = 6.017D 04 MAX Y = 7.055D 04 RANGE Y = 1.038D 04

```
          Y  = B(0) + B(1)X1 + B(2)X2 + B(3)X3 + B(4)X4 + B(5)X5 + B(6)X6
          Y  = EMPLOYMENT
          X1 = PRICE INDEX
          X2 = GROSS NATIONAL PRODUCT
          X3 = UNEMPLOYMENT
          X4 = SIZE OF ARMED FORCES
          X5 = NONINSTITIONAL POPULATION
          X6 = TIME IN YEARS
```

IND.VAR(I)	NAME	COEF.B(I)	S.E. COEF.	T-VALUE	R(I)SQRD	MIN X(I)	MAX X(I)	RANGE X(I)	REL.INF.X(I)
0		-3.48225863460D 06							
1	PRICE	1.50618722714D 01	8.49D 01	0.2	0.9926	8.300D 01	1.169D 02	3.390D 01	0.05
2	GNP	-3.58191792926D-02	3.35D-02	1.1	0.9994	2.343D 05	5.549D 05	3.206D 05	1.11
3	UNEMPY	-2.02022980382D 00	4.88D-01	4.1	0.9703	1.870D 03	4.806D 03	2.936D 03	0.57
4	ARMY	-1.03322686717D 00	2.14D-01	4.8	0.7214	1.456D 03	3.594D 03	2.138D 03	0.21
5	NONIST	-5.11041056534D-02	2.26D-01	0.2	0.9975	1.076D 05	1.301D 05	2.247D 04	0.11
6	TIME	1.82915146461D 03	4.55D 02	4.0	0.9987	1.947D 03	1.962D 03	1.500D 01	2.64

```
NO. OF OBSERVATIONS               16
NO. OF IND. VARIABLES              6
RESIDUAL DEGREES OF FREEDOM        9
F-VALUE                         330.3
RESIDUAL ROOT MEAN SQUARE       304.85407355
RESIDUAL MEAN SQUARE          92936.00615998
RESIDUAL SUM OF SQUARES      836424.05543986
TOTAL SUM OF SQUARES      185008825.99999990
MULT. CORREL. COEF. SQUARED      .9955
```

		----ORDERED BY COMPUTER INPUT----					----ORDERED BY RESIDUALS----		
IDENT.	OBSV.	WS DISTANCE	OBS. Y	FITTED Y	RESIDUAL	OBSV.	OBS. Y	FITTED Y	ORDERED RESID.	SEQ

IDENT.	OBSV.	WS DISTANCE	OBS. Y	FITTED Y	RESIDUAL	OBSV.	OBS. Y	FITTED Y	ORDERED RESID.	SEQ
JASA P	1947	2396.	60323.000	60055.660	267.340	1956	67857.000	67401.606	455.394	1
JASA R	1948	1799.	61122.000	61216.014	-94.014	1961	69331.000	68989.068	341.932	2
JASA E	1949	1345.	60171.000	60124.713	46.287	1951	63221.000	62911.285	309.715	3
JASA C	1950	889.	61187.000	61597.115	-410.115	1947	60323.000	60055.660	267.340	4
JASA I	1951	545.	63221.000	62911.285	309.715	1949	60171.000	60124.713	46.287	5
JASA S	1952	329.	63639.000	63888.311	-249.311	1955	66019.000	66004.695	14.305	6
JASA I	1953	175.	64989.000	65153.049	-164.049	1954	63761.000	63774.180	-13.180	7
JASA O	1954	30.	63761.000	63774.180	-13.180	1957	68169.000	68186.269	-17.269	8
JASA N	1955	16.	66019.000	66004.695	14.305	1958	66513.000	66552.055	-39.055	9
JASA	1956	102.	67857.000	67401.606	455.394	1960	69564.000	69649.671	-85.671	10
JASA T	1957	271.	68169.000	68186.269	-17.269	1948	61122.000	61216.014	-94.014	11
JASA E	1958	584.	66513.000	66552.055	-39.055	1959	68655.000	68810.550	-155.550	12
JASA S	1959	872.	68655.000	68810.550	-155.550	1953	64989.000	65153.049	-164.049	13
JASA T	1960	1297.	69564.000	69649.671	-85.671	1962	70551.000	70757.758	-206.758	14
JASA	1961	1874.	69331.000	68589.068	341.932	1952	63639.000	63888.311	-249.311	15
JASA	1962	2446.	70551.000	70757.758	-206.758	1950	61187.000	61597.115	-410.115	16

Figure 19

313

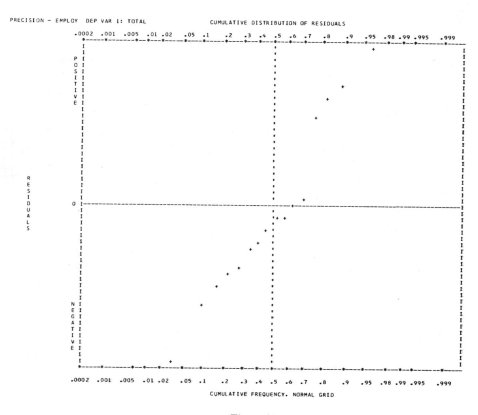

PRECISION — EMPLOY   DEP VAR 1: TOTAL   CUMULATIVE DISTRIBUTION OF RESIDUALS

Figure 20

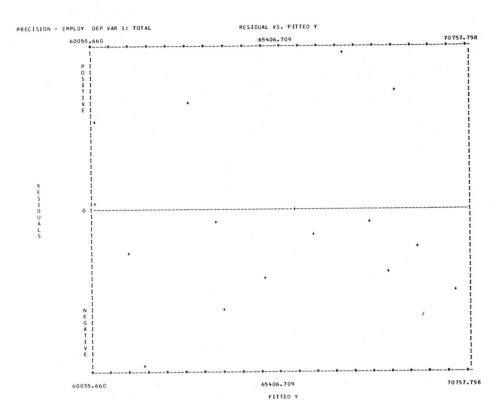

Figure 21

```
                                    CP VALUES FOR THE SELECTION OF VARIABLES

PRECISION - EMPLOY   DEP VAR 1: TOTAL        SEARCH NUMBER 1


NUMBER OF OBSERVATIONS                      16
NUMBER OF VARIABLES IN FULL EQUATION         6
NUMBER OF VARIABLES IN BASIC EQUATION        0
REMAINDER OF VARIABLES TO BE CONSIDERED      6


EQUATION  P   CP   VARIABLES IN EQUATION

          7  7.0   FULL EQUATION

                   NO BASIC SET OF VARIABLES

     1    5  4.6   BASIC SET PLUS      3       4       5       6
                                     UNEMPY   ARMY   NONIST   TIME

     2    6  6.1   BASIC SET PLUS      1       3       4       5       6
                                     PRICE   UNEMPY   ARMY   NONIST   TIME

     3    7  7.0   BASIC SET PLUS      1       2       3       4       5       6
                                     PRICE    GNP   UNEMPY   ARMY   NONIST   TIME

     4    6  5.0   BASIC SET PLUS      2       3       4       5       6
                                      GNP   UNEMPY   ARMY   NONIST   TIME

     5    5  3.2   BASIC SET PLUS      2       3       4       6
                                      GNP   UNEMPY   ARMY    TIME

     6    6  5.1   BASIC SET PLUS      1       2       3       4       6
                                     PRICE    GNP   UNEMPY   ARMY    TIME

SWEEP NUMBER  64,   DELTA Z(K,K) = -2.D-16


SEQ  EQUATION  P   ORDERED CP
 1       5     5      3.2
 2       1     5      4.6
 3       4     6      5.0
 4       6     6      5.1
 5       2     6      6.1
 6       3     7      7.0
```

Figure 22

316

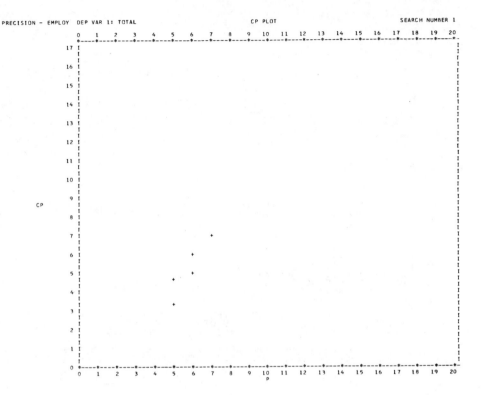

Figure 23

USER'S INSTRUCTIONS
LINEAR LEAST-SQUARES CURVE FITTING PROGRAM

ORDER OF CARDS FOR EACH PROBLEM

1 CONTROL CARD.
2 FORMAT CARD(S), IF ANY.
3 TRANSFORMATION CARD(S), IF ANY.
4 INFORMATION CARD(S) FOR PRINTOUT, IF ANY.
5 NAMES OF VARIABLES CARD(S) FOR PRINTOUT, IF ANY
6 CP FACTORIAL SEARCH CARD, IF ANY.
7 DATA CARDS
8 END CARD (END IN COL 1-3) THE NUMBER OF END CARDS "MUST"
    EQUAL THE NUMBER OF CARDS PER OBSERVATION.
9 ADDITIONAL CP FACTORIAL SEARCH CARDS, IF ANY.

NOTE: FORMAT, DATA AND END CARDS ARE NOT NEEDED IN SUBSEQUENT
    PROBLEMS IF SAME INPUT DATA ARE REUSED.

LINEAR LEAST-SQUARES CURVE-FITTING PROGRAM

GLOSSARY OF CONTROL CARD

COLUMN   ----DESCRIPTION----   (NOTE: BLANKS = 0)

1-18	PROBLEM IDENTIFICATION.
19-20	NUMBER OF INDEPENDENT VARIABLES READ IN.
21-22	NUMBER OF DEPENDENT VARIABLES READ IN.
24	E CAUSES RESIDUALS, Y AND FITTED Y TO TO BE LISTED WITH E FORMAT RATHER THAN F FORMAT.
25	1 DO TRANSFORMATIONS.
26	1 WEIGHT OBSERVATIONS.
27	1 LET B(0) = ZERO, OTHERWISE B(0) IS CALCULATED VALUE.
28	1 CHECK OBSERVATION NUMBER SEQUENCE.
29	1 CHECK SEQUENCE WITHIN OBSERVATION NUMBER.
30	1 DO BACK TRANSFORMATION OF DEPENDENT VARIABLE.
31	1 PUNCH DATA FOR CONFIDENCE INTERVAL PROGRAM,
32	0 (OR BLANK) LIST ALL INPUT DATA.
	1 LIST ONLY 1ST OBSERVATION.
	2 DO NOT LIST ANY INPUT DATA.
33	0 LIST TRANSFORMATIONS AND ALL TRANSFORMED DATA.
	1 LIST TRANSFORMATIONS AND 1ST TRANSFORMED OBSERVATION.
	2 DO NOT LIST TRANSFORMATIONS OR TRANSFORMED DATA.
34	1 DO NOT LIST SUMS OF VARIABLES.
35	1 DO NOT LIST RAW SUMS AND CROSS PRODUCTS WHEN B(0)=0.
36	1 DO NOT LIST RESIDUAL SUMS AND CROSS PRODUCTS.
37	1 DO NOT LIST MEANS AND STANDARD DEVIATIONS OF VARIABLES.
38	1 DO NOT LIST SIMPLE CORRELATION COEFFICIENTS.
39	1 DO NOT LIST INVERSE MATRIX.
40	1 DO NOT PRINT PLOTS OF RESIDUALS VS. FITTED Y.
	2 PLOT (A) RESIDUALS AND (B) COMPONENT-EFFECTS-PLUS-RESIDUALS VS. EACH INDEPENDENT VARIABLE.
	3 PLOT (A) RESIDUALS ONLY.
	4 PLOT (B) COMPONENT-EFFECTS-PLUS-RESIDUALS ONLY.
	5 PLOT (B) WITH INCREMENTS EXPANDED TO FILL EACH PLOT.
41	0 READ DATA WITH STANDARD FORMAT(A6,I4,I2,10F6.3).
	1 READ DATA WITH FORMAT TO BE READ, 1CARD,72COL.ASSUMED
	2 READ DATA WITH READ DATA (REDATA) SUBROUTINE.
42	0 DO NOT SAVE DATA FOR NEXT PROBLEM.
	1 SAVE DATA FOR NEXT PROBLEM.
	2 REUSE DATA FROM PREVIOUS PROBLEM.
45	NUMBER OF FORMAT CARDS IF GREATER THAN 1. (8 MAX.)
46-47	NUMBER OF INFORMATION CARDS TO BE READ FOR DISPLAY ON PRINTOUT IF DESIRED, 72 COL.EACH, 12 CARDS MAXIMUM.
48	0 DO NOT READ NAMES OF VARIABLES FOR DISPLAY ON PRINTOUT.
	1 READ NAMES OF VARIABLES FROM CARDS. (NAMES IN SAME POSITION AS VARIABLES ON TRANSFORMATION CARDS - 6 COL.EACH, 4X.) ONLY ACTIVE VARIABLES AFTER TRANSFORMATION ARE LISTED.
	2 REUSE NAMES OF VARIABLES FROM PREVIOUS PROBLEM.
49	1 SEARCH FOR CANDIDATE EQUATIONS VIA CP.
50	NUMBER OF CP FACTORIAL SEARCH CARDS TO BE READ PER DEPENDENT VARIABLE. KEYPUNCH IN ASCENDING ORDER THE IDENTIFING NUMBER OF THE VARIABLES (AFTER TRANSFOR-MATIONS) TO BE USED IN THAT SEARCH, 18(2X,I2) FORMAT. ALL OTHER VARIABLES WILL BE PLACED IN BASIC EQUATION.
51	1 CALCULATE COMPONENT EFFECT OF EACH VARIABLE ON EACH OBSERVATION AND ESTIMATE THE STANDARD DEVIATION FROM NEAR NEIGHBORS.
	2 MAKE ONLY NEAR-NEIGHBOR ESTIMATE.
52-53	NUMBER OF FILE IF DATA ARE READ FROM SEPERATE FILE, WITH OR WITHOUT AN END CARD.
54-55	NUMBER OF INDEPENDENT VARIABLES GREATER THAN THE 99 ALLOWED IN COLS 19-20.
56-58	TO REDUCE PRINTOUT WITH MANY OBSERVATIONS, NUMBER OF CENTRAL OBSERVATIONS AND RESIDUALS NOT TO BE PRINTED WITH E (FORMAT) IN COL 24.
59	TO CALCULATE FIT AND RESIDUALS OF DIFFERENT OBSERVATIONS:
	1 WRITE B(0) AND B(I)S ON RETAINED FILE IN ONE PROBLEM.
	2 READ B(0) AND B(I)S FROM THAT FILE IN THE NEXT PROBLEM WITH A NEW SET OF OBSERVATIONS.

319

Computer Nonlinear Least-Squares Curve Fitting Program

Abstract

This computer program allows the user to estimate the coefficients of a nonlinear equation such as $Y = A/(x + B)^2$ and $Y = Ax^B + C$—equations that are nonlinear in the coefficients. An iterative technique is used; the estimates at each iteration are obtained by Marquardt's maximum neighborhood method which combines the Gauss (Taylor series) method and the method of steepest descent.

Since numerous forms of equations can be used, the user must specify the form by providing a subroutine to compute the values of the equation's coefficients. In addition, the user must provide a control card, a format card for reading data, and estimates of the starting values of the coefficients. If desired, information cards and coefficient name cards can be read for display on the printout. Such displays are helpful to record the form of equation, the purpose of the run and any additional information that may help identify the printout in the future. Identification of the coefficients by name is particularly helpful when working with large or complex equations.

The output of the program is a printed report which includes a description of the problem, the starting values of the coefficients, the size of the incremental steps, a summary of each iteration and a summary of the final fit (in terms similar to those in the Linear Least-Squares Curve Fitting Program). The statistics calculated include the number of observations, the number of coefficients, the residual degrees of freedom, the maximum and minimum value of the dependent variable, as well as its range, the standard error and

t-value for each coefficient, the residual sum of squares, the residual mean square, and the residual root mean square.

Listings are made of the observed and fitted values of the dependent variable—both in the sequence in which observations were given to the computer, and in the order of the magnitude of the differences between the observed and fitted values. Plots are made to indicate (1) whether these differences are normally distributed and (2) how they are distributed over all the fitted values of the dependent variable.

Provisions are made to run multiple problems as well as different equations using the same data.

An example is given.

General Information

Background

This program was originally written by D. A. Meeter, University of Wisconsin, using D. W. Marquardt's Maximum Neighborhood Method (1963). The program was revised and programmed, using FORTRAN IV conventions, for IBM 360 System computers by F. S. Wood. The nomenclature, printout, and plots are consistent with the *Linear* Least-Squares Curve Fitting Program.

References

Marquardt, D. W., "An Algorithm for Least-Squares Estimation of Non-Linear Parameters," *J. Soc. Ind. App. Math.*, **11** 2 (June 1963), pp. 431–441.

Meeter, D. A., "Non-Linear Least Squares (GAUSHAUS)," University of Wisconsin Computing Center, 1964; program revised 1966.

Form of equation

Any that is nonlinear in the coefficients such as

$$Y = b_1 x_1{}^{b_2} + b_3 + b_4 x_2 + \cdots + b_K x_M$$

Computer Listing

where Y = the fitted value of the dependent variable,   FITTED Y
$x_1, x_2, \ldots$ = independent variables, and   IND. VAR.
$b_1, b_2, \ldots, b_i, \ldots, b_K$ = estimated coefficients,   COEF. B(I)

Note: the dependent and independent variable may be transformations of the observed values.

Restrictions

Variables
  Maximum (NVARX)*   80
  Minimum
    Dependent   1
    Independent   1
Observations
  Maximum (NOBMAX)*   170
  MINIMUM   One greater than the number of
                                   coefficients being estimated

* These restrictions may be altered by changing the values for NVARX and NOBMAX in the main program.

Input

Control card

Cols. 1–20. Problem Identification. These columns may contain any grouping of letters, numbers, and blanks.

Col. 21. Source of Data.
0 (or blank) Read observations from data cards.
1 Reuse data from previous problem.

Cols. 23–24. Number of Coefficients to Be Estimated.

Cols. 27–28. Number of Independent Variables to Be Read.

Col. 32. Number of the Equation to Be Used. The program, as written, allows the user to compile as many as five alternate equations at one time. Each control card designates which of the five is to be used.

Cols. 33–36. Starting Value of Lambda. Lambda is used as a multiplier to scale the space or size of the steps taken. If not specified, the starting value for lambda will be set at 0.1. Occasionally a value of 1 will be required to control the initial iterations.

Cols. 37–40. Value of Nu. Nu is used as a divider or a multiplier to change the size of lambda depending on whether the sum of squares of that iteration is relatively close or far from the value of the previous iteration. It is a measure of whether that iteration is far from or near the minimum sum of squares.

Cols. 43–44. Maximum Number of Iterations. If not specified, the maximum number of iterations is set at 20.

Cols. 45–48. Multiplier Used to Increment Value of Coefficients. This multiplier is used to increment the coefficients from one iteration to the next. If not specified, the value of the multiplier is set at 0.01.

Cols. 49–56. Magnitude of Sum of Squares Criterion. This is one of two criteria used to end converging iterations. If a value of 0.0001 is specified, the iterations will stop when there is a change of less than 0.0001 in the residual sum of squares from one iteration to the next. If the value is not specified, there will be *no* control by this criterion.

Cols. 57–64. Magnitude of Ratio of Coefficients Criterion. This is the second of two criteria used to end converging iterations. If a value of 0.001 is specified, the iterations will stop when there is a change of less than 0.001 in the ratio of *all* comparable coefficients from one iteration to the next. If a value is not specified, there will be *no* control by this criterion.

Cols. 65–66. Number of Information Card(s). If desired, the user can specify the number of 72 column information cards that are to be read for display on the computer printout. This information can include the form of the equation used, a statement of the purpose of the run or other pertinent information that will help identify the printout in the future. The information cards (12 max) follow the cards containing the starting values of the coefficients.

Col. 68. Names of Coefficients.
0 (or blank) Do not read names of coefficients for display on printout.
1 Read names of coefficients from cards for display on printout. The name of each coefficient is keypunched in the first 6 columns of a 10 column field, 7 per card (in the same positions as the corresponding numerical values keypunched on the starting coefficients cards). These coefficient name cards follow the information cards.

Format card

The format card indicates the fields in which information is to be found on the data cards. At least three fields are required; (1) identification of the observation, (2) the dependent variable, and (3) at least one independent variable. The format card is written using FORTRAN conventions; for example:

$$(A4, 6X, F10.2, 8F4.0/F4.2)$$

This indicates (*a*) the identification of the observation written in alphanumeric characters are in the first 4 columns, (*b*) skip six columns, (*c*) the dependent variable in columns 11–21 has a field of 10 with 2 numbers to the right of the decimal, (d) there are 8 independent variables in columns 22–53, each in a field of 4 columns with no number to the right of the decimal, and (e) the last independent variable is in the first 4 columns of the next card— so indicated by the slash. In this example there are two cards per observation so two END cards are required.

The format card is not needed in subsequent problems if the same data are reused.

Starting values of coefficients

The starting values (guesses) of the coefficients are keypunched on cards 10 columns per coefficient, 7 per card. These cards follow the format card.

Data cards

Any number of cards can be used per observation as long as it is so indicated on the format card. The first field of 4 columns contain the identification of the observation, the second field the dependent variable, and the third field the first independent variable. This field must be large enough to contain 999. in the END Card after the data. The program counts the number of observations.

As with the format card, the data cards and the END card(s) are not needed on subsequent problems if the data are reused.

Order of cards

First Problem
1. Control card
2. Format card
3. Starting coefficients card(s)
4. Information card(s), if desired
5. Names of coefficients card(s), if desired
6. Data cards
7. END card(s)—999. in location of 1st independent variable

Second Problem (Different Data)
Repeat 1 through 7 above

Third Problem (Same Data as in Second Problem)
1. Control card (1 in col. 21)
2. Starting coefficients card(s)
3. Information card(s), if desired
4. Names of coefficients card(s), if desired

Equation subroutine

The equations to be used are written in FORTRAN and placed in Subroutine MODEL1, MODEL2, MODEL3, MODEL4, or MODEL5 as follows:

Columns
4 5 6 7

```
    SUBROUTINE MODEL1 (NPROB, B, FY, NOB, NC, X, NVARX,
  1 NOBMAX)
    IMPLICIT REAL * 8(A-H, O-Z)*
    DIMENSION B(NVARX), FY(NOBMAX), X(NVARX, NOBMAX)
    DO 10 J = 1. NOB
```

Example of equation

FY(J) = B(1)*X(1,J)**B(2) + B(3)

```
1 0   CONTINUE
      RETURN
      END
```

Double precision programs only

Note: All equations are written so that the estimated coefficients are positive. A glossary of nomenclature follows:

B(I)   Array of coefficients indexed by I from 1 to NC
FY(J)   Fitted value of Y, array-indexed by J from 1 to NOB.
NC   Number of coefficients.
NOB   Number of observations.
NOBMAX   Maximum number of observations.
NPROB   Problem identification.
NVARX   Maximum number of variables.
X(M,J)   Double-indexed array of independent variables.
            M is index of the independent variable, and
            J is index of the observations from 1 to NOB

Example 1

The problem for this example is from E. S. Keeping, *Introduction to Statistical Inference*, Van Nostrand Company, Princeton, 1962, p. 354.

The energy, y, radiated from a carbon filament lamp per cm² per second was measured at six filament temperatures, x. The observed data are given in the following table, where y is the absolute temperature of the filament in thousands of degrees Kelvin.

Observation	y	x
1	2.138	1.309
2	3.421	1.471
3	3.597	1.490
4	4.340	1.565
5	4.882	1.611
6	5.660	1.680

The experimenter wishes to fit the following equation:

$$Y = Ax^B$$

The information needed for keypunching is as follows:

Control Card

Column	Entry	Item
1–20	EXAMPLE 1, LAMP HEAT	Problem Identification
21	0	Read Data from Cards
24	2	No. Coefficients
28	1	No. Independent Variables
32	1	No. of Equation
33–36	0.10	Starting Value of Lambda
37–40	10.0	Value of Nu
43–44	20	Max No. of Iterations
45–48	0.01	Multiplier
49–56	0.0	Sum of Squares Criterion
57–64	0.000001	Ratio of Coefficients Criterion
65–66	3	No. of Information Cards
68	1	Read Names of Coefficients

Format Card

(A4, 2X, 2F6.0, F1.0)

Starting values of coefficients card

From log–log plots of y versus x, A and B are approximately 0.7 and 4.0 respectively.

Columns

1	11
0.7	4.0

Information cards

Columns
1
Y = A(X)EXPONENT B, WHERE
Y = ENERGY RADIATED FROM CARBON FILAMENT / CMSQRD
    / SEC.
X = ABSOLUTE TEMPERATURE OF FILAMENT IN M DEGREES
    KELVIN.

Names of coefficients card

Columns
1   11
A   EXP. B

Data cards

		Columns	
Card	*4*	*11–16*	*17–22*
First	1	2.138	1.309
Last	6	5.660	1.680
END			999.

Equation in subroutine MODEL 1

Columns
1   7
     FY(J) = B(1)*X(1, J)**B(2)
The printout of this problem is shown in Figures 1–5.

```
NON-LINEAR ESTIMATION              EXAMPLE 1, LAMP HEAT

MODEL NUMBER                1

                     Y = A(X)EXPONENT B
                     WHERE Y = ENERGY RADIATED FROM CARBON FILAMENT / (CM)SQRD / SEC.
                             X = ABSOLUTE TEMPERATURE OF FILAMENT IN M DEGREES, KELVIN.

NUMBER OF COEFFICIENTS      2

STARTING LAMBA              0.010

STARTING NU                10.000

MAX NO. OF ITERATIONS      10


DATA FORMAT ( A4, 2X, 2F6.0, F1.0 )

OBSV. NO.  IDENT.  SEQ.   1-11-21   2-12-22   3-13-23   4-14-24   5-15-25   6-16-26   7-17-27
     1     GAUS     1      2.138     1.309
     2     HAUS     1      3.421     1.471
     3              1      3.597     1.490
     4     TEST     1      4.340     1.565
     5              1      4.882     1.611
     6     PROB     1      5.660     1.680
     7     LEM      1      0.0     999.000



NON-LINEAR ESTIMATION,        EXAMPLE 1, LAMP HEAT
NUMBER OF OBSERVATIONS      6
NUMBER OF COEFFICIENTS      2
STARTING LAMBA             0.010
STARTING NU               10.000
MAX NO. OF ITERATIONS     10


INITIAL VALUES OF THE COEFFICIENTS
     1          2

 7.25000E-01 4.00000E 00

PROPORTIONS USED IN CALCULATING DIFFERENCE QUOTIENTS
     1          2

 1.00000E-02 1.00000E-02

INITIAL SUM OF SQUARES:  1.4718E-02
```

Figure 1

328

```
                              ITERATION NO.    1

EIGENVALUES OF MOMENT MATRIX - PRELIMINARY ANALYSIS

  2.21967E 02 3.71613E-01

DETERMINANT:   2.0203E-02

ANGLE IN SCALED. COORD.=81.30 DEGREES

VALUES OF COEFFICIENTS
      1         2
  7.62911E-01 3.87580E 00

LAMBDA: 1.000E-03                    SUM OF SQUARES AFTER LEAST-SQUARES FIT:  4.4735E-03



                              ITERATION NO.    2

DETERMINANT:   1.8787E-02

ANGLE IN SCALED. COORD.=81.52 DEGREES

VALUES OF COEFFICIENTS
      1         2
  7.68783E-01 3.86069E 00

LAMBDA: 1.000E-04                    SUM OF SQUARES AFTER LEAST-SQUARES FIT:  4.3173E-03



                              ITERATION NO.    3

DETERMINANT:   1.8653E-02

ANGLE IN SCALED. COORD.=82.15 DEGREES




VALUES OF COEFFICIENTS
      1         2
  7.68917E-01 3.86026E 00

LAMBDA: 1.000E-05                    SUM OF SQUARES AFTER LEAST-SQUARES FIT:  4.3172E-03



                              ITERATION NO.

DETERMINANT:   1.8636E-02

ANGLE IN SCALED. COORD.=38.44 DEGREES

VALUES OF COEFFICIENTS
      1         2
  7.68916E-01 3.86026E 00

LAMBDA: 1.000E-06                    SUM OF SQUARES AFTER LEAST-SQUARES FIT:  4.3169E-03

ITERATION STOPS - RELATIVE CHANGE IN EACH COEFFICIENT LESS THAN:  1.0000E-05

CORRELATION MATRIX

DETERMINANT:   7.3236E 01
      1         2
1   1.0000
2  -0.9906    1.0000
```

Figure 2

```
                   NON-LINEAR LEAST-SQUARES CURVE FITTING PROGRAM

EXAMPLE 1, LAMP HEAT  DEP.VAR.: MIN Y = 2.138E 00  MAX Y =  5.660E 00  RANGE Y =  3.522E 00

            Y = A(X)EXPONENT B
         WHERE Y = ENERGY RADIATED FROM CARBON FILAMENT / (CM)SQRD / SEC.
               X = ABSOLUTE TEMPERATURE OF FILAMENT IN M DEGREES, KELVIN.

                                                     95% CONFIDENCE LIMITS
IND.VAR(I)   NAME   COEF.B(I)    S.E. COEF.  T-VALUE  LOWER          UPPER
    1          A   7.68916E-01   1.82E-02     42.4    7.19E-01     8.19E-01
    2       EXP. B 3.86026E 00   5.09E-02     75.9    3.72E 00     4.00E 00

NO. OF OBSERVATIONS              6
NO. OF COEFFICIENTS              2
RESIDUAL DEGREES OF FREEDOM      4
RESIDUAL ROOT MEAN SQUARE       0.03285139
RESIDUAL MEAN SQUARE            0.00107921
RESIDUAL SUM OF SQUARES         0.00431686

--------ORDERED BY COMPUTER INPUT----------        --------------ORDERED BY RESIDUALS----------------
OBS. NO.   OBS. Y   FITTED Y   RESIDUAL            OBS. NO.   OBS. Y   FITTED Y  ORDERED RESID.  SEQ.
GAUS       2.138     2.174     -0.036                        4.882     4.845      0.037          1
HAUS       3.421     3.411      0.010                        3.597     3.584      0.013          2
           3.597     3.584      0.013              HAUS      3.421     3.411      0.010          3
TEST       4.340     4.333      0.007              TEST      4.340     4.333      0.007          4
           4.882     4.845      0.037              GAUS      2.138     2.174     -0.036          5
PROB       5.660     5.697     -0.037              PROB      5.660     5.697     -0.037          6
```

Figure 3

329

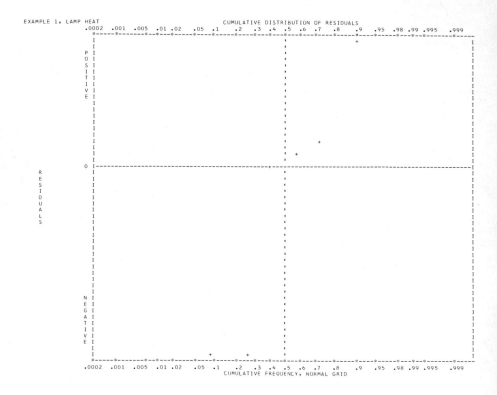

Figure 4

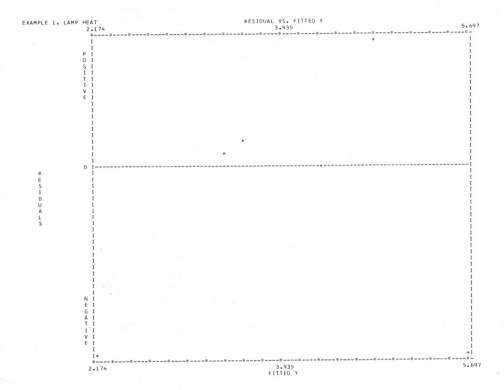

EXAMPLE 1, LAMP HEAT

RESIDUAL VS. FITTED Y

Figure 5

NON-LINEAR LEAST-SQUARES CURVE FITTING PROGRAM

USER'S INSTRUCTIONS

CONTROL CARD
    COL. 1-20 IDENTIFICATION OF PROBLEM, 5A4.
         21 0 OR BLANK, OBSERVATIONS READ FROM CARDS.
            1, REUSE DATA FROM PREVIOUS PROBLEM.
      23-24 NO. OF COEFFICIENTS TO BE ESTIMATED, I2.
      27-28 NO. OF INDEPENDENT VARIABLES TO BE READ IN, I2
      31-32 NO. OF THE EQUATION TO BE USED, I2.
      33-36 STARTING VALUE FOR LAMBDA, F4.2, (E.G. 0.1),
            USED AS A MULTIPLIER TO SCALE THE SPACE OR
            SIZE STEPS TAKEN.
      37-40 VALUE OF NU, F4.0, (E.G. 10.),
            DIVIDER AND MULTIPLIER TO CHANGE SIZE OF
            LAMBDA DEPENDING ON WHETHER SUM OF SQUARES OF
            ITERATION IS FAR OR NEAR VICINITY OF MINIMUM.
      43-44 MAX NUMBER OF ITERATIONS, I2, (E.G. 20).
      45-48 MULTIPLIER USED TO INCREMENT VALUE OF
            COEFFICIENTS, F4.3 (E.G. 0.01).

      NOTE: IF VALUES IN COLUMNS 33-48 ARE NOT DEFINED
            ON THE CONTROL CARD, THEIR LEVEL WILL BE SET
            AUTOMATICALLY TO THE ABOVE E.G. VALUES.

            CRITERIA FOR ENDING CONVERGING ITERATIONS.
      49-56 SUM OF SQUARES CRITERIA, F8.7 (E.G. 0.0001,
            A CHANGE OF LESS THAN 0.0001 IN THE RESIDUAL
            SUM OF SQUARES).
      57-64 RATIO OF COEFFICIENTS CRITERIA , F8.7,
            (E.G. 0.001 A CHANGE OF LESS THAN 0.001 IN
            THE RATIO OF ALL COMPARABLE COEFFICIENTS).

      NOTE: VALUES IN COLS.49-64 CAN BE SET AT 0.0 IF
            CONTROL OF EITHER OR BOTH IS NOT DESIRED.

      65-66 NO. OF INFORMATION CARDS TO BE READ FOR DISPLAY
            ON PRINTOUT IF DESIRED, 72 COL.EACH, 12 MAX.
         68 1 READ NAMES OF COEFFICIENTS FROM CARDS FOR
            DISPLAY ON PRINTOUT, 1ST 6 OF 10 COLS. /
            COEFFICIENT, 7 / CARD.

USER'S INSTRUCTIONS CONTINUED

FORMAT CARD)
    COL. 1-72 E.G. (A4, F6.0, (NOIND*F4.0)
              IDENT IDENTIFICATION OF OBSERVATION, A4
              Y(J) DEPENDENT VARIABLE, J-TH OBSERVATION, F6.0
              X(I,J) INDEPENDENT VARIABLE, I-TH VARIABLE, J-TH
                OBSERVATION, NOIND*(F4.0)

EXAMPLE OF SUBROUTINE MODEL

EQUATIONS TO BE USED WILL BE WRITTEN IN FORTRAN AND PLACED
IN SUBROUTINE MODEL1, 2, 3, 4, OR 5 AS FOLLOWS:

```
SUBROUTINE MODEL1 (NPROB, B, FY, NOB, NC, X, NVARX, NOBMAX)
DIMENSION  B(NVARX), FY(NOBMAX), X(NVARX,NOBMAX)
DO 10  J=1,NOB
FY(J)=B(1)*X(1,J)**B(2) + B(3)   --- EXAMPLE OF EQUATION
10 CONTINUE
RETURN
END
```
              NOTE: ALL EQUATIONS ARE TO BE WRITTEN SO THAT THE
                    COEFFICIENTS ESTIMATED AND CALCULATED WILL BE
                    POSITIVE.

ORDER OF CARDS FOR EACH PROBLEM

1 CONTROL CARD.
2 FORMAT CARD (72COL). E.G. (A4,F6.0,(NOIND*F6.0)) TO READ DATA.
3 STARTING VALUES (GUESSES) OF COEFFICIENTS(10COL/COEF, 7/CARD).
4 INFORMATION CARDS FOR PRINTOUT,IF ANY(72COL/CARD, 12 CARDS MAX).
5 NAMES OF COEFFICIENTS CARD(S) FOR PRINTOUT, IF ANY
   (1ST 6 OF 10 COLS. / COEFFICIENT, 7 / CARD).
6 DATA CARDS.
7 END CARD (999. IN LOCATION OF X(1) VARIABLE).
   NOTE: FORMAT, DATA AND END CARDS ARE NOT NEEDED IN SUBSEQUENT
         PROBLEMS IF SAME INPUT DATA ARE REUSED.

Bibliography

Acton, F. S., *The Analysis of Straight-Line Data*, Wiley, 1959.

Anderson, R. L., and T. A. Bancroft, *Statistical Theory in Research*, McGraw-Hill, 1952.

Anscombe, F. J., and T. W. Tukey, "The Examination and Analysis of Residuals," *Technometrics*, **5**, 2 (May 1963), pp. 141–160.

Barlett, M. S., "Fitting a Straight Line When Both Variables Are Subject to Error," *Biometrics*, September 1949, p. 207.

Behnken, D. W., "Estimation of Copolymer Reactivity Ratios: An Example of Non-Linear Estimation," *J. Polymer Sci.* Part A, **2** (1964), pp. 645–668.

Bennett, C. A., and N. L. Franklin, *Statistical Analysis in Chemistry and the Chemical Industry*, Wiley, 1954.

Berkson, J., "Are There Two Regressions?" *J. Amer. Statist. Assoc.*, **45**, (1950), pp. 164–180.

Blom, G., *Statistical Estimates and Transformed Beta Variables*, Wiley, 1958.

Bogue, R. H., "Calculation of the Compounds in Portland Cement," *Ind. Eng. Chem. (Anal. Ed.)*, **1**, 4 (October 15, 1929).

Booth, G. W., and T. I. Peterson, "Non-Linear Estimation," IBM Share Program No. 687, 1958.

Box, G. E. P., "Fitting Empirical Data," *Ann. N.Y. Acad. Sci.*, **86**, (May 1960), Article 3, pp. 792–816.

Box, G. E. P., and J. S. Hunter, "The 2^{k-p} Fractional Factorial Designs I and II," *Technometrics*, **3**, No. 3 (1961), pp. 311–351, and No. 4 (1961), pp. 449–458.

Brownlee, K. A., *Statistical Theory and Methodology in Science and Engineering*, Wiley, Second Edition, 1965.

Daniel, C., "Use of Half-Normal Plots in Interpreting Factorial Two-Level Experiments," *Technometrics*, **1**, No. 4 (1959), pp. 311–341.

Davies, O. L., Editor, *Statistical Methods in Research and Production*, Hafner, 1958.

Davies, O. L., Editor, *Design and Analysis of Industrial Experiments*, Hafner, 1960.

Dixon, W. J. and F. J. Massey, *Introduction to Statistical Analysis*, McGraw-Hill, 1951.

Draper, N. R., and H. Smith, *Applied Regression Analysis*, Wiley, 1966.

Fisher, R. A., and F. Yates, *Statistical Tables for Biological, Agricultural and Medical Research*, Hafner, 1953, pp. 86–87.

Gorman, J. W., and J. E. Hinman, "Simplex Lattice Designs for Multicomponent Systems," *Technometrics*, **4**, No. 4 (November 1962), pp. 463–487.

Gorman, J. W., and R. J. Toman, "Selection of Variables for Fitting Equations to Data," *Technometrics*, **8**, No. 1 (February 1966), pp. 27–51.

Gorman, J. W., "Fitting Equations to Mixture Data with Restraints on Composition," *J. of Quality Technology*, **2**, No. 4 (October 1970), pp. 186–194.

335

Hader, R. J., and A. H. E. Grandage, "Simple and Multiple Regression Analyses," *Experimental Designs in Industry* (edited by Chew), Wiley, 1956, pp. 109–137.

Hald, A., *Statistical Theory and Engineering Applications*, Wiley, 1952.

Hamaker, H. C., "On Multiple Regression Analysis," *Statist. Neerlandica*, **16**, (1962), pp. 31–56, (in English).

Hansen, W. C., "Potential Compound Composition of Portland Cement Clinker," *J. Materials*, **3**, No. 1 (1968).

Hartley, H. O., "The Modified Gauss-Newton Method for the Fitting of Non-Linear Regression Functions by Least-Squares," *Technometrics*, **3**, No. 2 (1961), pp. 269–280.

Hocking, R. R., and R. N. Leslie, "Selection of the Best Subset in Regression Analysis," *Technometrics*, **9**, No. 4 (November 1967), pp. 531–540.

Hoerl, A. E., "Fitting Curves to Data," *Chemical Business Handbook* (edited by J. H. Perry), McGraw-Hill, 1954, Section 20, pp. 55–77.

Keeping, E. S., *Introduction to Statistical Inference*, D. Van Nostrand, 1962.

Longley, J. W., "An Appraisal of Least-Squares Programs for the Electronic Computer from the Point of View of the User," *J. Amer. Statist. Assoc.* **62**, No. 319, September 1967.

Madansky, A., "The Fitting of Straight Lines When Both Variables Are Subject to Error," *J. Amer. Statist. Assoc.*, **54**, (1959), pp. 173–205.

Mallows, C. L., "Choosing Variables in a Linear Regression: A Graphical Aid," presented at the Central Regional Meeting of the Institute of Mathematical Statistics, Manhattan, Kansas, May 7–9, 1964.

Mandel, J., *The Statistical Analysis of Experimental Data*, Wiley, 1964.

Marquardt, D. W., "Solution of Non-Linear Chemical Engineering Models," *Chem. Eng. Progr.*, **55**, No. 6 (June 1959), pp. 65–70.

Marquardt, D. W., and T. Baumeister, "Least-Squares Estimation of Non-Linear Parameters," IBM Share Program No. 1428, 1962.

Marquardt, D. W., "An Algorithm for Least-Squares Estimation of Non-Linear Parameters," *J. Soc. Ind. Appl. Math.*, **11**, No. 2 (June 1963), pp. 431–441.

Meeter, D. A., "Non-Linear Least Squares (GAUSHAUS)," University of Wisconsin Computing Center, 1964; program revised 1966.

Moriceau, J., "Une Method Statistique d'Evaluation des Temps Operationels," *Rev. Statistique Appliquee*, **2**, No. 3 (1954), pp. 57–74.

National Bureau of Standards, *Fractional Factorial Experimental Designs for Factors at Two Levels*, Applied Mathematics Series 48, 1962.

Oosterhoff, J., "On the Selection of Independent Variables in a Regression Equation," Stiching Mathematisch Centrum, 2 Boerhaavestraat 49, Amsterdam, Report s319 (VP 23), December 1963 (in English).

Owens, D. B., *Statistical Tables*, Wesley, 1962, pp. 373–382.

Ralston, A., and H. S. Wilf, Editors, *Mathematical Methods for Digital Computers*, Wiley, 1960, pp. 191–203.

Rand Corporation, *A Million Random Digits with 100,000 Normal Deviates*, The Free Press, 1955.

Scheffé, H., "Experiments with Mixtures," *J. Roy. Statist. Soc.*, Series B, **20**, (1958), pp. 344–360.

Scheffé, H., *The Analysis of Variance*, Wiley, 1961.

Wood, F. S., "The Use of Individual Effects and Residuals in Fitting Equations to Data," *Technometrics*, *15*, No. 4 (1973), pp. 677-695.

Woods, H., H. H. Steinour, and H. R. Starke, "The Heat Evolved by Cement During Hardening," *Eng. News-Record*, October 6, 1932, pp. 404–407, and October 13, 1932, pp. 435–437.

Woods, H., H. H. Steinour, and H. R. Starke, "Effect of Composition of Portland Cement on Heat Evolved During Hardening," *Ind. Eng. Chem.*, **24**, No. 11 (November 1932), pp. 1207–1214.

Woods, H., H. H. Steinour, and H. R. Starke, "Heat Evolved by Cement in Relation to Strength," *Eng. News-Record*, April 6, 1933, pp. 431–433.

Yates, F., *The Design and Analysis of Factorial Experiments*, Technical Communication No. 35, Harpenden, England: Imperial Bureau of Soil Science, 1935.

Index